IET MATERIALS, CIRCUITS AND DEVICES SERIES 61

Asynchronous Circuit Applications

Other volumes in this series:

Volume 1	**Advances in High-Power Fiber and Diode Laser Engineering** Ivan Divliansky (Editor)
Volume 2	**Analogue IC Design: The current-mode approach** C. Toumazou, F.J. Lidgey and D.G. Haigh (Editors)
Volume 3	**Analogue–Digital ASICs: Circuit techniques, design tools and applications** R.S. Soin, F. Maloberti and J. France (Editors)
Volume 4	**Algorithmic and Knowledge-based CAD for VLSI** G.E. Taylor and G. Russell (Editors)
Volume 5	**Switched Currents: An analogue technique for digital technology** C. Toumazou, J.B.C. Hughes and N.C. Battersby (Editors)
Volume 6	**High-frequency Circuit Engineering** F. Nibler *et al.*
Volume 8	**Low-Power High-Frequency Microelectronics: A unified approach** G. Machado (Editor)
Volume 9	**VLSI Testing: Digital and mixed analogue/digital techniques** S.L. Hurst
Volume 10	**Distributed Feedback Semiconductor Lasers** J.E. Carroll, J.E.A. Whiteaway and R.G.S. Plumb
Volume 11	**Selected Topics in Advanced Solid State and Fibre Optic Sensors** S.M. Vaezi-Nejad (Editor)
Volume 12	**Strained Silicon Heterostructures: Materials and devices** C.K. Maiti, N.B. Chakrabarti and S.K. Ray
Volume 13	**RFIC and MMIC Design and Technology** I.D. Robertson and S. Lucyzyn (Editors)
Volume 14	**Design of High Frequency Integrated Analogue Filters** Y. Sun (Editor)
Volume 15	**Foundations of Digital Signal Processing: Theory, algorithms and hardware design** P. Gaydecki
Volume 16	**Wireless Communications Circuits and Systems** Y. Sun (Editor)
Volume 17	**The Switching Function: Analysis of power electronic circuits** C. Marouchos
Volume 18	**System on Chip: Next generation electronics** B. Al-Hashimi (Editor)
Volume 19	**Test and Diagnosis of Analogue, Mixed-signal and RF Integrated Circuits: The system on chip approach** Y. Sun (Editor)
Volume 20	**Low Power and Low Voltage Circuit Design with the FGMOS Transistor** E. Rodriguez-Villegas
Volume 21	**Technology Computer Aided Design for Si, SiGe and GaAs Integrated Circuits** C.K. Maiti and G.A. Armstrong
Volume 22	**Nanotechnologies** M. Wautelet *et al.*
Volume 23	**Understandable Electric Circuits** M. Wang
Volume 24	**Fundamentals of Electromagnetic Levitation: Engineering sustainability through efficiency** A.J. Sangster
Volume 25	**Optical MEMS for Chemical Analysis and Biomedicine** H. Jiang (Editor)
Volume 26	**High Speed Data Converters** A.M.A. Ali
Volume 27	**Nano-Scaled Semiconductor Devices** E.A. Gutiérrez-D (Editor)
Volume 28	**Security and Privacy for Big Data, Cloud Computing and Applications** L. Wang, W. Ren, K.R. Choo and F. Xhafa (Editors)
Volume 29	**Nano-CMOS and Post-CMOS Electronics: Devices and modelling** Saraju P. Mohanty and Ashok Srivastava
Volume 30	**Nano-CMOS and Post-CMOS Electronics: Circuits and design** Saraju P. Mohanty and Ashok Srivastava
Volume 32	**Oscillator Circuits: Frontiers in design, analysis and applications** Y. Nishio (Editor)
Volume 33	**High Frequency MOSFET Gate Drivers** Z. Zhang and Y. Liu
Volume 34	**RF and Microwave Module Level Design and Integration** M. Almalkawi
Volume 35	**Design of Terahertz CMOS Integrated Circuits for High-Speed Wireless Communication** M. Fujishima and S. Amakawa
Volume 38	**System Design with Memristor Technologies** L. Guckert and E.E. Swartzlander Jr
Volume 39	**Functionality-Enhanced Devices: An alternative to Moore's law** P.-E. Gaillardon (Editor)
Volume 40	**Digitally Enhanced Mixed Signal Systems** C. Jabbour, P. Desgreys and D. Dallett (Editors)
Volume 43	**Negative Group Delay Devices: From concepts to applications** B. Ravelo (Editor)
Volume 45	**Characterisation and Control of Defects in Semiconductors** F. Tuomisto (Editor)
Volume 47	**Understandable Electric Circuits: Key concepts, 2nd Edition** M. Wang
Volume 53	**VLSI Architectures for Future Video Coding** M. Martina (Editor)
Volume 58	**Magnetorheological Materials and Their Applications** S. Choi and W. Li (Editors)
Volume 60	**IP Core Protection and Hardware-Assisted Security for Consumer Electronics** A. Sengupta and S. Mohanty
Volume 67	**Frontiers in Securing IP Cores: Forensic detective control and obfuscation techniques** A Sengupta
Volume 68	**High Quality Liquid Crystal Displays and Smart Devices: Vol. 1 and Vol. 2** S. Ishihara, S. Kobayashi and Y. Ukai (Editors)
Volume 69	**Fibre Bragg Gratings in Harsh and Space Environments: Principles and applications** B. Aïssa, E.I. Haddad, R.V. Kruzelecky and W.R. Jamroz
Volume 70	**Self-Healing Materials: From fundamental concepts to advanced space and electronics applications, 2nd Edition** B. Aïssa, E.I. Haddad, R.V. Kruzelecky and W.R. Jamroz
Volume 71	**Radio Frequency and Microwave Power Amplifiers: Vol. 1 and Vol. 2** A. Grebennikov (Editor)
Volume 73	**VLSI and Post-CMOS Electronics Volume 1: VLSI and Post-CMOS electronics and Volume 2: Materials, devices and interconnects** R. Dhiman and R. Chandel (Editors)

Asynchronous Circuit Applications

Edited by
Jia Di and Scott C. Smith

The Institution of Engineering and Technology

Published by The Institution of Engineering and Technology, London, United Kingdom

The Institution of Engineering and Technology is registered as a Charity in England & Wales (no. 211014) and Scotland (no. SC038698).

First published 2019

The Institution of Engineering and Technology
Michael Faraday House
Six Hills Way, Stevenage
Herts, SG1 2AY, United Kingdom

www.theiet.org

British Library Cataloguing in Publication Data
A catalogue record for this product is available from the British Library

ISBN 978-1-78561-817-8 (hardback)
ISBN 978-1-78561-818-5 (PDF)

Typeset in India by MPS Limited
Printed in the UK by CPI Group (UK) Ltd, Croydon

Contents

About the editors

Dr. Jia Di received B.S. and M.S. degrees from Tsinghua University, China, in 1997 and 2000, respectively. He completed his Ph.D. in Electrical Engineering at the University of Central Florida in 2004. He joined the Computer Science and Computer Engineering Department of the University of Arkansas in Fall 2004, where he is now a Professor and 21st Century Research Leadership Chair. His research area mainly focuses on asynchronous integrated circuit design for different applications and hardware security. His Trustable Logic Circuit Design Lab has been well sponsored by various federal agencies and industry. He has published one book and over 100 papers on technical journals and conferences. He also holds 5 U.S. patents. He is a senior member of IEEE, an eminent member of Tau Beta Pi, and an elected member of the National Academy of Inventors.

Dr. Scott C. Smith is Department Chair and Professor of Electrical Engineering and Computer Science at Texas A&M University Kingsville. Prior to this, he was a Professor of Electrical and Computer Engineering at North Dakota State University from 2013 to 2019, including 4 years as Department Chair. Before that, he was a Tenured Associate Professor of Electrical Engineering at the University of Arkansas from 2007 to 2013, including 2 years as Interim Associate Department Head; and prior to that, he was an Assistant Professor at University of Missouri – Rolla (now called Missouri University of Science and Technology) from 2001 to 2007, where he was promoted to Associate Professor and tenured effective September 2007. He received his Ph.D. degree in Computer Engineering from the University of Central Florida in 2001, the MSEE degree from the University of Missouri – Columbia in 1998, and BSEE and BS Computer Engineering degrees from the University of Missouri – Columbia in 1996. His expertise includes the areas of Computer Architecture, Embedded Systems, Digital Logic, FPGAs, Asynchronous Logic, NULL Convention Logic, CAD Tools for Digital Design, Computer Arithmetic, VHDL, VLSI, Secure/Trustable Hardware, Wireless Sensor Networks, Robotics, and Cyber-Physical Systems, where he has published 24 refereed journal papers, 71 refereed conference papers, 8 U.S. patents, 1 co-authored and another co-edited book, and 4 additional book chapters. He has co-founded 3 tech start-up companies. He is a senior member of IEEE and a member of the National Academy of Inventors, Sigma Xi, IEEE-HKN, and Tau Beta Pi.

Chapter 1

Introduction

Jia Di[1] and Scott C. Smith[2]

Our world is essentially asynchronous, where time is continuous, and natural influences (e.g., temperature, humidity, illuminance, etc.) can change at any time, not only at predefined increments. Responses of animals and plants are event-driven, not dictated by specific time intervals. When computer scientists and electronic circuit developers create abstractions of our world to design computers and electronic circuits, various approximations are applied. As illustrated in Table 1.1, compared to their analog counterparts, digital design paradigms utilize discrete values to represent "their world." Between the two digital paradigms, synchronous logic incorporates a further approximation, representing time as a discrete series of events, by utilizing a periodic clock to signal when change can occur. Asynchronous logic, on the other hand, is a more natural event-driven approach that does not rely on an approximation for time (i.e., there is no synchronizing clock signal).

This fundamental difference between asynchronous and synchronous logic paradigms grants each with unique design tradeoffs. Table 1.2 lists a subset of such tradeoffs. Across all levels of design abstraction (e.g., architectural description, gate-level netlists, transistor-level schematics, and physical layouts), these tradeoffs are translated into various circuit design considerations, such as active/leakage power consumption, propagation delay, throughput, area/size, reliability/robustness, modularity/reusability, noise/emission, design complexity, design automation, etc., which are critical for system architects and circuit designers.

1.1 Overview of asynchronous circuits

The theory of asynchronous logic (i.e., circuits being self-timed instead of being externally timed by a periodic clock signal, like synchronous circuits) was first proposed in the 1950s. Since then, many research and development activities on asynchronous circuits have been carried out by both industry and academia, resulting in numerous asynchronous design paradigms being invented and demonstrated in silicon. Asynchronous circuits can be grouped into two broad implementation types,

[1]Computer Science and Computer Engineering Department, University of Arkansas, Fayetteville, AR, USA
[2]Department of Electrical Engineering and Computer Science, Texas A&M University-Kingsville, Kingsville, TX, USA

Table 1.1 Illustration of different computation paradigms

	Discrete time	**Continuous time**
Discrete value	Digital, synchronous logic	Digital, asynchronous logic
Continuous value	Switched-capacitor analog	General analog computation

Table 1.2 Subset of tradeoffs between asynchronous and synchronous logic paradigms

Asynchronous	**Synchronous**
Continuous time computation	**Discrete** time computation
Local handshaking/self-timed control	**Global clock** control
Observed delay is the **average** of possible circuit paths, for some paradigms	Observed delay is the **maximum** overall possible circuit paths
Local switching from **data-driven** computation typically yields **lower power** operation	**Global** activity from **clock-driven** computation requires careful **clock gating** for low power operation
Throughput determined solely by **device speed**, for some paradigms	Throughput determined by **device speed** and additional **operating margins**
Data must be **encoded** for some paradigms, which requires **additional wires** (e.g., two wires per bit)	**Unencoded** data acceptable (i.e., one wire per bit)

bounded delay (BD) and quasi-delay insensitive (QDI), each with numerous different implementation paradigms. Bounded delay circuits typically utilize a bundled data representation, where data requires one wire per bit (same as synchronous circuits), and includes one additional wire to signal when a group of data wires, referred to as a bundle, is valid. QDI circuits, on the other hand, encode data validity along with the actual data being transmitted, and therefore require more than one wire per bit. An overview of both QDI and BD circuits is provided below.

A typical data encoding for QDI circuits is dual-rail logic, which is a 1-hot encoding requiring two wires per bit, where $(D^1 = 0, D^0 = 1)$ = DATA0, $(D^1 = 1, D^0 = 0)$ = DATA1, $(D^1 = 0, D^0 = 0)$ = absence of DATA, also referred to as the NULL or spacer state, and $(D^1 = 1, D^0 = 1)$ is an invalid state that will not occur in a properly functioning circuit. Other data encodings are also sometimes utilized, including quad-rail logic (i.e., 1-hot encoding scheme utilizing four wires to represent 2 bits of data) [1], other 1-hot encodings that require more than two wires per bit [1], and encodings that rely on rail transitions to represent data [2]. These encodings allow the QDI circuit to know when its data are valid without referencing time, and based on this knowledge, to generate handshaking signals to convey this to other parts of the circuit.

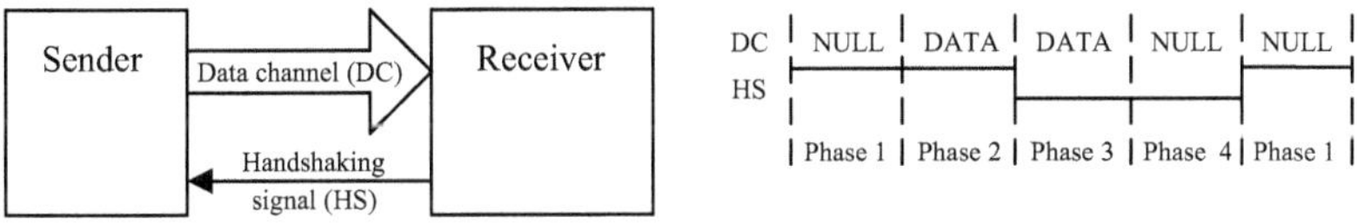

Figure 1.1 Four-phase handshaking protocol

Typical QDI circuits alternate between valid DATA states (i.e., all data signals are DATA) and the NULL state (i.e., all data signals are NULL), and utilize a 4-phase handshaking protocol to communicate with neighboring circuits, although some QDI paradigms utilize two-phase handshaking (e.g., [2]). Four-phase handshaking requires a single handshaking signal between sender and receiver to request and acknowledge data transmissions between the two, as shown in Figure 1.1. In Phase 1, the data channel is in the NULL state, and the receiver requests data by asserting the handshaking signal. In Phase 2, data are sent by the sender after receiving the handshaking request by setting the data channel to DATA. In Phase 3, the receiver gets the data and acknowledges this by deasserting the handshaking signal. In Phase 4, the sender resets the data channel back to NULL after receiving the handshaking acknowledgment. After the receiver sees NULL on the data channel, it can request the next data by asserting the handshaking signal, which is Phase 1 again.

In the general case where a sender transmits to more than 1 receiver, the sender must ensure that all receivers acknowledge the DATA/NULL transmission before sending the subsequent NULL/DATA transmission. This is accomplished by utilizing completion logic consisting of C-elements [3] to conjoin handshaking signals from multiple receivers. A C-element operates as follows: when all inputs are asserted the output is asserted, and when all inputs are deasserted the output is deassserted; otherwise, the output does not change (i.e., a C-element contains hysteresis state-holding capability). Completion logic can be implemented utilizing either bit-wise or full-word completion. Bit-wise completion only sends the completion signal from receiver *b* back to each sender whose output is sent to receiver *b*; whereas full-word completion partitions a circuit into stages, and combines all receiver handshaking outputs for a stage into a single signal, which is used as the handshaking input to all senders for that stage. Figure 1.2 depicts both bit-wise and full-word completion for an example with two senders and three receivers.

There are a variety of different QDI paradigms that utilize the typical dual-rail logic and 4-phase handshaking, which vary in terms of combinational logic (C/L) implementation, and partitioning of C/L and registration/latching functionality. Two commonly used QDI paradigms are pre-charge half buffer (PCHB) [4] and NULL Convention Logic (NCL) [5]. PCHB combines C/L and registration/latching into a single gate structure, which yields a very fine-grained pipeline, while NCL separates C/L and registration/latching functionality, resulting in a coarser-grained pipeline. For feedback loops containing N DATA tokens, at least $2N + 1$ asynchronous registers/latches are required to prevent deadlock.

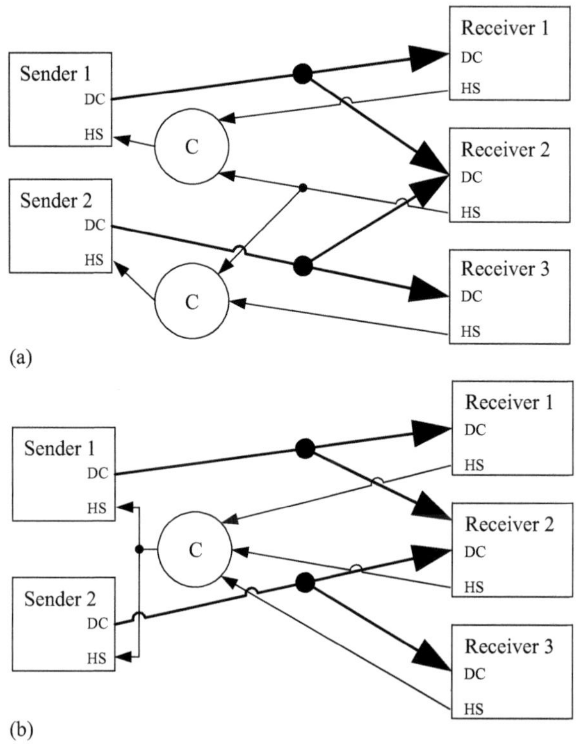

Figure 1.2 (a) Bit-wise completion and (b) full-word completion

A PCHB gate has dual-rail data inputs and outputs (e.g., X and Y, and F, respectively, for the NAND2 example in Figure 1.3), and a handshaking input and output, R_{ack} and L_{ack}, respectively. The set functions, F^0 and F^1, are implemented to achieve the particular gate functionality. The 2-input NOR gates connected to both inputs' rails and the outputs' rails detect when a dual-rail signal is either DATA or NULL; and the C-element connects these completion detection signals to generate the gate's acknowledge signal, L_{ack}. The weak inverter arrangement is used to hold the output DATA until precharged back to NULL to attain delay-insensitivity. When L_{ack} is asserted, signifying request-for-data (*rfd*), the inputs will eventually become DATA; and when L_{ack} is deasserted, signifying request-for-NULL (*rfn*), the inputs will eventually become NULL. The function evaluates, and the output becomes DATA whenever both L_{ack} and R_{ack} are *rfd* and one or both of the X and Y inputs are DATA. If R_{ack} is *rfd* and L_{ack} is *rfn*, or vice versa, the state is held by the weak inverters. When L_{ack} and R_{ack} are both *rfn*, the output is precharged back to NULL. Whenever the inputs and outputs are all DATA, L_{ack} changes to *rfn*; and when the inputs and output are all NULL, L_{ack} changes to *rfd*.

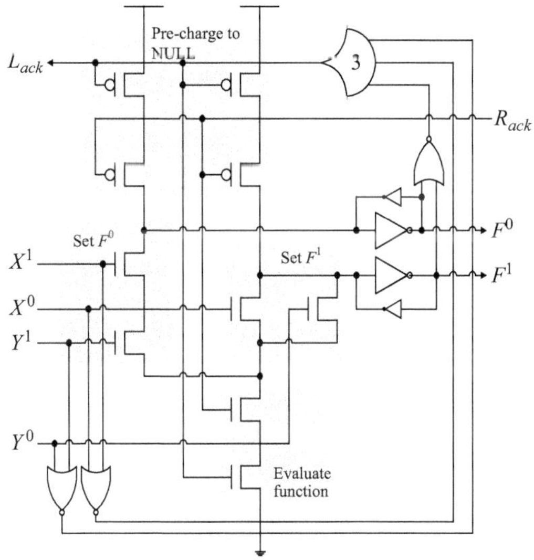

Figure 1.3 PCHB NAND2 gate

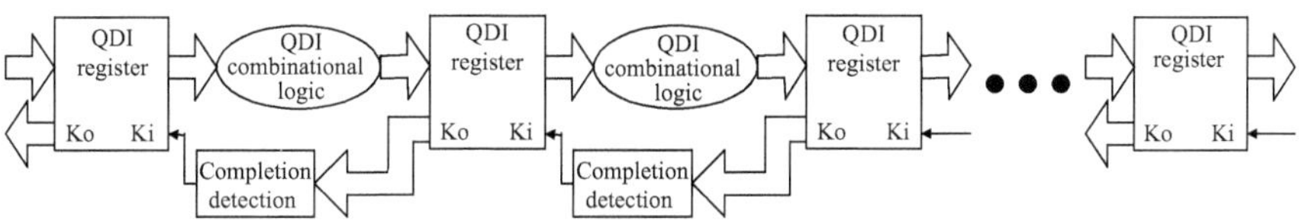

Figure 1.4 NCL system framework: shown using full-word completion (i.e., single K_i signal per stage)

The framework for NCL systems consists of QDI combinational logic sandwiched between QDI registers/latches, as shown in Figure 1.4. This is similar to synchronous systems; however, input wavefronts are controlled by local handshaking signals and completion detection instead of by a global clock signal.

NCL circuits are comprised of 27 fundamental gates, shown in Table 1.3, which constitute the set of all functions consisting of four or fewer variables, where each rail of a multirail data signal is considered a separate variable. The primary type of NCL gate, shown in Figure 1.5, is a THmn threshold gate, where $1 \leq m \leq n$. THmn gates have n inputs and at least m of the n inputs must be asserted before the output will become asserted. In a THmn gate, each of the n inputs is connected to the rounded portion of the gate; the output emanates from the pointed end of the gate; and the gate's threshold value, m, is written inside of the gate. Another type of threshold gate is referred to as a weighted threshold gate, denoted as THmnW$w_1 w_2 \ldots w_R$. Weighted threshold gates have an integer value, $m \geq w_R > 1$, applied to *inputR*. Here $1 \leq R < n$; where n is the number of inputs; m is the gate's threshold; and

Table 1.3 27 Fundamental NCL gates

NCL gate	Set function
TH12	A + B
TH22	AB
TH13	A + B + C
TH23	AB + AC + BC
TH33	ABC
TH23w2	A + BC
TH33w2	AB + AC
TH14	A + B + C + D
TH24	AB + AC + AD + BC + BD + CD
TH34	ABC + ABD + ACD + BCD
TH44	ABCD
TH24w2	A + BC + BD + CD
TH34w2	AB + AC + AD + BCD
TH44w2	ABC + ABD + ACD
TH34w3	A + BCD
TH44w3	AB + AC + AD
TH24w22	A + B + CD
TH34w22	AB + AC + AD + BC + BD
TH44w22	AB + ACD + BCD
TH54w22	ABC + ABD
TH34w32	A + BC + BD
TH54w32	AB + ACD
TH44w322	AB + AC + AD + BC
TH54w322	AB + AC + BCD
THxor0	AB + CD
THand0	AB + BC + AD
TH24comp	AC + BC + AD + BD

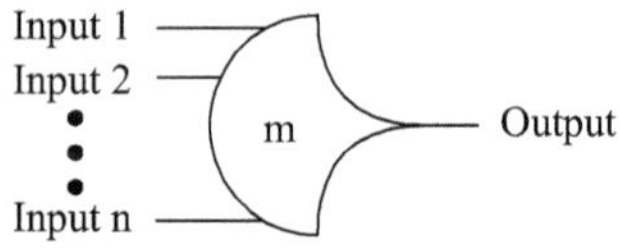

Figure 1.5 THmn NCL gate

w_1, w_2, ..., w_R, each >1, are the integer weights of *input1*, *input2*, ... , *inputR*, respectively. For example, consider the TH34w2 gate, whose $n = 4$ inputs are labeled *A*, *B*, *C*, and *D*, shown in Figure 1.6. The weight of input *A* is therefore 2. Since the gate's threshold, *m*, is 3, this implies that in order for the output to be asserted, either inputs *B*, *C*, and *D* must all be asserted, or input *A* must be asserted along with any other input, *B*, *C*, or *D*.

NCL gates are designed with *hysteresis* state-holding capability, such that after the output is asserted, or set, all inputs must be deasserted before the output will be deasserted. Hysteresis helps ensure a complete transition of inputs back to NULL

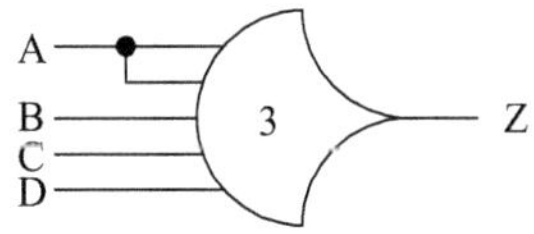

Figure 1.6 TH34w2 NCL gate: $Z = AB + AC + AD + BCD$

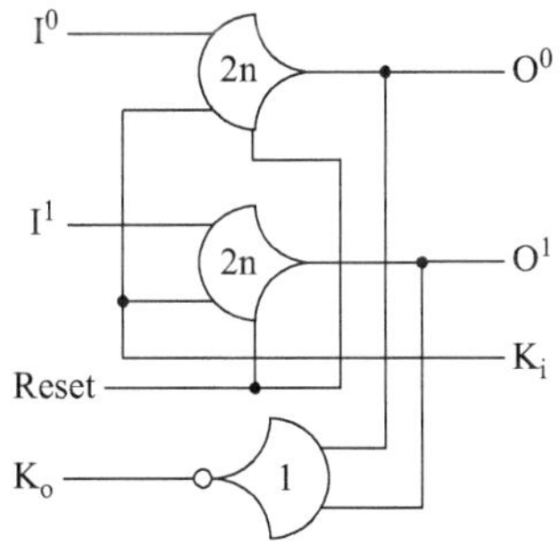

Figure 1.7 QDI register/latch shown as reset to NULL

before asserting the output associated with the next wavefront of input data. Therefore, a THnn gate is equivalent to an n-input C-element; and a TH1n gate is equivalent to an n-input OR gate. NCL threshold gates may also include a *reset* input to initialize the output. Circuit diagrams designate resettable gates by either a *d* or an *n* appearing inside the gate, along with the gate's threshold. *d* denotes the gate as being reset to logic 1; *n*, to logic 0.

These resettable gates are used to design a QDI register/latch, as shown in Figure 1.7, which is then replicated *N* times to form an N-bit register stage. These registers consist of TH22 gates that pass a DATA value at the input only when K_i (equivalent to PCHB R_{ack}) is *rfd*, and likewise, pass NULL only when K_i is *rfn*. They also contain a NOR gate to generate K_o (equivalent to PCHB L_{ack}), which is *rfn* when the register output is DATA and *rfd* when the register output is NULL. The register in Figure 1.7 is reset to NULL, since both TH22 gates are reset to logic 0; however, it could instead be reset to a DATA value by replacing exactly one of the TH22n gates with a TH22d gate.

NCL C/L must be designed to be both input-complete and observable. Input-completeness requires that all outputs of a C/L circuit may not transition from NULL to DATA until all inputs have transitioned from NULL to DATA, and that all outputs of a C/L circuit may not transition from DATA to NULL until all inputs have transitioned from DATA to NULL. In circuits with multiple outputs, it is acceptable according to Seitz's "weak conditions" of delay-insensitive signaling, for some of the outputs to transition without having a complete input set present, as long as all outputs cannot transition before all inputs arrive. Observability requires every gate transition to be observable at the output, which means that every gate that transitions is necessary to transition at least one output.

NCL C/L can also be relaxed [6] by replacing select NCL gates with hysteresis with their Boolean equivalent (i.e., same *set* function, but no hysteresis, such that the gate output is logic 0 whenever the gate's *set* function is false), while still maintaining input-completeness and observability. When NCL circuits are constructed using only threshold gates with hysteresis, ensuring input-completeness and observability of the NULL to DATA transition guarantees input-completeness and observability of the DATA to NULL transition, since gate hysteresis ensures that a gate output cannot transition to 0 until all its inputs transition to 0. However, this is not the case for relaxed NCL circuits, which are comprised of both threshold gates with hysteresis and Boolean gates.

One commonly used BD paradigm is micropipelines [7], which utilizes a two-phase bundled data protocol requiring two handshaking signals per stage, *Req*, which signals when its corresponding data bundle is valid, and *Ack*, which acknowledges a data transmission, as shown in Figure 1.8. When the sender is ready to transmit, it sets its *N* data output wires to their correct Boolean values, and then toggles its *Req* output (1) to signal that a new transmission has been initiated. The *Req* wire includes a delay element, which must be at least as long as the worse-case path through that stage's C/L to ensure that the *Req* transition does not arrive at the receiver before all *M* data input wires to the receiver are their correct value (2). After the *Req* transition arrives (i.e., Req_R toggles), the receiver can latch its *M* data inputs, after which it toggles its *Ack* output (3) to notify the sender that it can transmit the next data bundle (4). As shown in the signal waveforms in Figure 1.8, two-phase handshaking is utilized, where a $0 \rightarrow 1$ transition on *Req*/*Ack* is treated the same as a $1 \rightarrow 0$ transition. A detailed review of other types of BD circuits is included at the beginning of Chapter 7.

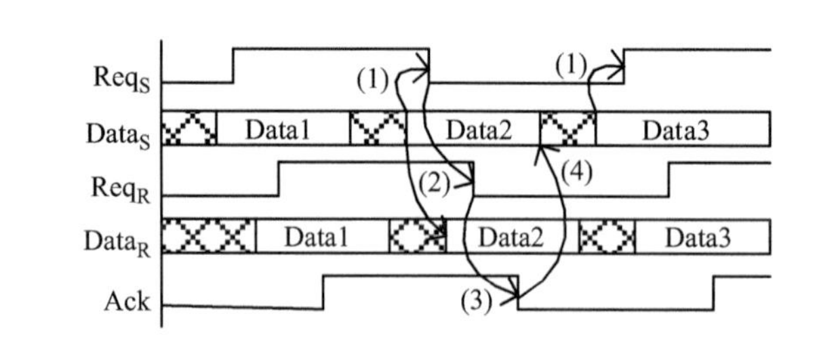

Figure 1.8 Two-phase micropipeline handshaking protocol

1.2 Advantages of asynchronous circuits

Most asynchronous paradigms share a number of common advantages as listed below. This is by no means a complete list, just some general advantages, while specific asynchronous paradigms may have other more specific benefits.

- *Flexible timing requirement*—Asynchronous circuits do not utilize a clock for timing/synchronization. Instead, handshaking protocols control and coordinate the circuit's behavior, which allows for more timing flexibility. For example, in BD asynchronous circuits, as long as the propagation delay of each independent pipeline stage is shorter than the predetermined delay bound for that particular stage, the circuit will function properly. The timing requirement for QDI circuits is even more flexible, as data and control are encoded together, and completion detection is used to generate the handshaking signals rather than relying on a predetermined delay for control. While timing closure has become an increasingly difficult task in synchronous circuit design, it is much simpler for asynchronous circuits. This flexible timing requirement feature is one of the most important advantages of asynchronous circuits, and the reason for several of the other following benefits.
- *Robust operation*—Synchronous circuits are susceptible to delay fluctuations in circuit elements due to process/voltage/temperature (PVT) variabilities. Since these variabilities are inevitable, especially process variation as transistor size continues to shrink, ensuring reliable operation for large complex synchronous circuits is a challenge. Asynchronous circuits, on the other hand, are much more robust with respect to PVT variabilities due to their flexible timing requirements, since the induced delay fluctuations are automatically tolerated by the handshaking protocols, thereby guaranteeing proper circuit behavior.
- *Improved performance*—In a synchronous circuit, all pipeline stages are coordinated by the same clock, whose period is required to be longer than the worst-case delay of any stage. In contrast to this worst-case performance, QDI asynchronous circuits yield data-dependent, average-case performance. When new data arrives, each pipeline stage finishes its computation as quickly as possible; and after finishing its computation, each stage is ready to pass the result to its next stage and get new data from its previous stage. However, since the delay of each pipeline stage is dependent on the specific data pattern being processed, a given pipeline stage may complete its computation earlier or later than its neighboring stages, in which case it will automatically wait to send or receive data, as needed, guaranteed by its handshaking protocol. Hence, there is no need to scale performance to accommodate the slowest stage, resulting in improved, average-case performance; however, this is somewhat offset by the additional performance overhead of resetting a stage to NULL before processing each subsequent data wavefront. QDI paradigms that incorporate both handshaking and C/L into a single gate structure, such as PCHB, result in a very finely grained pipeline, yielding high performance. BD asynchronous

circuits can also yield high performance since they utilize bundled data and therefore do not require the return to NULL overhead.

- *High energy efficiency*—In synchronous circuits, unless specifically gated, the clock toggles continuously, typically at a high rate of speed; and therefore, all clock tree components continuously switch, which consumes dynamic power, even when useful work is not being performed. Additionally, all flip-flops have internal gates that transition every clock edge, even if the flip-flop output does not change. On the other hand, asynchronous circuits are naturally event-driven, which can be thought of as automatic clock gating. While waiting on new data to process after completing all previous computations, asynchronous circuits inherently remain idle, such that no transitions occur. Additionally, during operation, only those circuit components necessary for that specific computation task switch, while other circuit components remain idle.
- *High modularity and scalability*—As the design complexity of modern systems-on-chip (SoCs) continues to increase, designers need to integrate many existing intellectual properties (IPs) together and verify the SoC's functionality in a short period time in order to meet time-to-market. This requires each IP to have a well-defined interface, easy migratability across different process nodes, and accurate timing information for various system specifications (e.g., temperature). While such requirements are very difficult for synchronous IPs and SoCs, it is fairly straightforward for asynchronous systems. Due to their timing flexibility, asynchronous IPs migrate more easily between process nodes, and the resulting timing fluctuations have little impact on correctness of their behaviors. For SoC integration, only minor timing analysis and circuit tweaking are needed to optimize performance. Therefore, asynchronous circuits can be more easily scaled up to form larger systems.
- *Low noise and emission*—In synchronous circuits, the high-frequency clock signal causes substantial electromagnetic interference (EMI) emission spikes, especially at the clock frequency fundamental, which can become an issue for surrounding circuits. Additionally, the concentrated switching activity at the clock edge generates a large amount of electrical noise, which can corrupt adjacent wires. On the other hand, asynchronous circuit switching activity is much more distributed and relies on localized handshaking signals instead of a periodic global clock, which leads to much lower noise and a much flatter EMI emission spectrum without a large spike, making asynchronous circuits easier to integrate with other circuit and system components.

1.3 Overview of asynchronous circuit applications

Despite the multitude of asynchronous circuit advantages discussed above, synchronous circuits have dominated the semiconductor industry. There are several reasons for this, but most stem from the underlying fact that until recently, synchronous circuits were good enough to design most next-generation ICs. This has led to a substantial investment in developing EDA tools for designing synchronous circuits over the past 50+ years, and little effort toward developing similar commercial asynchronous EDA tools. Additionally, the synchronous paradigm is traditionally what's taught in Electrical Engineering, Computer Engineering, and Computer Science curricula, such that the vast majority of IC designers have little understanding of asynchronous circuits, and what their advantages and tradeoffs are compared to synchronous circuits. Therefore, in the past, asynchronous circuits were primarily utilized for niche markets and in the research domain.

However, as transistor size continues to decrease, asynchronous circuits are being looked to by industry to solve power dissipation and process variability issues associated with today's ever-decreasing feature size. In 2003, the International Technology Roadmap for Semiconductors (ITRS) predicted a gradual industry shift from synchronous to asynchronous design styles to help mitigate power, robustness to process variation, and timing issues, as process technology continued to shrink [8]. The 2005 ITRS edition predicted asynchronous circuits to comprise 22% of the industry by 2013 [9], which was confirmed as 20% in the most recent 2013 ITRS [10]. Looking forward, ITRS predicts asynchronous circuit usage to continue to grow, accounting for a little over 50% of logic in the multibillion dollar semiconductor industry by 2027 [10].

There exist many applications in which asynchronous circuits clearly outperform their synchronous counterparts with unmatched benefits, leveraging their advantages discussed previously. One example is the surging wave of neuromorphic computing for artificial intelligence applications. Since neurons are event-driven, an asynchronous implementation is the natural choice for these devices. In 2014, the pioneering IBM TrueNorth neuromorphic processor, comprised of approximately 1 million neurons and 268 million synapses, totaling 5.4 billion transistors, was a fully asynchronous system that achieved a power consumption of 70 mW for real-time operation, which translates to 46 billion synaptic operations per second per watt [11].

The subsequent chapters present a number of applications for asynchronous circuits, each accompanied with the corresponding asynchronous circuit design theory, sample circuit implementations, results, and analysis. Chapter 2 discusses dynamic voltage scaling techniques for on-demand power consumption in asynchronous circuits. Chapter 3 presents a method for balancing power consumption and performance by incorporating parallelism in asynchronous data processing platforms. Chapter 4 discusses asynchronous circuit design for robust operation when utilizing an ultra-low supply voltage. Chapter 5 presents the design of an event-driven asynchronous circuit for interfacing with analog circuits in mixed-signal systems. Chapter 6 discusses utilizing asynchronous circuits to interface with sensors. Chapter 7 presents a design methodology for high-speed (multi-Tera bit per second) self-timed circuits utilizing bundled data. Chapter 8 details a globally asynchronous locally synchronous (GALS) network-on-chip (NoC) architecture that combines synchronous and asynchronous design styles, utilizing asynchronous communication to interface a multitude of synchronous circuits, each controlled by their own separate clock. Chapter 9 discusses the design of an asynchronous field programmable gate array (FPGA), which yields improved performance and reduced power. Chapters 10 and 11 detail asynchronous circuit design techniques for robust operation in extreme temperatures and high radiation environments, respectively. Chapter 12 presents a technique for designing secure asynchronous circuits to mitigate side-channel attacks. Chapter 13 discusses an asynchronous control mechanism for superconductive circuits. Finally, asynchronous EDA tools are detailed in Chapters 14 and 15, namely an NCL synthesis tool (Uncle) and an NCL verification tool, respectively.

References

[1] S. C. Smith and J. Di, "Designing asynchronous circuits using NULL convention logic (NCL)," *Synthesis Lectures on Digital Circuits and Systems*, Vol. 4/1, 2009.

[2] D. H. Linder and J. H. Harden, "Phased logic: supporting the synchronous design paradigm with delay-insensitive circuitry," *IEEE Transactions on Computers*, Vol. 45/9, pp. 1031–1044, 1996.

[3] D. E. Muller, "Asynchronous logics and application to information processing," in *Switching Theory in Space Technology*, Stanford University Press, pp. 289–297, 1963.

[4] A. J. Martin and M. Nystrom, "Asynchronous techniques for system-on chip design", *Proceedings of the IEEE*, Vol. 94/6, pp. 1089–1120, 2006.

[5] K. M. Fant and S. A. Brandt, "NULL convention logic: a complete and consistent logic for asynchronous digital circuit synthesis," *International Conference on Application Specific Systems, Architectures, and Processors*, pp. 261–273, 1996.

[6] C. Jeong and S. M. Nowick, "Optimization of robust asynchronous circuits by local input completeness relaxation," *Asia and South Pacific Design Automation Conference*, pp. 622–627, 2007.

[7] I. E. Sutherland, "Micropipelines," *Communications of the ACM*, Vol. 32/6, pp. 720–738, 1989.

[8] https://www.dropbox.com/sh/0ce36nq4118wiag/AACZ1MVxbt8GBSPlla7-FoMda?dl=0&preview=Design2003.pdf [Accessed April 2019].

[9] https://www.dropbox.com/sh/2urwqghq1gzk511/AADuZE5F68lz2DYGpA3TspSna?dl=0&preview=Design.pdf [Accessed April 2019].

[10] https://www.dropbox.com/sh/6xq737bg6pww9gq/AACQWcdHLffUeVloszVY6Bkla?dl=0&preview=2013_Design-v3.pdf [Accessed April 2019].

[11] P. A. Merolla, J. V. Arthur, R. Alvarez-Icaza, *et al.*, "A million spiking-neuron integrated circuit with a scalable communication network and interface," *Science*, Vol. 345/6197, p. 668, 2014.

Chapter 2

Asynchronous circuits for dynamic voltage scaling

Kwen-Siong Chong[1], *Tong Lin*[1], *Weng-Geng Ho*[1], *Bah-Hwee Gwee*[2] *and Joseph S. Chang*[2]

2.1 Introduction

Dynamic voltage scaling (DVS) [1] refers to the scaling of the magnitude of the voltage of supply rail, V_{DD}, to provide a means of power/speed trade-off. Specifically, for high speed demands (accompanied by high power dissipation), V_{DD} is "dialed-up" and conversely "dialed-down" when the demand for speed is modest (accompanied by low power dissipation). The full-range V_{DD} in DVS spans three voltage regimes:

(a) Usual operating voltage regime with high speed: nominal voltage;
(b) Low operating voltage regime with medium speed: near-threshold (near-V_t) voltage; and
(c) Very low operating voltage regime with very slow speed: sub-threshold (sub-V_t) voltage.

To appreciate the implications of the various operating voltage regimes, Figure 2.1 plots our simulation results of the delay and total power dissipation of a CMOS inverter (@130 nm) versus V_{DD} @50 kHz switching rate. The delay herein is defined as the sum of high-to-low (t_{HL}) and low-to-high (t_{LH}) switching delays, where the low and high levels are defined as 10% and 90% V_{DD}, respectively. The process option of RVT (regular-V_t; $|V_t| \approx 0.4$ V) is considered, and for sake of easy comparison, the plots are normalized to the RVT inverter @nominal $V_{DD} = 1.2$ V.

It can be observed from Figure 2.1 that by reducing V_{DD} from nominal to near-/sub-V_t, the total power dissipation (both dynamic and static) of an inverter is substantially reduced. For example, when V_{DD} is scaled from nominal $V_{DD} = 1.2$ V to deep sub-V_t, $V_{DD} = 0.15$ V, the total power dissipation of the inverter based on the RVT process is reduced by ~51×. The effect of scaling V_{DD} is even more dramatic

[1]Temasek Laboratories, Nanyang Technological University, Singapore, Singapore
[2]VIRTUS, IC Design Centre of Excellence, School of Electrical and Electronic Engineering, Nanyang Technological University, Singapore, Singapore

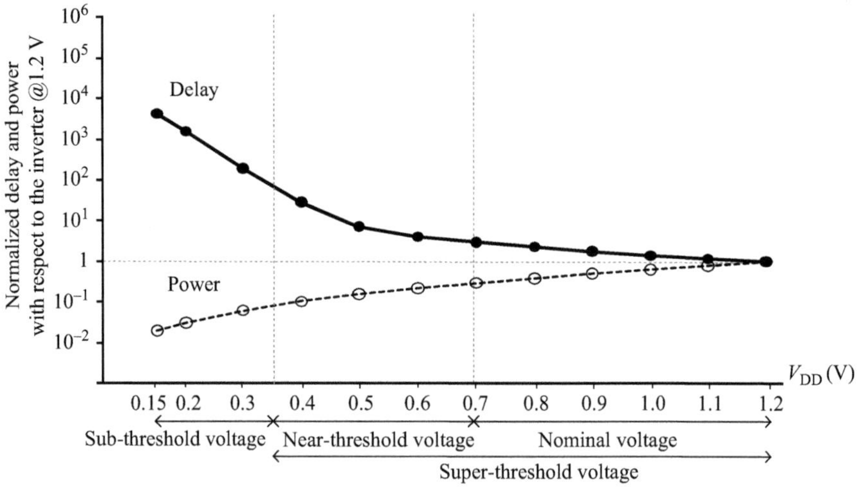

Figure 2.1 Delay and power characteristics of inverters (@50 kHz) 130 nm CMOS, normalized to characteristics @1.2 V

to delay, particularly in near-/sub-V_t. For example, for V_{DD} scaled from 1.2 V to 0.15 V, the delay of the RVT inverter is ~4,262× longer. Of specific interest, it has been analytically shown that the theoretical lowest power dissipation point is in the extreme sub-V_t voltage regime [2]. It is also well documented that in terms of the DVS, the most energy efficient point is in the sub-V_t voltage regime [3] and this point is not necessarily at the lowest possible voltage operation. Put differently, it would be advantageous to determine this most energy-efficient point for a given system and this would be the lowest practical voltage in the range of the DVS (as reducing the voltage further would undesirably incur both higher energy and slower speed).

To enable DVS, process-voltage-temperature (PVT) variations in circuit conditions in an integrated circuit (IC) are generally considered as some of the most challenging aspects of circuit design as these circuit conditions often dictate if a given circuit is operating error-free, particularly when it is desired to optimize the circuit design. The PVT variations become even more variable as the operating voltage is reduced and the characterizations therein become ambiguous to the point of extreme/intractable when the V_{DD} is near the sub-V_t voltage $V_{DD} \ll V_t$. To accommodate the PVT variations in the practical realization of current-art complex digital near-V_t systems, the approaches employed include strict operating environments (e.g., expensive highly controlled fabrication processes and electrical conditions), transistor upsizing [4], current-mode approaches [5], adaptive body biasing [6], double-gate MOSFET [7], self-calibration techniques [8], redundancy/duplication circuitry [9], and, to a large extent, "pessimistic" designs with large delay safety margins (even with the aforesaid approaches fully or in part adopted). The large delay safety margins allowed for would typically include the worst-case delay, including clock skew, setup-time, and hold-time for registers, etc. Consequently, the design of such systems for operation robustness based on the synchronous design philosophy [10] (where a global clock is used) for near-V_t and sub-V_t operations would be challenging [11] and/or such systems may be unnecessarily slower than warranted. Nonetheless, because a complete profile for PVT variations is ambiguous particularly in the sub-V_t voltage regime, the synchronous design philosophy is unable to practically guarantee robust error-free operation. Furthermore, the yield of synchronous circuits for sub-V_t voltage operation could be low, and their reliability issues cannot be assumed. In the context of our project, the full-range of DVS is often not practically realizable by the synchronous design philosophy.

Conversely, the alternative to the synchronous design philosophy is the somewhat esoteric asynchronous design philosophy [12] where handshake signals (instead of the global clock) are used to ascertain error-free synchronization. In Figure 2.2, we depict the synchronous and asynchronous design philosophies where the asynchronous design philosophy can be well-leveraged for robust DVS operation. In Figure 2.2, the lower level classifications within the asynchronous design philosophy are depicted. In the second row, there are three general asynchronous timing approaches: bundled-data (BD), quasi-delay-insensitive (QDI)/speed-independent (SI)/timed-pipeline (TP) and delay-insensitive. Delay-insensitive has been well analyzed to be impractical [13] for realization. Bundled-data circuits are, to some extent, similar to synchronous circuits in the sense that its operation relies on a bounded delay assumption for gates and wires. As the bounded delay assumption may be unmatched/insufficient due to the virtually intractable PVT variations, the ensuing circuits are not robust for sub-V_t operation. QDI/SI/TP circuits are conveniently grouped together here because they innately detect their computation timing—their specific timing assumptions have been well-documented elsewhere [14]. Of significance, QDI circuits can accept arbitrary wire delays save the isochronic forks [15] which are assumed to have the same wire branch delays, a condition that can be satisfied in practice. The QDI asynchronous approach hence offers the most practical approach [16] for sub-V_t operation. In the third row, we classify asynchronous circuits based on their pipeline granularity—one is based on the block-level pipeline structure with data-control-decomposition [17], and the other based on the gate-level pipeline structure with Integrated-Latch [18]. In the last row, there are various asynchronous cell design approaches. For the block-level pipeline structures, they include precharged-static-logic (PCSL) [19], null convention logic (NCL) [20], delay-insensitive-minterm-synthesis (DIMS) [21] and direct static logic implementation (DSLI) [22]. For the gate-level pipeline structures, they include sense-amplifier half buffer (SAHB) [23], PS0 [24], LP2/1 [25], sense-amplifier pass-transistor-logic (SAPTL) [26], single-track asynchronous pulse logic (STAPL) [27], single-track full buffer (STFB) [28] and precharge half buffer (PCHB) [29].

This chapter presents a few practical asynchronous realizations, based on the QDI/pseudo-QDI timing assumptions, suitable for full range of DVS. Both block-level and gate-level QDI asynchronous circuits and their associated applications

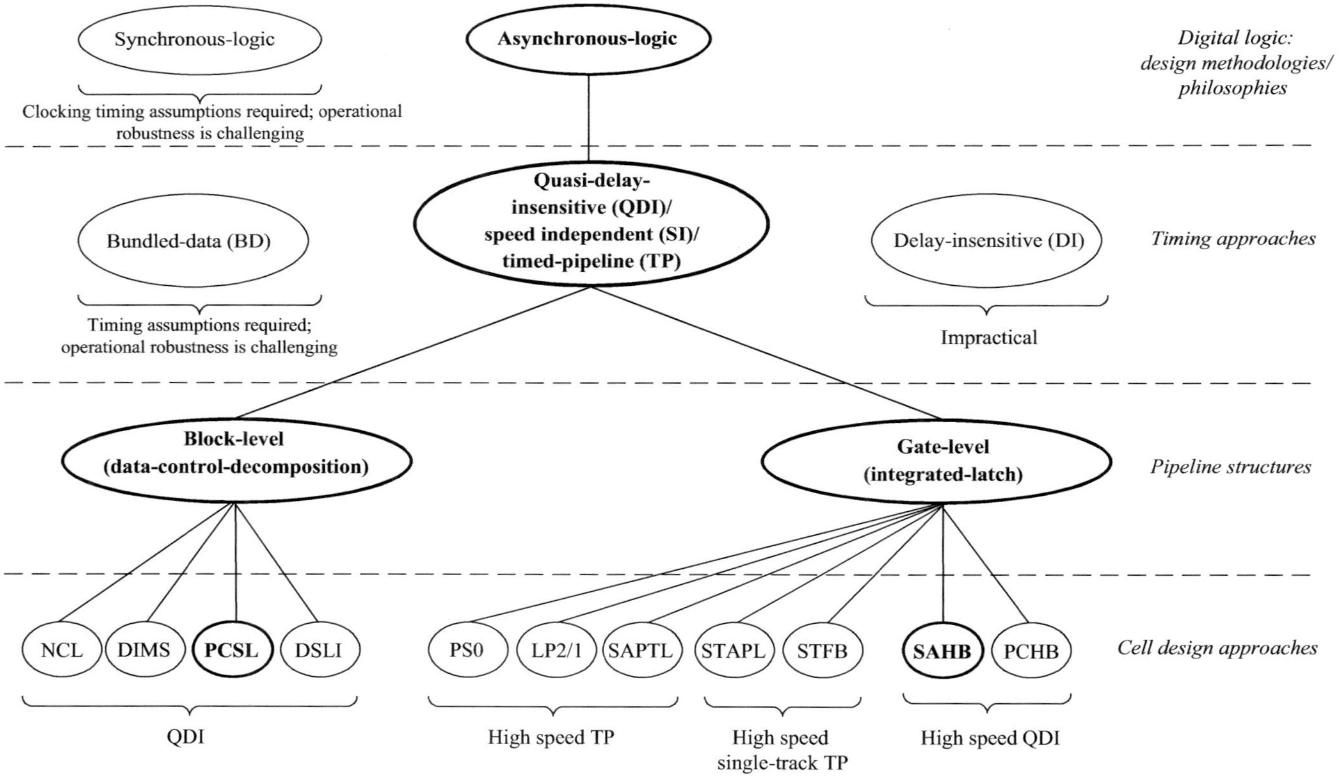

Figure 2.2 General classification of digital-logic circuits—the lower level classifications are under the asynchronous design philosophy

will be described. For asynchronous cell design approaches, we present our PCSL for block-level asynchronous circuits and our SAHB for gate-level asynchronous circuits. The chapter is organized as follows. Section 2.2 presents the block-level asynchronous circuits. Section 2.3 presents the gate-level asynchronous circuits. Section 2.4 concludes this chapter.

2.2 Block-level asynchronous circuits

In this sub-section, we first introduce a full-range DVS scheme embodying the block-level QDI asynchronous circuits with the aim to minimize power/energy under prevailing conditions. We term this adaptive V_{DD} scaling system sub-threshold self-adaptive V_{DD} scaling, or SSAVS for short. The SSAVS system is exemplified by means of an implementation of a frequency response masking (FRM) filter bank (FB) design. To reduce the hardware, power, and delay overheads of conventional QDI asynchronous cells, we present our "precharged-static-logic" (PCSL) cells for realizing the dual-rail datapath for block-level QDI asynchronous circuits. To further reduce the hardware and power overheads of the standardized asynchronous QDI protocol, we present a simplification to said protocol, and coin this simplified protocol as "pseudo-QDI." For this proposed protocol, although requiring a timing assumption, we show that the timing assumption is easily satisfied in practical digital circuits and systems.

2.2.1 Quasi-delay-insensitive (QDI) sub-threshold self-adaptive V_{DD} scaling (SSAVS)

2.2.1.1 SSAVS system design

We demonstrate an SSAVS system by means of an 8 × 8-bit quad-channel FRM FB [30] realized in the block-level QDI asynchronous pipeline structure. This asynchronous FRM FB was part of a wireless sensor network (WSN) node [19]. This WSN node was expected to operate in extreme environmental conditions including temperature range from −55°C to +125°C and with modest yet wide range of workload requirements—it needs to process input samples at a varying rate from 0.1 kSamples/s (kS/s) to 100 kS/s. Further, as a standalone device operating on batteries, it needs to preserve power/energy whenever possible. These stringent requirements call for adaptive V_{DD} scaling system, in particular, full-range dynamic V_{DD} scaling into sub-V_t region to maximally preserve power/energy yet meeting workload requirement and prevailing conditions.

Figure 2.3 depicts the proposed SSAVS system including a power management module embodying the SSAVS controller and its associated adjustable V_{DD} means (a Buck DC-DC Converter), and the 8 × 8-bit quad-channel QDI FRM FB. There are two V_{DD} voltage rails in the overall proposed SSAVS system: a fixed V_{DD_NOM} = 1.2 V and a variable V_{DD_ADJ} whose sub-V_t voltage typically ranges from 150 mV to 400 mV. For ease of illustration, the specific V_{DD} rail is shown in parenthesis for the supply rails and for signals of the various modules. A constantly-powered simple processor (@V_{DD_NOM}) is used to monitor sensor input

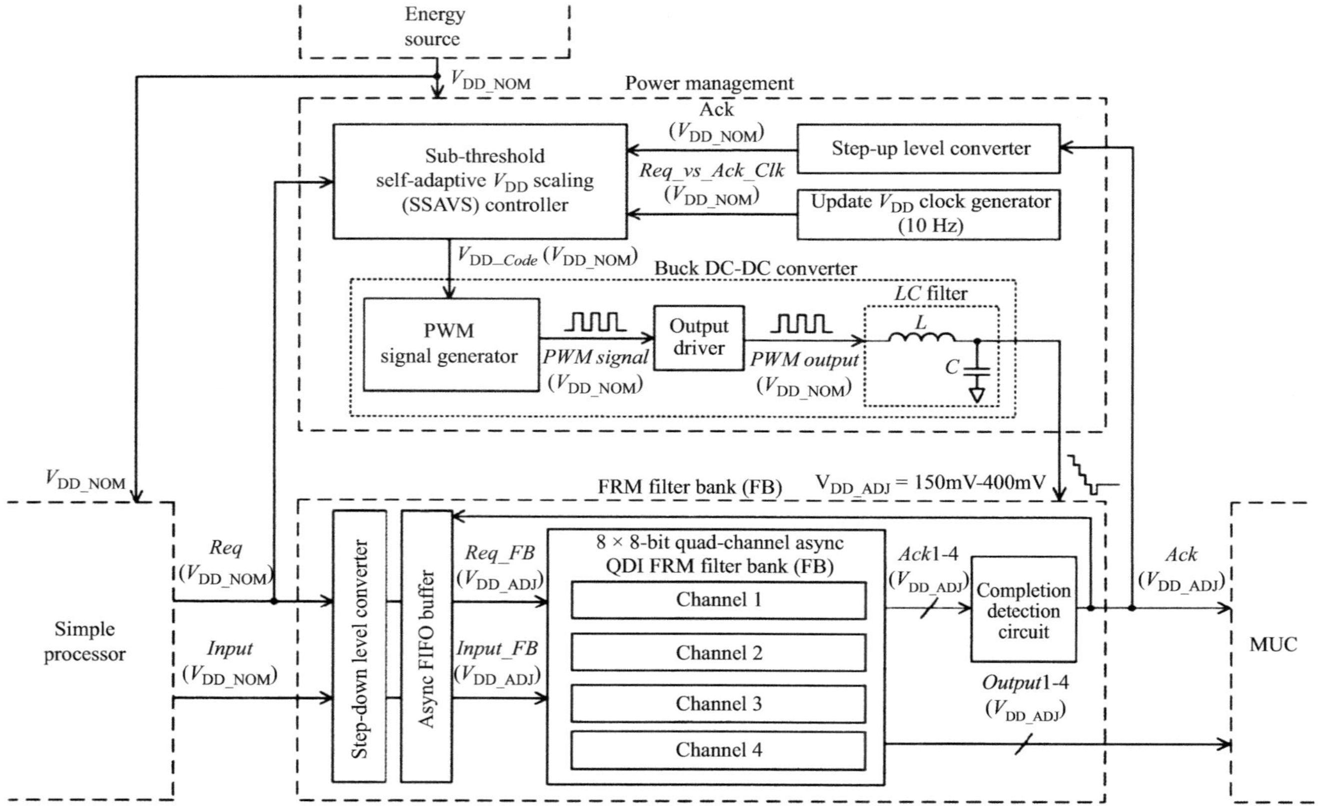

Figure 2.3 Overall structure of the proposed SSAVS system with an asynchronous QDI FRM filter bank (FB) as part of a WSN node [19] realized in the block-level asynchronous pipeline; $V_{DD\_NOM} = 1.2$ V, V_{DD_ADJ} *ranges from 150 mV to 400 mV*

of the WSN and ascertain if the input is possibly useful, and if it is, the WSN goes into active mode where the simple processor signals the power management module to energize the asynchronous FRM FB via V_{DD_ADJ}.

In Figure 2.3, the voltage of *Input* (useful input samples) and of *Req* signals is first adjusted from V_{DD_NOM} = 1.2 V to V_{DD_ADJ} by the Step-Down Level Converter, and are thereafter buffered by the Async FIFO Buffer (depth of 50) before input (*Input_FB* and *Req_FB*) to the asynchronous FRM FB. The FB outputs (*Output*1-4) and their associated *Ack* (combined from *Ack*1-4 via the Completion Detection Circuit) are output to a microcontroller unit (MCU) for further processing. *Ack* is also fed back to the Async FIFO Buffer. The *Req* and *Ack* signals are input to the power management module, and *Ack* is stepped up from V_{DD_ADJ} to V_{DD_NOM}. The SSAVS controller within the power management module monitors the number of *Req* and *Ack* signals in each *Req_vs_Ack_Clk* period (an adjustable 10 Hz clock generated by the Update V_{DD} Clock Generator for a target throughput of <1 kS/s). The V_{DD}*_Code* is a 5-bit code that sets one of 24 voltage levels (in the Buck DC-DC Converter) ranging from "00000" = 50 mV to "10111" = 1.2 V (in 50 mV steps) for V_{DD_ADJ}.

Figure 2.4 graphically depicts an example of the self-adjustment of V_{DD_ADJ}. When the WSN node is first initiated, the SSAVS controller outputs V_{DD}*_Code* = "10111," equivalently V_{DD_ADJ} = 1.2 V, and the speed of the FB would far exceed the required computation. In this scenario, the number of FB *Ack* clocks will be equal to the number of *Req* clocks in each *Req_vs_Ack_Clk* period. In the next *Req_vs_Ack_Clk* period, the SSAVS controller will subsequently decrement V_{DD}*_Code* by 1 bit to "10110" and V_{DD_ADJ} correspondingly reduces by

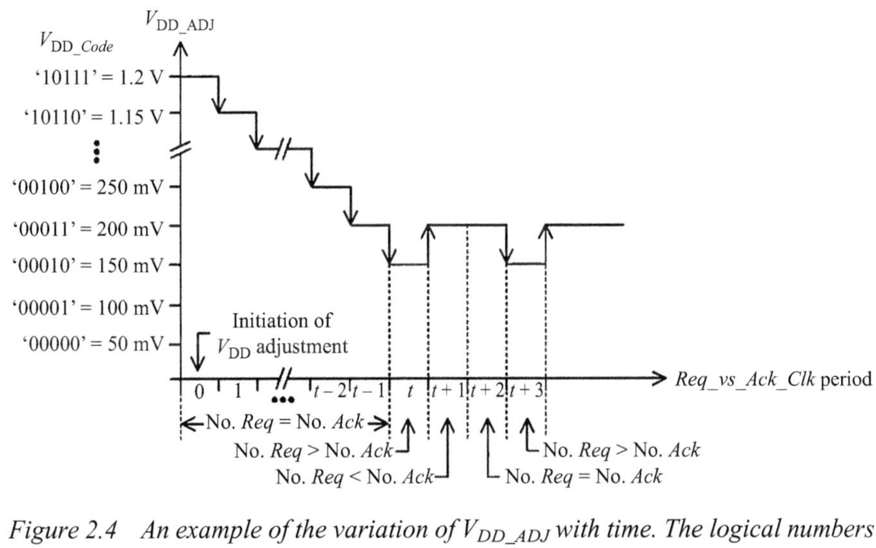

Figure 2.4 An example of the variation of V_{DD_ADJ} with time. The logical numbers on the ordinate are V_{DD}_Code and their corresponding DC voltages (V_{DD_ADJ})

50 mV to 1.15 V. The process continues where V_{DD}_*Code* is continuously decremented as with the voltage of V_{DD_ADJ} commensurably reduced. Eventually, at period *t* in Figure 2.4, V_{DD}_*Code* is decremented to "00010," equivalently $V_{DD\_ADJ} = 150$ mV. This is the juncture where the speed of the FRM FB is just slightly slower than the *Input* data rate for the prevailing conditions—the number of *Req* clocks hence exceeds the number of *Ack* clocks in one *Req_vs_Ack_Clk* period.

Although the speed of the FRM FB is slightly too slow, no error occurs because the unconsumed inputs are stored in the async FIFO buffer (Figure 2.3). In the next period, $t + 1$, the SSAVS controller reacts accordingly by incrementing V_{DD}_*Code* by 1 bit to "00011" and the corresponding V_{DD_ADJ} increased by 50 mV to 200 mV. With V_{DD_ADJ} increased, the speed of the FRM FB now slightly exceeds the required computation and the unconsumed inputs stored in the FIFO buffer (*Input_FB*) are in turn computed at a slightly faster rate than the *Input* data rate. Consequently, the number of *Req* clocks is now less than the number of *Ack* clocks and at the end of this $t + 1$ period, all unconsumed inputs in the FIFO may have been cleared; if not, the voltage of V_{DD_ADJ} remains (or increased further) in the next time period(s). If cleared, in the next period $t + 2$, the number of *Req* clocks again equals to the number of *Ack* clocks (as in time periods preceding *t*). This is the same scenario where the FB, as a consequence of the slightly raised V_{DD_ADJ}, is capable of computing faster than the *Input* data rate. In the next period $t + 3$, the scenario is that as in period *t*, and the operation repeats accordingly. Table 2.1 summarizes the three operational conditions.

In short, the voltage of V_{DD_ADJ} of the FB is in situ adaptively self-adjusted to be as low as possible (within 50 mV) to meet the throughput for the prevailing operating conditions, and on average, the voltage of V_{DD_ADJ} is slightly higher than the actual required minimum. Hence, the FB is ultra-low-power and highly power-efficient. Note that the overheads for this self-adjusting V_{DD} are very modest (a counter) and the circuit operation is uninterrupted whilst V_{DD} transitions.

2.2.1.2 Precharged-static-logic (PCSL)

A crucial design step in realizing block-level QDI asynchronous pipelines is the design of dual-rail logic cells in the datapath block of the asynchronous pipelines.

Table 2.1 Operation of the SSAVS controller

Operational conditions	SSAVS controller action
(i) Operation is too fast (No. *Req* clocks = No. *Ack* clocks)	Decrement V_{DD}_*Code* and decrease V_{DD_ADJ} (e.g., period < *t* in Figure 2.4)
(ii) Operation is too slow (No. *Req* clocks > No. *Ack* clocks)	Increment V_{DD}_*Code* and increase V_{DD_ADJ} (e.g., period *t* in Figure 2.4)
(iii) Operation is slightly faster than required & the FIFO buffer is being cleared (No. *Req* clocks < No. *Ack* clocks)	Maintain V_{DD}_*Code* and maintain V_{DD_ADJ} (e.g., period $t + 1$ in Figure 2.4)

In view of the need for sub-V_t operation, it is imperative to adopt circuits based on the static logic family to mitigate the effects of critical transistor sizing [31]. To mitigate the hardware, power, and delay overheads of reported QDI asynchronous cells, we propose "precharged-static-logic" (PCSL) cells [22].

Figure 2.5(a) depicts the basic architecture of PCSL. This basic architecture comprises an Inverting Static-Logic Cell, three transistors (for output precharging during the reset phase/evaluation during the computation phase), and two inverters (for output buffering). The outputs are *Q.T* (Output True) and *Q.F* (Output False). In PCSL cells, when *Req* is "0," both outputs are "0." On the other hand, when *Req* is "1" (indicating that an operation is ready) and when the input signals are valid, the operation commences and an ensuing output is obtained. The architecture of the PCSL cell involves an integration of the subcircuit associated with the *Req* signal and a buffer (to each output) into the standard static-logic library cell (redesigned for dual-rail asynchronous), thereby sharing of (common) transistors. This reduces the number of transistors, resulting in simultaneous lower power/energy dissipation, faster speed and smaller IC area (see Table 2.2). On the basis of this architecture, Figures 2.5(b)–(g) depict the schematic of six basic PCSL cells (all with 3-transistor limit in any stack, this is to mitigate the effect of I_{on}/I_{off} degradation [32] in sub-V_t operation).

To depict the hardware advantage of the proposed PCSL logic style, the 2-input AND/NAND gate in Figure 2.5(b) can be compared to the same gate realized by three reported static-logic QDI styles in Figures 2.6(a)–(c): (a) DIMS style [21], (b) NCL with complex gates [33] (denoted NCL1) and (c) NCL with fast-reset complex gates [34] (denoted NCL2). On the basis of simulations (@130 nm CMOS), Table 2.2 benchmarks energy-per-operation (E_{per}), delay and IC area of the aforesaid six basic cells of the various styles. The competing cells are normalized to the PCSL cells whose actual values are shown within parentheses. The average attributes are tabulated in the last row.

It is apparent from Table 2.2 that the cells embodying the proposed PCSL logic style feature the lowest E_{per}, save the simple AND/NAND and OR/NOR gates of NCL1. On average, E_{per} of cells embodying the reported DIMS, NCL1 and NCL2 logic styles is significantly higher: 4.0×, 1.6×, and 1.9×, respectively. It is also apparent that the cells embodying the proposed PCSL logic style feature the shortest delay (the sum of two components, t_{LH} (computation phase) and t_{HL} (reset phase), averaged over all input combinations), save the simple AND/NAND and OR/NOR gates of NCL1. On average, the reported DIMS, NCL1 and NCL2 cells are significantly slower: 4.1×, 1.8× and 1.9×, respectively. It is also apparent that the cells embodying the proposed PCSL logic style require the smallest IC area; the layouts are based on the standard-cell approach where the cell height is fixed at 4 μm and the cell width is in multiples of 0.4 μm. On average, the IC area required for cells embodying the reported DIMS, NCL1 and NCL2 logic styles is significantly larger: 4.7×, 2.6×, and 2.7×, respectively; from a perspective of dual-rail asynchronous and (single-rail) synchronous circuits, the smaller IC area is worthwhile because the IC area overhead of the former is somewhat mitigated. In short, cells embodying the proposed PCSL logic style simultaneously exhibit the lowest E_{per}, shortest delay, and smallest IC area.

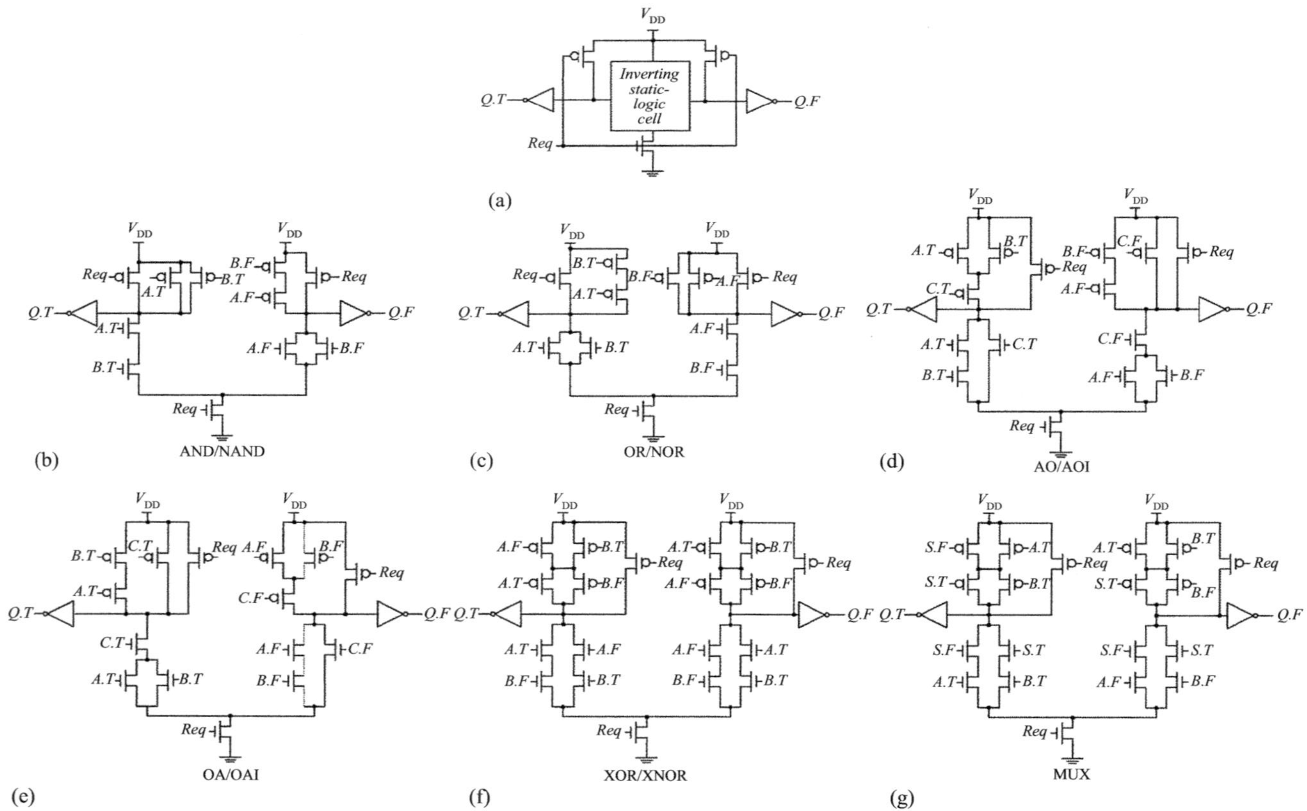

Figure 2.5 *(a) Proposed precharged-static-logic (PCSL) architecture and six basic cells embodying the proposed PCSL dual-rail QDI logic style, (b) 2-input AND/NAND gate, (c) 2-input OR/NOR gate, (d) 3-input AO/AOI gate, (e) 3-input OA/OAI gate, (f) 2-input XOR/XNOR gate, and (g) 2-input MUX*

Table 2.2 Energy-per-operation (E_{per}), delay and IC area of dual-rail library cells embodying various logic styles @V_{DD} = 150 mV and 130 nm CMOS process

	E_{per} (normalized to PCSL (10^{-18} J))				Delay (normalized to PCSL (ns))				IC area (normalized to PCSL (μm^2))			
	PCSL	DIMS	NCL1	NCL2	PCSL	DIMS	NCL1	NCL2	PCSL	DIMS	NCL1	NCL2
AND/NAND	1 (185)	3.2	0.9	1.2	1 (663)	3.2	1.0	1.2	1 (21)	4.5	1.6	1.6
OR/NOR	1 (185)	3.2	0.9	1.3	1 (661)	3.2	1.0	1.2	1 (21)	4.2	1.5	1.5
AO/AOI	1 (229)	6.8	2.0	2.2	1 (696)	7.0	2.2	2.3	1 (26)	6.8	3.1	3.3
OA/OAI	1 (229)	6.9	2.0	2.2	1 (696)	7.0	2.2	2.3	1 (26)	6.8	3.1	3.3
XOR/XNOR	1 (330)	1.8	2.0	2.1	1 (997)	2.1	2.3	2.3	1 (34)	2.9	3.1	3.3
MUX	1 (326)	1.8	2.0	2.1	1 (1,021)	2.1	2.3	2.3	1 (32)	3.1	3.2	3.4
Average	1 (247)	4.0	1.6	1.9	1 (789)	4.1	1.8	1.9	1 (27)	4.7	2.6	2.7

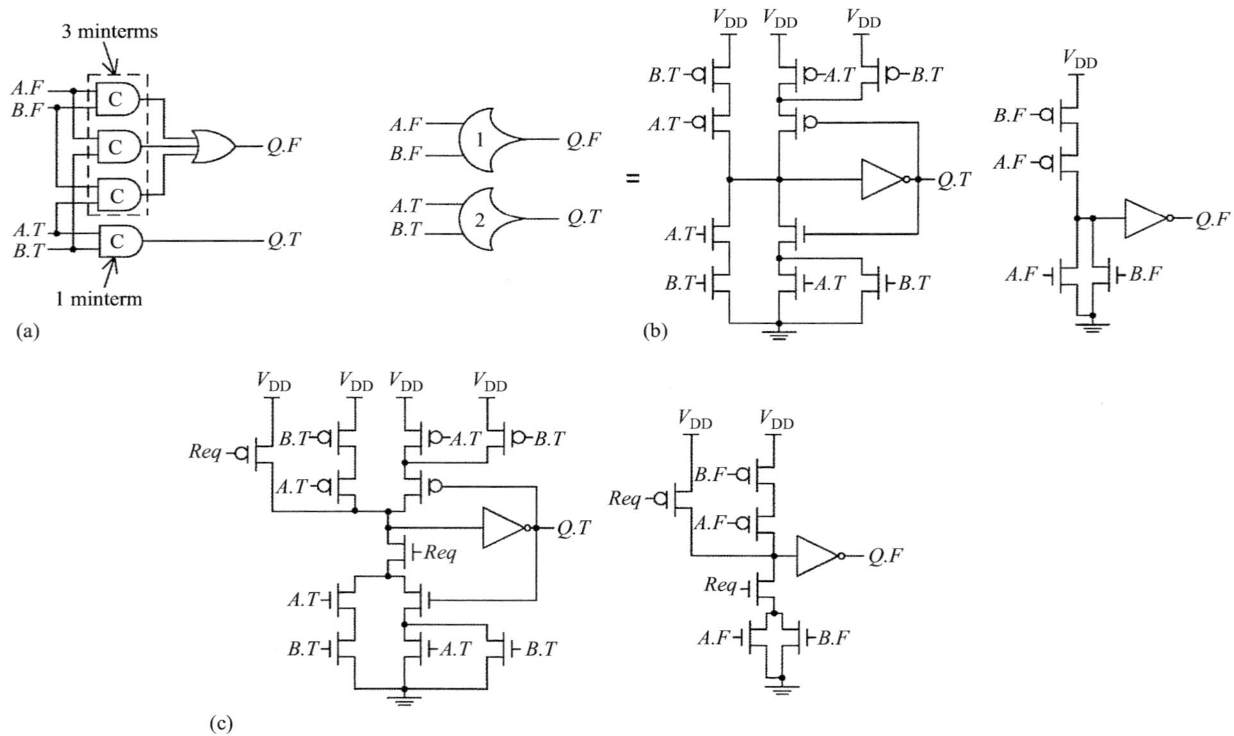

Figure 2.6 *Reported dual-rail AND/NAND circuit designs: (a) delay-insensitive-minterm-synthesis (DIMS), (b) NULL-convention-logic (NCL) with complex gates (NCL1), and (c) NCL with fast-reset complex gates (NCL2)*

2.2.1.3 Block-level QDI asynchronous FRM FB

A semicustom design flow for block-level QDI asynchronous pipeline is adopted, where the front-end is designed using an assortment of in-house design tools and commercial synthesis tools based on a flow similar to NCL-X [33]. The back-end implementation, on the other hand, is based on commercial electronic design automation (EDA) tools with our customized library cells (including the proposed PCSL). Each FB channel is independent and Figure 2.7 depicts the block diagram of one FB channel embodying an FIR filter realizing the FRM algorithm. As the throughput requirement of the intended WSN is somewhat modest, a serial implementation is adopted, where each FB channel comprises an async read/write controller, an 8 × 8-bit coefficient memory, an 8 × 8-bit data memory, an 8-bit PCSL multiplier and a 20-bit PCSL adder. To preserve the QDI protocol and proper asynchronous handshaking, datapath completion detection (DCD) and latch completion detection (LCD) circuits are included with Muller C-elements (denoted by a gate symbol with "C") [33]. All asynchronous dual-rail latches in the datapath are initialized to an "empty" value except for Latch 3 which is used to hold the accumulated product and is initialized to a valid "0."

The *Input_FB* data and *Req_FB* clock from the Async FIFO Buffer (see Figure 2.3) are input to each FB channel. The Async Read/Write Controller in Figure 2.7 first initiates a write operation by providing a valid memory address on *Data_Addr* and asserting *Write_Req* to write the *Input_FB* data into the 8 × 8-Bit Data Memory. Upon write completion, the Async Read/Write Controller subsequently initiates the first read operation for the multiply-accumulate (MAC) operation from both the 8 × 8-Bit Data Memory and the 8 × 8-Bit Coefficient Memory by providing them with valid memory addresses on *Data_Addr* and *Coeff_Addr*, and then asserting *Read_Req*. The input data and its corresponding coefficient are respectively read out to Latch 1 and Latch 2, and subsequently multiplied by the 8-Bit PCSL Multiplier. The multiplication product is captured by Latch 4 and sign-extended to 20 bits to accommodate potential overflow. The 20-Bit PCSL Adder is used to add this product to the accumulated product stored in Latch 3. The result of the adder is looped back to Latch 3, thereby updating its value and completing the first MAC operation. The MAC operation repeats until the last tap of the filter. When *Output* (one of *Output*1-4 in Figure 2.3) is finally computed, the Async Read/Write Controller of each channel will assert its *Ack* clock to indicate completion. The overall *Ack* clock is output to the Async FIFO Buffer which subsequently resets *Input_FB* and de-asserts the *Req_FB* clock. This in turn resets all FB channels and the system is now ready to process the next input data from the FIFO.

2.2.1.4 Circuit realization and measurement results

We will first demonstrate the robustness of the proposed asynchronous FB to PVT variations, particularly large V_{DD} variations, on the basis of physical measurements on prototype ICs (@130 nm CMOS) embodying the SSAVS system and the FB. Figure 2.8 depicts the die microphotograph (left) and its layout (right) of the prototype IC. The asynchronous FB embodying 4 channels occupies an

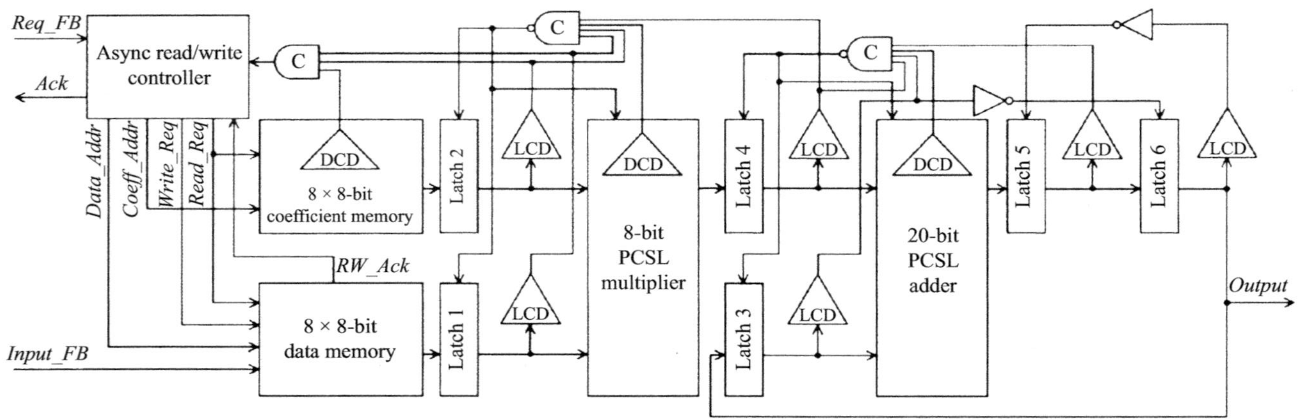

Figure 2.7 Block diagram of one channel of the 8 × 8-bit quad-channel QDI async FRM FB

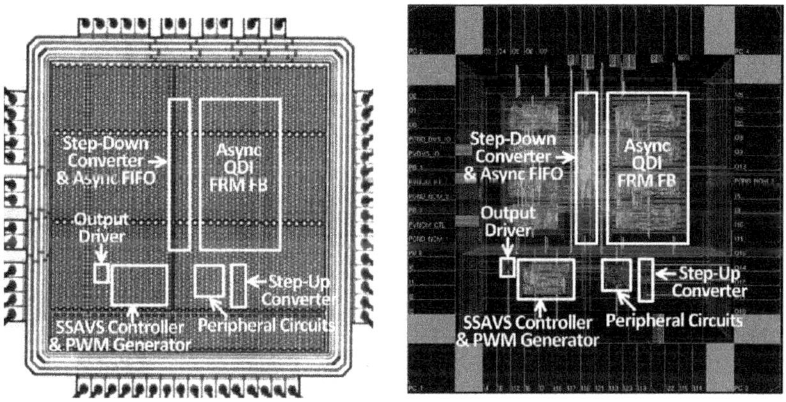

Figure 2.8 Die microphotograph (left) and layout (right) of the fabricated test-chip embodying the proposed SSAVS system with QDI asynchronous FRM filter bank

IC area of ~0.18 mm^2. All 30 prototype ICs tested were fully functional for $V_{DD} \geq 130$ mV ($|V_t| \approx 400$ mV), and this in some sense corroborates the robustness of the design. The functionality was verified by sampling the input data (generated from a pattern generator) and comparing the ensuing output data (by means of a logic analyzer) with that expected.

Consider first the robustness of the proposed asynchronous FB against PVT variations, in this case, V_{DD} varying at 1 kHz between 150 mV and 300 mV as shown in the top trace of Figure 2.9. Under this "harsh" V_{DD} condition, the asynchronous FB operates without error as verified by the *Ack* signal (and by means of a logic analyzer), depicted as the bottom trace in Figure 2.9. It can be appreciated that as V_{DD} can be varied widely without error and since the FB operation is uninterrupted, the asynchronous FB readily lends itself to being self-adjusted using the SSAVS system to the lowest voltage possible that meets the throughput for the prevailing conditions.

Consider now two examples of the SSAVS system that demonstrate its in situ self-adjusting V_{DD}. In the first example, the operation of the SSAVS system earlier delineated in Figure 2.4 is now physically depicted in Figure 2.10(a) with the top and bottom traces being V_{DD_ADJ} and *Ack*, respectively. Figure 2.10(b) depicts the second example where in addition to V_{DD_ADJ} self-adjusting to the throughput rate, it also self-adjusts to the prevailing conditions. In the top trace of Figure 2.10(b), the prototype IC is subjected to a sudden temperature drop (by means of freezer spray onto the package thereof) at some juncture, and V_{DD_ADJ} self-adjusts by first increasing to between 200 mV and 250 mV, and thereafter to between 250 mV and 300 mV as the cold permeates the IC package. Although not shown here, the converse is obtained when the prototype IC is subjected to heat, for example from a hot air gun—V_{DD_ADJ} reduces and finally toggles between two lower voltage levels.

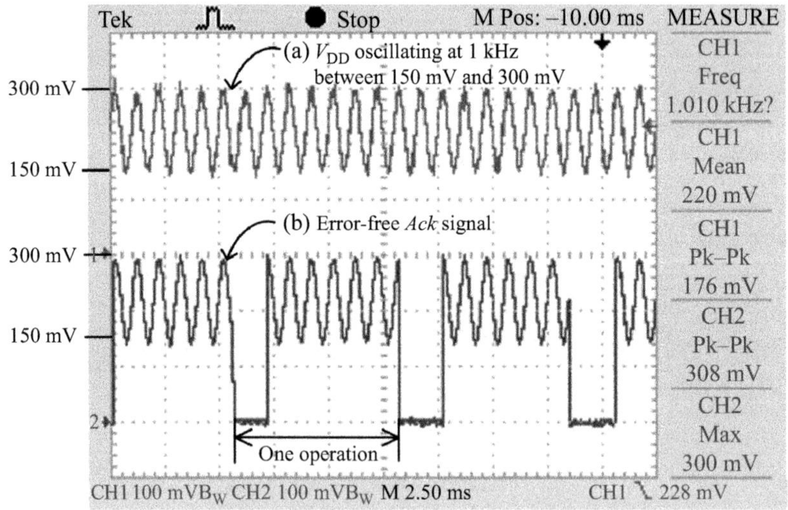

Figure 2.9 (a) High V_{DD} variations @ 1 kHz, 150 mV–300 mV and (b) error-free response (Ack signal) from the proposed asynchronous QDI FRM filter bank

2.2.2 *Pseudo-quasi-delay-insensitive sub-threshold self-adaptive V_{DD} scaling (SSAVS)*

Although QDI asynchronous circuits are highly robust, the costs (i.e., power/energy overheads) associated with QDI are high. In this sub-section, we present an alternative QDI approach, coined "pseudo-QDI" [35], for SSAVS with the objective of reduced power/energy overheads compared to the standardized QDI, yet retaining robustness. The approach comprises a simplified asynchronous 4-phase pipeline structure (see Figure 2.11(b)) with our proposed PCSL dual-rail logic cells delineated earlier. The salient difference between the proposed pseudo-QDI pipeline and a standardized QDI pipeline (henceforth termed "True-QDI") is the removal of the DCD while preserving the LCD. This simplified technique places an additional timing requirement on the reset cycle of the 4-phase asynchronous operation—specifically that certain internal nodes must reset before the next cycle of evaluation commences, in part facilitated by the fast-reset nature of our proposed PCSL cells. We show that this timing requirement can be easily satisfied, thereby ensuring robust operation even under severe PVT variations in sub-V_t region (see measurement results on prototype ICs in Section 2.2.2.3).

2.2.2.1 Asynchronous pseudo-QDI realization approach

Consider first the design of a true-QDI pipeline embodying our proposed PCSL cells that provide for sub-V_t operation. To preserve its delay-insensitivity attribute (save the fundamental isochronic fork assumption [13]), the QDI pipeline needs to address the issues of "input completeness" [36] (where all inputs need to be

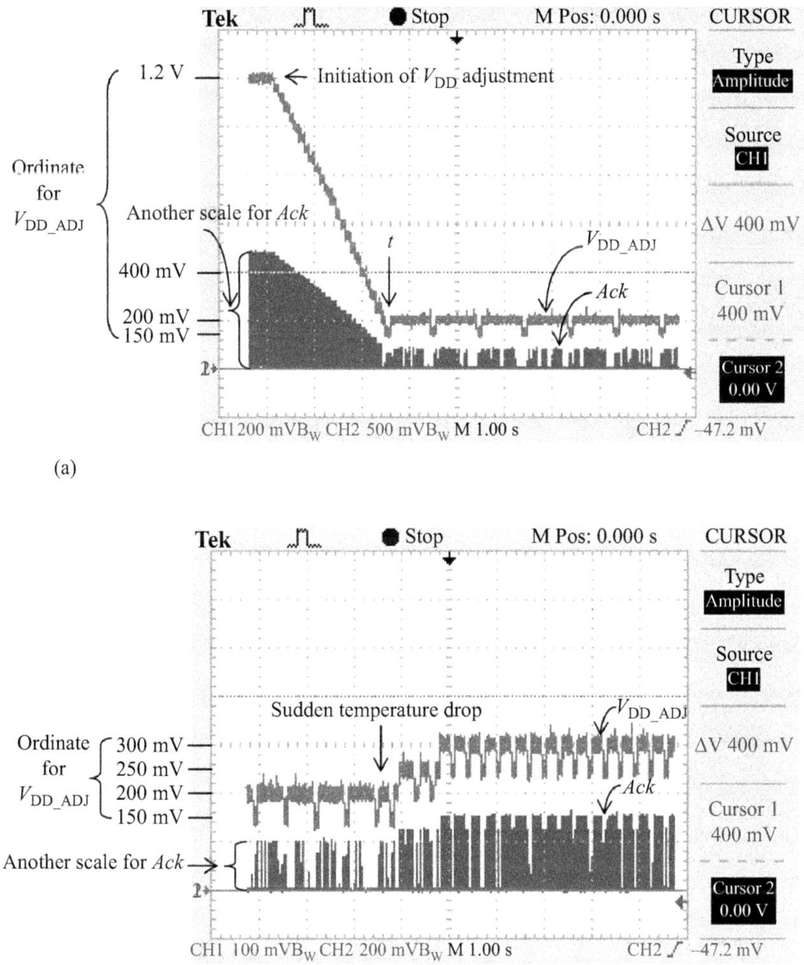

Figure 2.10 Example of the captured waveforms depicting (a) self-adjustment of V_{DD_ADJ} and Ack from the asynchronous QDI FRM filter bank and (b) self-adjustment of V_{DD_ADJ} and Ack under sudden temperature drop

acknowledged before a new pipeline operation commences) and "gate orphan" [36] (where an internal gate is enabled to switch its output but the switching is masked from the observable outputs of the entire circuit). To address these two issues, either the NCL-X pipeline structure [33] or the NCL-D pipeline structure [37] may be used. We adopt the former because it occupies a much smaller area due to its relatively simple realization [33] of datapaths where a functional circuit can first be synthesized (using a (single-rail) standard synthesis tool), followed by a single-rail to dual-rail conversion.

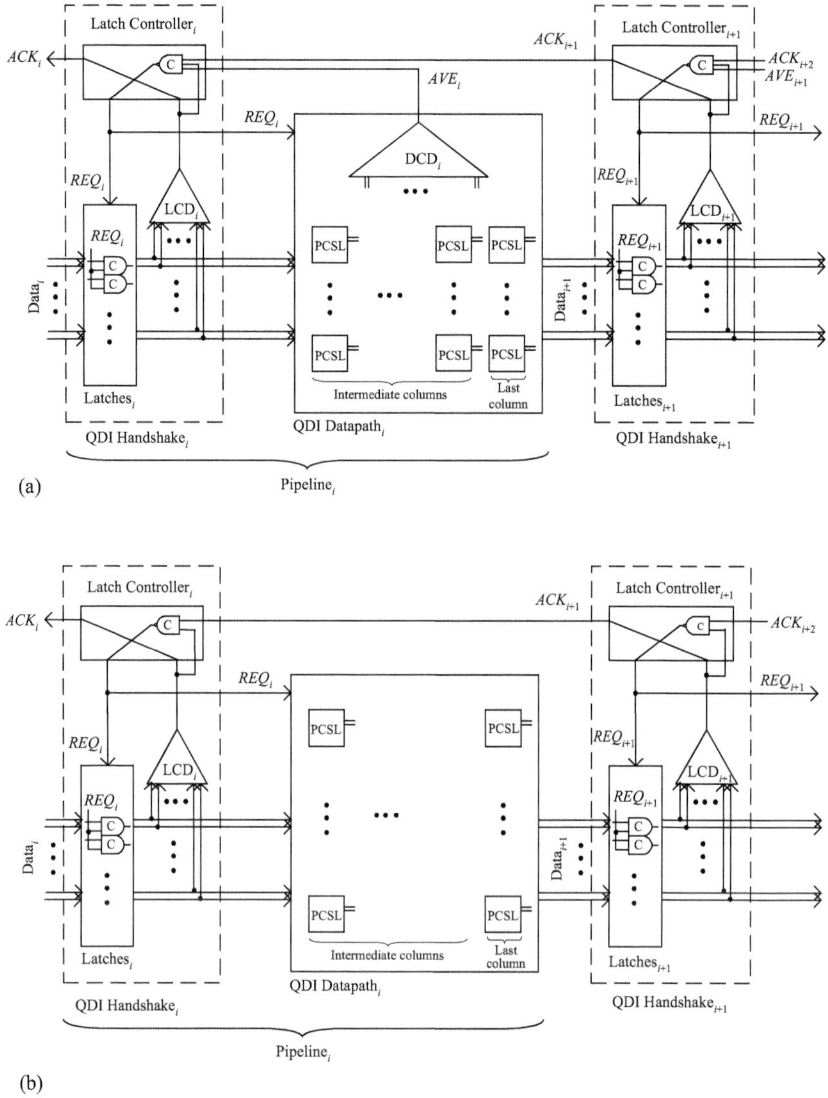

Figure 2.11 (a) The conventional asynchronous true-QDI pipeline and (b) our proposed asynchronous pseudo-QDI pipeline embodying the PCSL cells

Figure 2.11(a) depicts the adopted asynchronous true-QDI pipeline stage (ith stage) comprising a QDI Handshake$_i$ (consisting of a Latch Controller$_i$, a Latches$_i$ and a latch completion detection (LCD$_i$)) and an asynchronous QDI Datapath$_i$; a QDI Handshake$_{i+1}$ is also shown for ease of illustration. The QDI Handshake$_i$ controls the asynchronous QDI Datapath$_i$ according to a sequence of predefined handshake signals. Initially, $ACK_{i+1} = 0$ and $REQ_i = 1$, indicating that (dual-rail) Latches_i are transparent and are waiting for valid Data_i. When Data_i is all valid, LCD_i will check the data and acknowledge Latches_{i-1} (not shown) of the preceding pipeline. $ACK_i = 1$ also acknowledges Latch Controller_i to ensure the input completeness of the pipeline. The valid Data_i will trigger QDI Datapath_i for computation. Once the output (Data_{i+1}) is valid and is stored in Latches_{i+1} (if $REQ_{i+1} = 1$), LCD_{i+1} will acknowledge Latch Controller_i. To address the gate orphan issues (if any), all outputs of the dual-rail PCSL circuits in the intermediate columns have to be checked by a datapath completion detection (DCD_i in Figure 2.11(a)) before the intermediate detection signal, AVE_i, can be asserted. Latch Controller_i will thereafter de-assert REQ_i to reset the PCSL circuits, and both AVE_i and ACK_{i+1} will likewise reset to "0." Once Data_i becomes empty, ACK_i is de-asserted to "0," and LCD_i will revert REQ_i back to its initial condition (REQ_i = "1"), awaiting Data_i to be valid again. This asynchronous pipeline (with DCD_i) fully satisfies the QDI protocol, hence "true-QDI" as described earlier.

It is well-established [33] that the area and energy overheads of DCD_i are large especially if the complexity of the functional circuits in QDI Datapath_i is high. Nevertheless, the delay overhead of DCD_i is largely insignificant as DCD_i executes in parallel with the functional circuits (and with QDI Handshake_{i+1}).

To alleviate the area and energy overheads of DCD_i, DCD_i may be removed in the pipeline. We denote this asynchronous modality as "pseudo-QDI" where an implicit timing condition is required to satisfy the QDI signal protocol. As the *REQ* signal is already integrated into our PCSL circuits, they immediately lead themselves to the pseudo-QDI pipeline depicted in Figure 2.11(b). The pseudo-QDI pipeline operates exactly as its true-QDI counterpart except that the Latch Controller_i no longer waits for the assertion and de-assertion of AVE_i as in the true-QDI pipeline. Note that as long as an implicit timing condition is abided by (see Section 2.2.2.2 below), the robustness of the pseudo-QDI pipeline is not compromised.

2.2.2.2 Timing analysis on the proposed pseudo-QDI realization approach

Consider now the delay properties in the pseudo-QDI pipeline depicted in Figure 2.11(b) by considering two scenarios:

(a) QDI Datapath_i embodying only one level (column) of PCSL circuits—a fine-grain gate-level pipeline where every circuit is pipelined, and
(b) QDI Datapath_i embodying multiple levels (columns) of PCSL circuits—a coarse-grain block-level pipeline where many circuits are collectively grouped to form a pipeline.

For brevity in the analysis, $t_{\uparrow\text{cycle}}$ is denoted as the forward cycle time ($REQ_i{+} \rightarrow REQ_i{-}$ for valid Data_i sent to Pipeline_i until Latches_i is closed), and $t_{\downarrow\text{cycle}}$ the reset cycle time ($REQ_i{-} \rightarrow REQ_i{+}$ for empty Data_i sent to Pipeline_i until Latches_i is reopened for the next operation). The cycle delay $t_{\text{cycle}} = t_{\uparrow\text{cycle}} + t_{\downarrow\text{cycle}}$ is an indication of the speed of the asynchronous pipeline.

For scenario (a), the inputs of the PCSL circuits are checked and acknowledged by LCD_i, and their outputs are subsequently checked and acknowledged by LCD_{i+1} (of the next pipeline stage). In this scenario, the QDI property is preserved, and the pipeline operation is robust.

For scenario (b), an implicit delay assumption arises for the $t_{\downarrow\text{cycle}}$ path when $REQ_i- \rightarrow REQ_i+$; there is no delay assumption for the $t_{\uparrow\text{cycle}}$ path. This implicit delay assumption arises because LCD_{i+1} can only check the primary outputs of QDI Datapath$_i$ at the last column, but not the intermediate output signals of the PCSL circuits (at the intermediate columns) where a "gate orphan" may exist. We formulate the necessary implicit timing condition in (2.1) for error-free operation:

$$\max\left(t_{\downarrow\left(\text{PSCL}_{\text{col}\neq\text{last}}\right)}\right) < t_{\downarrow\text{cycle}}$$
$$< \max\left[\left(t_{\downarrow\text{LCD}_i} + t_{\downarrow\text{Latches}_i}\right), \left(t_{\downarrow(\text{PCSL}_{\text{col}=\text{last}})} + t_{\downarrow\text{Latches}_{i+1}} + t_{\downarrow\text{LCD}_{i+1}}\right)\right] + t_{\downarrow\text{LC}_i} \tag{2.1}$$

where

- $t_{\downarrow\left(\text{PCSL}_{\text{col}\neq\text{last}}\right)}$ is the reset delay for the PCSL circuits at the intermediate columns,
- $t_{\downarrow(\text{PCSL}_{\text{col}=\text{last}})} =$ is the reset delay for the PCSL circuits at the last column,
- $t_{\downarrow\text{Latches}_i}$ is the reset delay for Latches$_i$,
- $t_{\downarrow\text{Latches}_{i+1}}$ is the reset delay for Latches$_{i+1}$,
- $t_{\downarrow\text{LCD}_i}$ is the reset delays for LCD_i,
- $t_{\downarrow\text{LCD}_{i+1}}$ is the reset delays for LCD_{i+1} and
- $t_{\downarrow\text{LC}_i}$ is the reset delay for Latch Controller$_i$.

From the viewpoint of the pipeline schematic, ideally $t_{\downarrow(PCSL_{\text{col}\neq\text{last}})} \cong t_{\downarrow(PCSL_{\text{col}=\text{last}})}$ when REQ_i switches from "1" to "0" for the reset phase, where all PCSL circuits are simultaneously reset. In general, this implicit timing assumption is easily satisfied—specifically, as long as the ratio of $t_{\downarrow\text{cycle}}/t_{\downarrow(PCSL_{\text{col}\neq\text{last}})} > 1$ under all possible PVT variations, the pseudo-QDI pipeline remains robust.

2.2.2.3 Circuit realization and measurement results

On the basis of the true-QDI and our proposed pseudo-QDI approaches, we designed and monolithically realized two asynchronous quad-channel FRM filter banks (@130 nm CMOS). The true-QDI filter bank was the same embodied in the SSAVS system delineated in Section 2.2.1. On the basis of measurements from the prototype ICs, both filter banks were fully functional for $V_{\text{DD}} > 130$ mV. Further, as shown in Figure 2.12(a) both filter banks were also fully functional for extreme V_{DD} variations, and fully functional for wide temperature variations (not shown)—thereby depicting their robustness under severe sub-V_{t} PVT variations.

Figure 2.12(b) benchmarks the measured E_{per} of the two asynchronous filter banks in sub-V_{t}, depicting the ~40% lower E_{per} advantage of the proposed pseudo-QDI

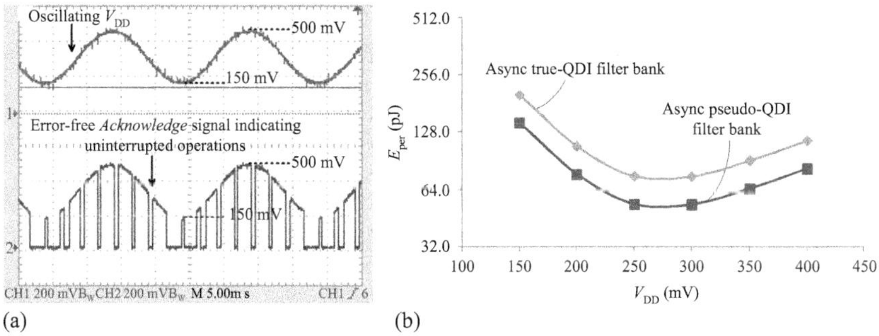

Figure 2.12 *(a) Robust sub-V_t operation of the fabricated pseudo-QDI filter bank under large V_{DD} variations and (b) measured energy/operation (E_{per}) of the asynchronous filter banks*

filter bank over its true-QDI counterpart. The proposed pseudo-QDI filter bank further features ~1.34 × smaller IC area advantage over its true-QDI counterpart.

In summary, we have described our proposed alternative QDI—the pseudo-QDI approach—for simultaneous lower E_{per} and smaller IC area than the standardized true-QDI, yet robust in sub-V_{t} (appropriate for SSAVS) and under extreme PVT variations.

2.3 Gate-level asynchronous circuits

In this sub-section, we present the SAHB—a new QDI cell design approach for gate-level asynchronous circuits—with emphases on high operational robustness, high speed, and low energy dissipation. We further describe an asynchronous QDI pipeline adder example embodying SAHB for full-range DVS operation.

2.3.1 Sense-amplifier half buffer (SAHB)

Figure 2.13 depicts the generic interface signals for the proposed dual-rail SAHB cell template. The data inputs are $Data_{in}$ and $nData_{in}$ and the data outputs are *Q.T*/*Q.F* and *nQ.T*/*nQ.F*. The left-channel handshake outputs are L_{ack} and nL_{ack}, and the right-channel handshake inputs are R_{ack} and nR_{ack}. $nData_{in}$, *nQ.T*, *nQ.F*, nL_{ack} and nR_{ack} are logical-complementary signals to the primary input/output signals of $Data_{in}$, *Q.T*, *Q.F*, L_{ack} and R_{ack}, respectively. For the sake of brevity, we will only use the primary input/output signals to delineate the operations of an SAHB cell. The SAHB cell strictly abides by the asynchronous 4-phase (4ϕ) handshake protocol—having two alternate operation sequences, evaluation and reset. Initially, L_{ack} and R_{ack} are reset to “0” and both $Data_{in}$ and *Q.T*/*Q.F* are empty, that is both of the rails in each signal are “0.” During the evaluation sequence, when $Data_{in}$ is valid (i.e., one of the rails in each signal is “1”) and R_{ack} is “0,” *Q.T*/*Q.F* is evaluated and latched, and L_{ack} is asserted to “1” to indicate the validity of the output. During the reset sequence, when $Data_{in}$ is empty and R_{ack} is “1,” *Q.T*/*Q.F* will then

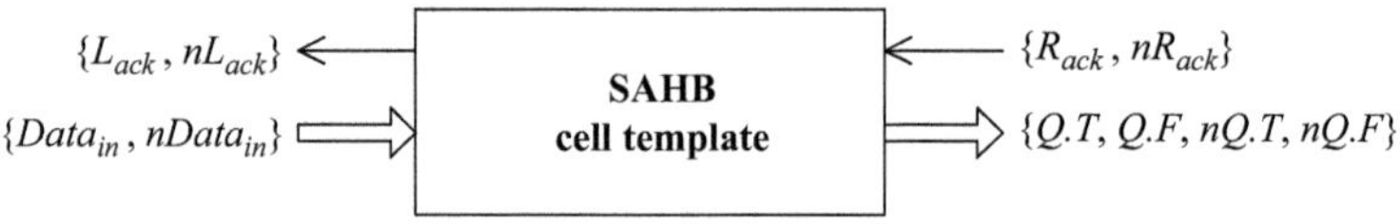

Figure 2.13 SAHB cell template

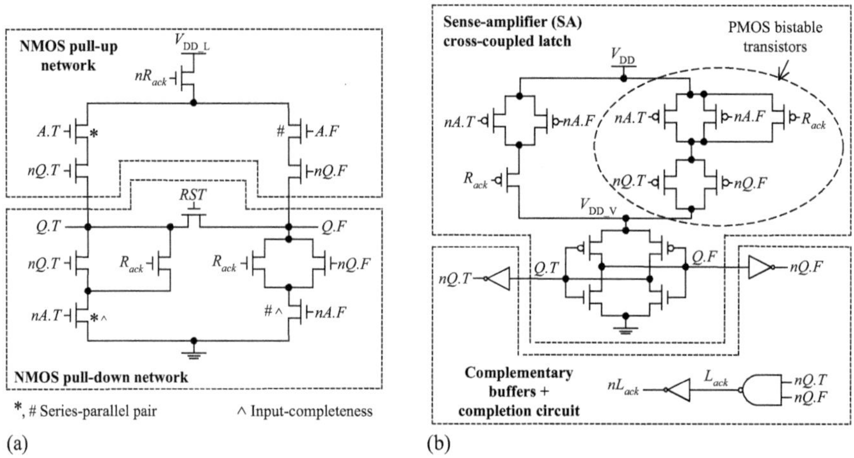

Figure 2.14 Circuit schematic of a buffer cell embodying SAHB: (a) evaluation block powered by V_{DD_L} and (b) SA block powered by V_{DD}

be empty and L_{ack} is de-asserted to "0." Subsequently, the SAHB cell is ready for next operation.

An SAHB cell comprises two building blocks—an evaluation block powered by V_{DD_L} and a sense-amplifier (SA) block powered by V_{DD}. Figures 2.14(a) and (b) depict the respective circuit schematic of an evaluation block and an SA block of a buffer cell embodying SAHB; the various sub-blocks are shown within the dotted blocks. The power lines V_{DD_L} and V_{DD} can be the same or different voltages [38]. The NMOS transistor with *RST* is optional for cell initialization.

In Figure 2.14(a), the evaluation block comprises an NMOS pull-up network and an NMOS pull-down network to respectively evaluate and reset the dual-rail output *Q.T*/*Q.F*. Of particular interest, the NMOS pull-up network features low parasitic capacitance (lower than the usual PMOS pull-up network whose transistor sizing is often 2× larger than that of the NMOS). From a structural view, the ensuing signals function as follows. Consider first the NMOS pull-up network where *Q.T*/*Q.F* is evaluated based on the data input (i.e., *A.T*/*A.F*) and nR_{ack} serves as an evaluation flow-control signal. The NMOS pull-up network realizes the buffer logic function. To reduce the short-circuit current, *nQ.T*/*nQ.F* will disconnect the evaluation function when *Q.T*/*Q.F* is evaluated.

Consider now the NMOS pull-down network where *Q.T*/*Q.F* is reset depending on the data input. For a single-input buffer cell depicted in Figure 2.14, the transistor configuration of *A.T*/*A.F* in the pull-up network is a series-parallel topology to the transistor configuration of *nA.T*/*nA.F* in the pull-down network. For examples of 2-input and 3-input cells, Figure 2.15 depicts 2-input AND/NAND, 2-input XOR/XNOR and 3-input AO/AOI cells. Their series-parallel pairs are marked with * and # for the *Q.T* path and *Q.F* path, respectively. In Figures 2.14 and 2.15, R_{ack} serves as the reset flow-control signal, connecting in series the data input (transistors marked with ^) for input-completeness [23].

In Figure 2.14(b), the SA block comprises an SA cross-coupled latch, complementary buffers and a completion circuit. The SA cross-coupled latch amplifies and latches *Q.T*/*Q.F*. The complementary buffers and completion circuit generate the complementary output signals (*nQ.T*/*nQ.F*) and the left-channel handshake signals (L_{ack}/nL_{ack}), respectively. From a structural view, the cross-coupled inverters serve as an amplifier where in the reset phase, both *Q.T* and *Q.F* are "0" and V_{DD_V} is floating. During the evaluation phase, *Q.T* and *Q.F* will develop a small voltage difference and when V_{DD_V} is connected to V_{DD}, the cross-coupled inverters will amplify (in a positive feedback mechanism) the voltage difference between *Q.T* and *Q.F*. To realize the input-completeness feature, the top left branch in the SA cross-coupled latch detects if all inputs (i.e., *nA.T* and *nA.F*) are ready and $R_{ack} = 0$. For bistable operation, the top right branch (within the dotted oblong circle) in the SA cross-coupled latch holds the output until all inputs are empty and $R_{ack} = 1$.

Initially, *A.T*, *A.F*, R_{ack} and L_{ack} are "0," and *nA.T*, *nA.F*, nR_{ack} and nL_{ack} are "1." During the evaluation phase, for example when *A.F* = "1" (*nA.F* = "0"), the voltage at node *Q.F* is partially charged to V_{DD_L} by the NMOS pull-up network in the evaluation block, and *Q.T* remains as "0" (via the NMOS pull-down network). As the input is now valid, the SA cross-coupled latch is turned on by connecting the virtual supply V_{DD_V} to V_{DD}, and amplifies *Q.F* to "1." *Q.F* is thereafter latched (together with the PMOS bistable transistors and the cross-coupled inverters) and *nQ.F* becomes "0" (to disconnect the node *Q.F* from the V_{DD_L} in the evaluation block to prevent any short-circuit current). L_{ack} is asserted to "1" (nL_{ack} = "0") to indicate the validity of the dual-rail output. During the reset phase, the input is empty (*nA.T* and *nA.F* are "1") and R_{ack} = "1," the dual-rail output becomes empty, and L_{ack} is de-asserted to "0." At this juncture, the SA block is ready for a new operation. Note that both the evaluation block and SA block are tightly coupled to reduce the number of switching nodes, thereby enhancing the speed and reducing the power dissipation. Furthermore, as both the evaluation and SA blocks operate in static-logic style, their transistor sizings are not critical.

Figures 2.15(a)–(c) depict the circuit schematic of three basic SAHB library cells: 2-input AND/NAND, 2-input XOR/XNOR and 3-input AOI/AOI cells. Similar to the buffer cell, the structure of the evaluation block and SA block of these cells are constructed based on their logic functions and input signals. These library cells will be used for benchmarking and for realizing the 64-bit SAHB pipeline adder.

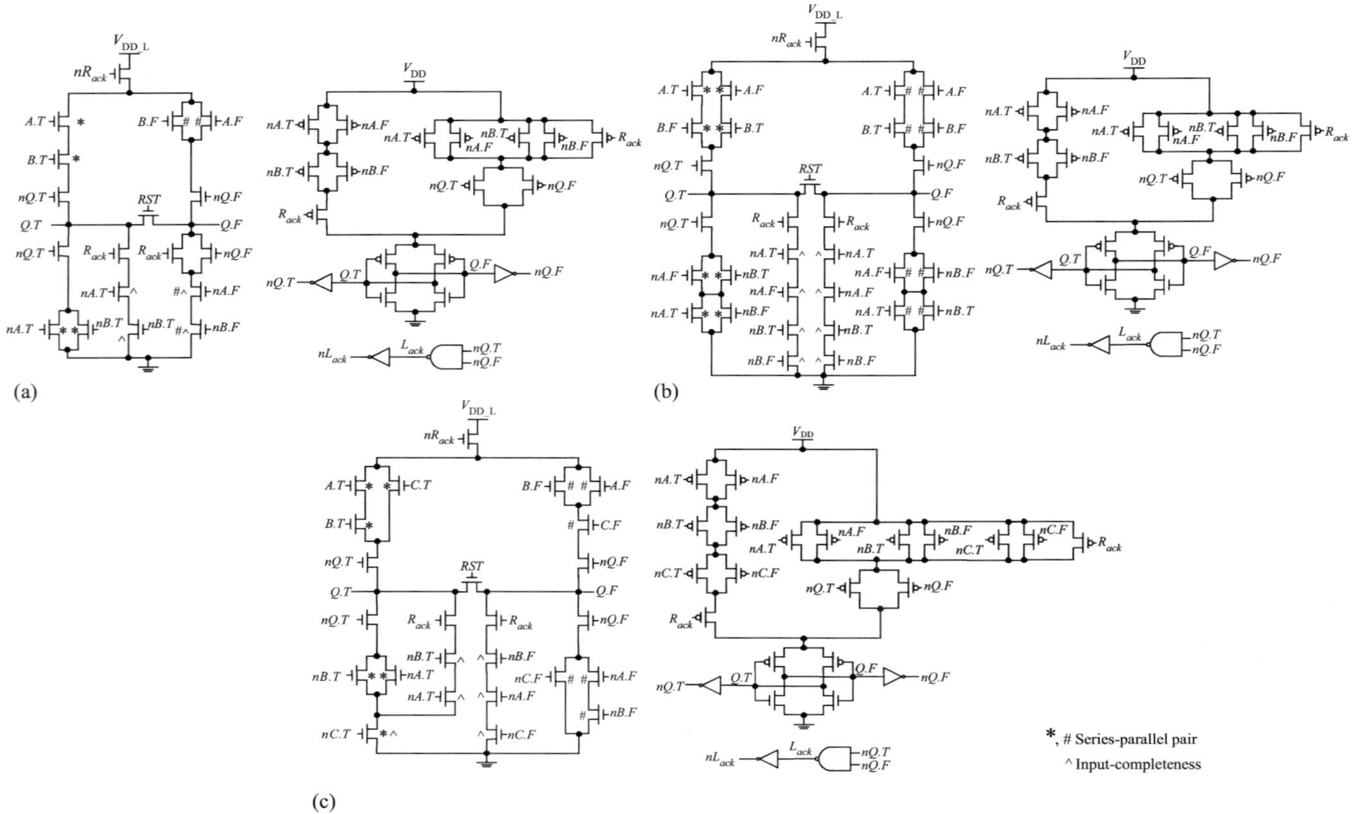

Figure 2.15 Dual-rail SAHB library cells: (a) 2-input AND/NAND, (b) 2-input XOR/XNOR, and (c) 3-input AO/AOI

2.3.2 Design example: Kogge–Stone (KS) 64-bit adder embodying SAHB

We now present the evaluation on a 64-bit Kogge–Stone (KS) pipeline adder embodying the SAHB cell design approach. We further perform the DVS operation on the adder and measure the results at various V_{DD}.

The KS SAHB adder IC was fabricated using ST microelectronics' (STM) 65 nm CMOS general purpose standard threshold voltage (GP-SVT) process whose $V_{tn} = 0.35$, $V_{tp} = -0.35$ V @$V_{DD} = 1$ V. Figures 2.16(a) and (b) respectively depict the microphotograph and the layout of the SAHB adder with the test structure. The core area of the KS SAHB adder is 306 μm × 209 μm.

All 20 KS SAHB adder prototype ICs were measured and were found to be fully functional. Of these 20 ICs, 5 and 15 ICs are functional for $V_{DD} \geq 0.25$ V and $V_{DD} \geq 0.3$ V, respectively. It is interesting to note that our design at subthreshold voltage features higher speed operation compared to some reported subthreshold designs. For example, based on the same 65 nm CMOS process, a recently reported 32-bit subthreshold KS adder [39] operates at 3 MHz @$V_{DD} = 300$ mV whereas our SAHB KS 64-bit adder design operates at higher speed of 3.76 MHz for the same V_{DD}.

Figure 2.17(a) depicts the V_{DD} (0.25 V) and output time-domain waveform for one of said five KS SAHB ICs. As these ICs were fully functional for V_{DD} ranging from sub-V_t voltage (0.3 V) → near-V_t voltage → nominal voltage (1.0 V), our SAHB approach is applicable for full-range DVS [1]. By comparison, the reported QDI and TP designs (PCHB, PS0, etc.) would likely be more applicable only to half-range DVS (i.e., near-V_t voltage → nominal voltage). This is because these reported designs adopt dynamic-logic style, where the cross-coupled inverters (in the integrated-latch) are not functionally robust in the sub-V_t voltage regime.

Consider now the operational robustness of our SAHB adder against V_{DD} variation for an in situ self-adaptive V_{DD} system [19] where V_{DD} is automatically adjusted such that the minimum V_{DD} voltage is applied—the intention is the lowest power operation for the given prevailing condition. The top and bottom traces of

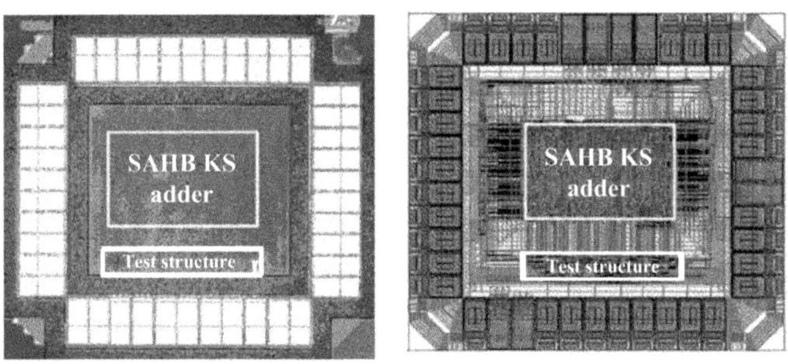

Figure 2.16 The 64-bit SAHB KS pipeline adder: (a) microphotograph and (b) layout view

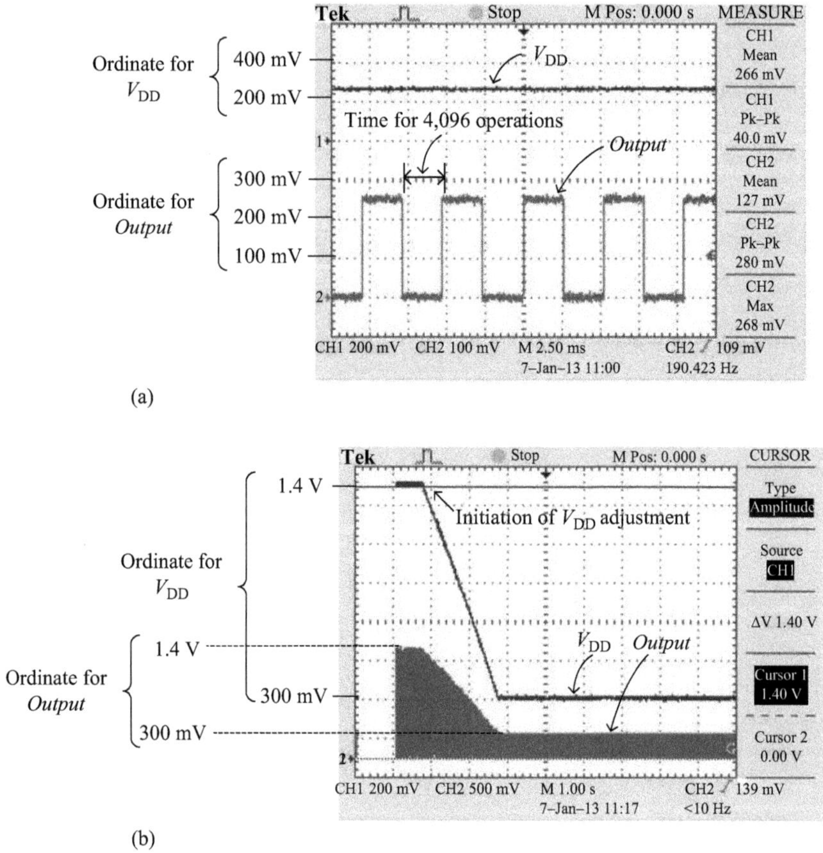

Figure 2.17 Signal waveforms of the SAHB adder operations: (a) sub-threshold (V_{DD} ~ 0.25 V) and (b) FDVS (V_{DD} from 1.4 V to 0.3 V)

Figure 2.17(b), respectively, depict the real-time varying V_{DD} (from 1.4 V to 0.3 V) and the generated output. It can be seen that even when V_{DD} is varied widely, the operation is uninterrupted and error-free. On this basis, circuits embodying our SAHB cell design approach are advantageous for power/speed trade-off through voltage scaling with low transition/recovery time [38,40].

2.4 Conclusions

We have presented the appropriateness of the QDI (and pseudo-QDI) asynchronous-logic design approach to realize circuits and systems suitable for full-range DVS (from the nominal voltage ↔ near-V_t voltage ↔ sub-V_t voltage regions). Both block-level and gate-level pipeline structures have been presented.

Using the block-level pipeline structure, we have presented an SSAVS system embodying block-level QDI asynchronous pipelines for a WSN with the objective of lowest possible power operation for the prevailing throughput and circuit conditions—V_{DD} adjusted to within 50 mV of the minimum voltage, yet high operational robustness with minimal overheads. High robustness has been achieved by adopting the asynchronous QDI protocols, and the embodiment of our proposed PCSL. A reduced-overhead design has further been shown by adopting the asynchronous pseudo-QDI protocols, and the embodiment of PCSL. Using the gate-level pipeline structure, we have presented our proposed SAHB cell design approach and evaluated an asynchronous QDI KS pipeline adder embodying SAHB for full-range DVS operation.

In summary, we show that QDI (and pseudo-QDI) asynchronous-logic, coupled with either PCSL or SAHB cell design approaches, provides a low-cost high-reliability solution for circuits and systems exclusively designed for error-free DVS.

References

[1] Chang J. S., Gwee B.-H., Chong K.-S. *Asynchronous-logic circuit for full dynamic voltage control*. US Patent US8791717B2, July 2014

[2] Ma W.-H., Kao J. C., Sathe V. S., Papaefthymion M. C. "187MHz subthreshold-supply charge recovery FIR." *IEEE J. Solid-State Circuits*, 2010, vol. 45(4), pp. 793–803

[3] Chang K.-L., Gwee B.-H., Chang J. S., Chong K.-S. "Synchronous-logic and asynchronous-logic 8051 microcontroller cores for realizing internet of things: a comparative study on dynamic voltage scaling and variation effects." *IEEE J. Emerg. Sel. Top. Circuits Syst.*, 2013, vol. 3(1), pp. 23–34

[4] Zhai B., Hanson S., Blaauw D., Sylvester D. "A variation-tolerant sub-200mV 6-T subthreshold SRAM." *IEEE J. Solid-State Circuits*, 2008, vol. 43(10), pp. 2338–2348

[5] Tajalli A., Alioto M., Leblebici Y. "Improving power-delay performance of ultra-low power subthreshold SCL circuits." *IEEE Trans. Circuits Syst. II*, 2009, vol. 56(2), pp. 127–131

[6] Jayakumar N., Khatri S. P. "A variation-tolerant sub-threshold design approach." *Proceedings of the 42nd Design Automation Conference (DAC)*, Anaheim, CA, USA, June 2005, pp. 716–719

[7] Hisamoto D., Lee W.-C., Kedzierski J., *et al*. "FinFET-a self-aligned double-gate MOSFET scalable to 20 nm." *IEEE Trans. Electron Devices*, 2000, vol. 47(12), pp. 2320–2325

[8] Wilson W. B., Un-Ku M., Lakshmikumar K. R., Liang D. "A CMOS self-calibrating frequency synthesizer." *IEEE J. Solid-State Circuits*, 2000, vol. 35(10), pp. 1437–1444

[9] Chang S.-C., Hsieh C.-T., Wu K.-C. "Re-synthesis for delay variation tolerance." *Proceedings of the 41st Design Automation Conf. (DAC)*, California, USA, June 2004, pp. 814–819

[10] Chong K.-S., Chang K.-L., Gwee B.-H., Chang J. S. "Synchronous-logic and globally-asynchronous-locally-synchronous (GALS) acoustic digital signal processors." *IEEE J. Solid-State Circuits*, 2012, vol. 47(3), pp. 769–780
[11] Raychowdhury A., Paul B. C., Bhunia S., Roy K. "Computing with subthreshold leakage: device/circuit/architecture co-design for ultralow-power subthreshold operation." *IEEE Trans. VLSI Syst.*, 2005, vol. 13(11), pp. 1213–1224
[12] Sparsø J., Furber S. *Principle of Asynchronous Circuit Design: A System Perspective*. Norwell, MA: Kluwer Academic, 2001
[13] Martin A. J. "The limitations to delay-insensitivity in asynchronous circuits." In *Proceedings of the 6th MIT Conf. on Advanced Research in VLSI*, 1990, pp. 263–278
[14] Beerel P. A., Ozdag R. O., Ferretti M. *A Designer's Guide to Asynchronous VLSI*. Cambridge: Cambridge University Press, 2010
[15] Martin A. J., Nystrom M. "Asynchronous techniques for system-on-chip designs." *Proceedings of the IEEE*, 2006, vol. 96(6), pp. 1104–1115
[16] Zhou R., Chong K.-S., Gwee B.-H. Chang J. S. "A low overhead quasi-delay-insensitive (QDI) asynchronous data path synthesis based on microcell-interleaving genetic algorithm (MIGA)." *IEEE Trans. Comput. Aid. Design Int. Circuits Syst.*, 2014, vol. 33(7), pp. 989–1002
[17] Nowick S. M., Singh M. "Asynchronous design – part 2: systems and methodologies." *IEEE Design Test*, 2015, vol. 32(3), pp. 19–28
[18] Golani P., Beerel P. A. "Area-efficient asynchronous multilevel single-track pipeline template." *IEEE Trans. VLSI Syst.*, 2014, vol. 22(4), pp. 838–849
[19] Lin T., Chong K.-S., Chang J. S., Gwee B.-H. "An ultra-low power asynchronous-logic in-situ self-adaptive V_{DD} system for wireless sensor networks." *IEEE J. Solid-State Circuits*, 2013, vol. 48(2), pp. 573–586
[20] Jorgenson R. D., Sorensen L., Leet D., Hagedom M. S., Lamb D. R., Fridell T. H., *et al.* "Ultra low-power operation in subthreshold regimes applying clockless logic." *Proceedings of the IEEE*, 2010, vol. 98(2), pp. 299–314
[21] Sparsø J., Staunstrup J., Dantzer-Sorensen M., "Design of delay insensitive circuits using multi-ring structures." *Proceedings of the European Design Automation Conference*, 1992, pp. 7–10
[22] Chang J. S., Gwee B.-H., Chong K.-S. *Digital Asynchronous-Logic: Dynamic Voltage Control*, Final Technical Report for DARPA Project, HR0011-09-2-0006, 2010
[23] Chong K.-S., Ho W.-G., Lin T., Gwee B.-H., Chang J. S. "Sense-amplifier half-buffer (SAHB): a low power high-performance asynchronous-logic QDI cell template." *IEEE Trans. VLSI Syst.*, 2017, vol. 25(2), pp. 402–415
[24] Williams T. E., Horowitz M. A. "A zero-overhead self-timed 160-ns 54-b CMOS divider." *IEEE J. Solid-State Circuits*, 1991, vol. 26(11), pp. 1651–1661

[25] Singh M., Nowick S. M. "The design of high-throughput asynchronous dynamic pipelines: lookahead pipelines." *IEEE Trans. VLSI Syst.*, 2007, vol. 15(11), pp. 1256–1269

[26] Liu T.-T., Alarcon L. P., Person M. D., Rabaey J. M. "Asynchronous computing in sense amplified-based pass transistor logic." *IEEE Trans. VLSI Syst.*, 2009, vol. 17(7), pp. 883 892

[27] Nystrom M., Ou E., Martin A. J. "An eight-bit divider implementation in asynchronous pulse logic." *Proceedings of the 10th IEEE International Symposium Asynchronous Circuits Systems (ASYNC)*, Crete, Greece, May 2004, pp. 19–23

[28] Ferretti M., Beerel P. A. "High performance asynchronous design using Single-Track Full-Buffer standard cells." *IEEE J. Solid-State Circuits*, 2006, vol. 41(6), pp. 1444–1454

[29] Martin A. J., Lines A., Manohar R., Nystrom M., Penzes P., Southworth R., *et al.* "The design of an asynchronous MIPS R3000 microprocessor." *Proceedings of the 17th Conf. Advance Research in VLSI*, Ann Arbor, USA, 1997, pp. 164–181

[30] Lim Y. C. "Frequency response masking approach for the synthesis of sharp linear phase digital filters." *IEEE Trans. Circuits Syst.*, 1986, vol. 33(4), pp. 357–364

[31] Rabaey J. *Low Power Design Essentials*. Springer Publishing Company, 2009

[32] Calhoun B. H., Wang A., Chandrakasan A. "Device sizing for minimum energy operation in subthreshold circuits." *Proceedings of the IEEE 2004 Custom Integrated Circuits Conference*, Orlando, USA, October 2004, pp. 95–98

[33] Kondratyev A., Lwin K. "Design of asynchronous circuits using synchronous CAD tools." *IEEE Design Test Comput.*, 2002, vol. 19(4), pp. 107–117

[34] Cortadella J., Kondratyev A., Lavagno L., Sotiriou C. "Coping with the variability of combinational logic delays." *Proceedings of the IEEE International Conference on Computer Design (ICCD)*, San Jose, USA, October 2004, pp. 505–508

[35] Lin T., Chong K.-S., Chang J. S., Gwee B.-H., Shu W. "A robust asynchronous approach for realizing ultra-low power digital self-adaptive V_{DD} scaling system." *Proceedings of the IEEE Sub-threshold Microelectronics Conference (SubVT)*, Waltham, USA, October 2012, pp. 1–3

[36] Smith S. C., Di J. *Designing Asynchronous Circuits using NULL Convention Logic (NCL)*. Morgan & Claypool, 2009

[37] Fant K. M., Bandt S. A. "Null conventional logic: a complete and consistent logic for asynchronous digital circuit synthesis." *Proceedings of the International Conference on Application-Specific Systems, Architectures and Processors (ASAP)*, Chicago, USA, August 1996, pp. 261–273

[38] Chang J. S., Gwee B.-H., Chong K.-S. *Digital cell.* US Patent US8994406B2, March 2015

[39] Fuketa H., Hashimoto M., Mitsuyama Y., Onoye T. "Adaptive performance compensation with in-situ timing error predictive sensors for subthreshold circuits." *IEEE Trans. VLSI Syst.*, 2012, vol. 20(2), pp. 333–343

[40] Ho W.-G., Chong K.-S., Ne K. Z. L., Gwee B.-H., Chang J. S. "Asynchronous-logic QDI quad-rail sense-amplifier half-buffer approach for NoC router design." *IEEE Trans. VLSI Syst.*, 2018, vol. 26(1), pp. 196–200

Chapter 3

Power-performance balancing of asynchronous circuits

Liang Men[1] *and Chien-Wei Lo*[1]

Handshaking protocols eliminate timing violations in the circuit, but the drawback is performance degrading. In the asynchronous design using NULL conventional logic (NCL) [1], each DATA/NULL cycle generates and propagates a feedback signal before the next cycle starts. Depending on the bit width and completion logic, the propagation delay could be twice of its synchronous counterpart [2]. Early complete detection and sleep mechanism are introduced into the multithreshold NCL (MTNCL) logic to reduce propagation delay in the research work [3,4]. This chapter focuses on the architecture design of the asynchronous circuit for power and performance optimization. Asynchronous pipelining features are applied to the parallel platforms for dynamic voltage scaling (DVS) and concurrent data processing. Advanced topics, including fine-grain core states control and heterogeneous architecture, are introduced in the second part of this chapter. In general, those design methodologies serve as reference to the asynchronous circuit application and its tape-out.

3.1 Pipelining the asynchronous design

Pipelining is the common concept in improving system throughput. When it is used in synchronous circuit, registers are inserted to the signal propagation path enabling multiple operations in single clock cycle. Timing, including setup/hold time and clock skew, determines the up boundary of the pipeline speedup. Asynchronous circuit uses handshaking and DATA/NULL cycles for signal processing. It is free of timing violation. However, asynchronous pipeline breaks the handshaking loop and leads to longer propagation delay if it is not carefully designed. In Section 3.1.1, an asynchronous carry-save-adder (CSA)-based multiplier using MTNCL architecture demonstrates how to balance the asynchronous pipeline for best speedup. Another pitfall on asynchronous pipelining is the dependency between multiple paths. The initial state of each path needs to be calibrated for

[1]Computer Science and Computer Engineering, University of Arkansas, Fayetteville, AR, USA

DATA cycle matching. A finite impulse response (FIR) example is shown in Section 3.1.2.

3.1.1 Pipeline balancing

Throughput is bounded by the worst propagation delay among the pipeline stages, so balancing the pipeline is critical for the best performance. In the synchronous design, the propagation delay is the forward path from one register to the next. While the propagation delay in the equivalent asynchronous circuit could be doubled because the asynchronous circuit not only has the forward path for signal processing but also the backward path for complete detection. Specifically, in the MTNCL circuit where sleep signal is used to generate the NULL cycle, the sleep buffer is a considerable contributor to the delay. Take the CSA in MTNCL as an example, Figure 3.1 shows the structure before pipelining, the DATA-to-DATA cycle T_{dd} is (3.1):

$$T_{dd} = T_{\text{comp\_datain}} + 2 \times T_{\text{sleep\_buf}} + T_{\text{comb}} + T_{\text{comp\_dataout}} \tag{3.1}$$

And the estimated pipeline throughput is shown in (3.2):

$$\text{Throughput} = \frac{1}{T_{dd}} \tag{3.2}$$

The throughput of the generic multiplier can be improved by adopting more pipeline stages. For the Boolean design, inserting registers in the critical path to divide the propagation delay evenly doubles the throughput. The same strategy is applied to the MTNCL architecture as shown in Figure 3.2. From (3.1), the T_{dd} of the MTNCL pipeline is not only determined by the delay of the combination logic. For the two pipeline stages in Figure 3.2, $T_{\text{comp\_datain}}$, $T_{\text{comp\_dataout}}$ and T_{comb} are the same. But the combination logic in stage 1 is much larger than the combination logic in stage 2. After buffering the sleep signal, $T_{\text{sleep\_buf1}}$ will be larger than $T_{\text{sleep\_buf2}}$. Since the circuit throughput is constrained by the maximum T_{dd} in the pipeline stages, the throughput of the two pipelined architecture is deteriorated as

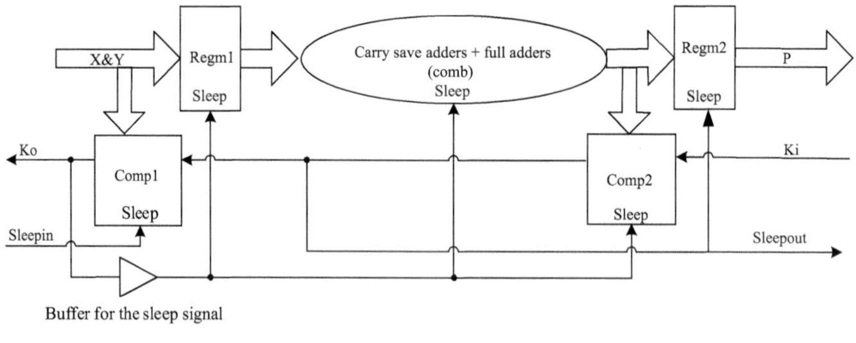

Figure 3.1 Nonpipelined carry save multiplier in MTNCL

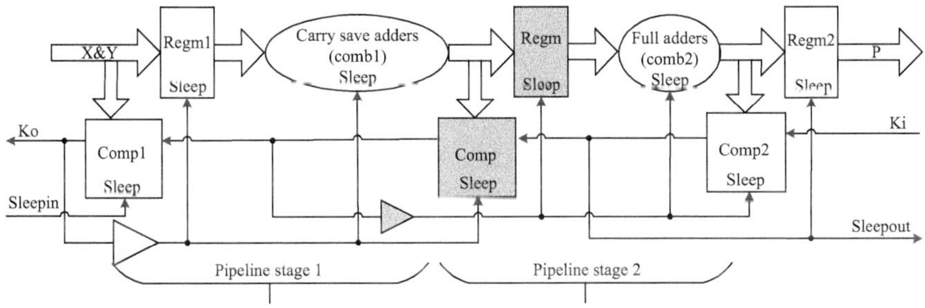

Figure 3.2 Pipelined carry save multiplier in MTNCL

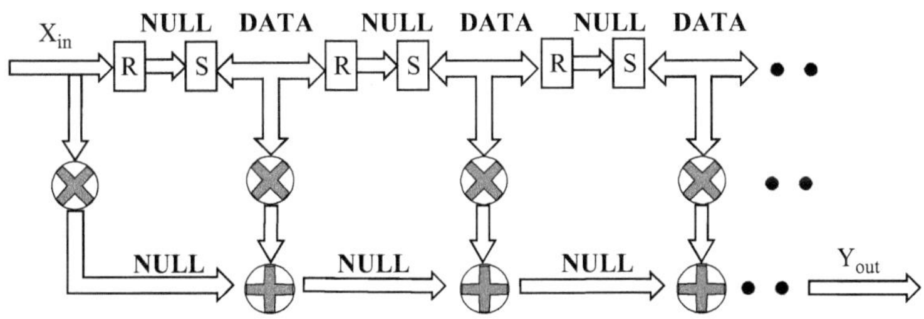

Figure 3.3 Initial states of the MTNCL FIR filter

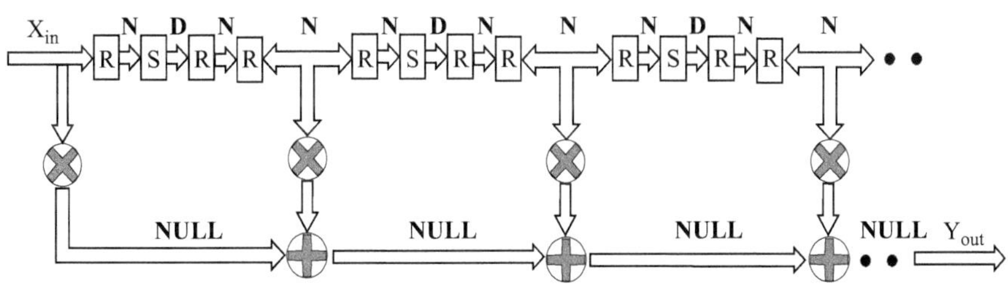

Figure 3.4 Throughput optimization of the MTNCL FIR filter

the number of input bits scale up. When partitioning the asynchronous design, not only the propagation delay need to be considered, the bit width and logic size, which determines the delay of the completion detection logic and the sleep buffer, can dominates the total delay and deteriorates pipelined performance.

3.1.2 Pipeline dependency

Pipelining asynchronous circuits with multiple data paths can be more complex. Take the FIR design as an example, all the individual components, including the shifter register, the adders and the multipliers, compose a tap-generic FIR filter with fixed 8-bit input. There are two pipelines in Figure 3.3, the bottom one convolutes the input data and the top one shifts the input data.

For the two pipelines structure, after reset, the data path in the bottom ones are all in "NULL" cycle. While the data path in the top pipeline is reset to "DATA" and "NULL" patterns for it was designed as the pattern delay shift register. The bottom pipeline is considered as "empty" and the top pipeline as already "full" after reset. The DATA can propagate through an "empty" pipeline but need to extrude a DATA to enter a "full" pipeline. When the first data enters the pipelines, it propagates through the bottom pipeline but is blocked at the first register in the top pipeline. After propagation delay of the bottom pipeline, the top pipeline can move forward, and those two pipelines will be able to take in next data. So, the throughput of this architecture is the reciprocal of the latency, rather than the maximum T_{dd} of the pipeline stages.

Although all the adders and multipliers in the FIR are optimized for throughput, the performance of the FIR is not improved because of the DATA congestion at the top pipeline. To improve the throughput caused by the latency of the circuit, multiple pipelined stages with NULL cycle initialization are implemented in the top pipeline, as shown in Figure 3.4. After reset, the top pipeline has the same number of "NULL" cycles as the bottom one, thus the DATA in the top pipeline can move forward after internal data comes in.

3.2 The parallel architecture and its control scheme

The speed of the asynchronous circuit can be further increased with parallelism. A parallel architecture is designed by putting multiple homogenous cores together in Figure 3.5, where the input data are processed in round-robin order. When the first data come, it is processed by the first core; the second data goes to the second core, and then the third and fourth, respectively. When the fifth data come, it waits until the first core is free. The maximum speedup of the platform is four times of the single core. Besides the computing units, the peripheral circuit includes demultiplexer and input sequence generator are designed to dispatch input data as well as the multiplexer and output sequence generator to guarantee the proper data exiting the platform. Besides the benefit of performance improvement, the unique features of the asynchronous pipeline can balance the power and performance of the platform with DVS.

3.2.1 DVS for the homogeneous platform

Dynamic voltage scaling has great potential to improve the energy efficiency of the multicore asynchronous platform when the data input rate is low. As the self-timed circuit can tolerate wider supply voltage drop than synchronous one, the benefit of dynamic voltage scaling is significant in asynchronous domain. The homogeneous platform is divided into two voltage domains. The peripheral circuit, including demultiplexer, multiplexer and input/output sequence generators, is working at maximum voltage supply; so, the input data can be dispatched to the internal cores

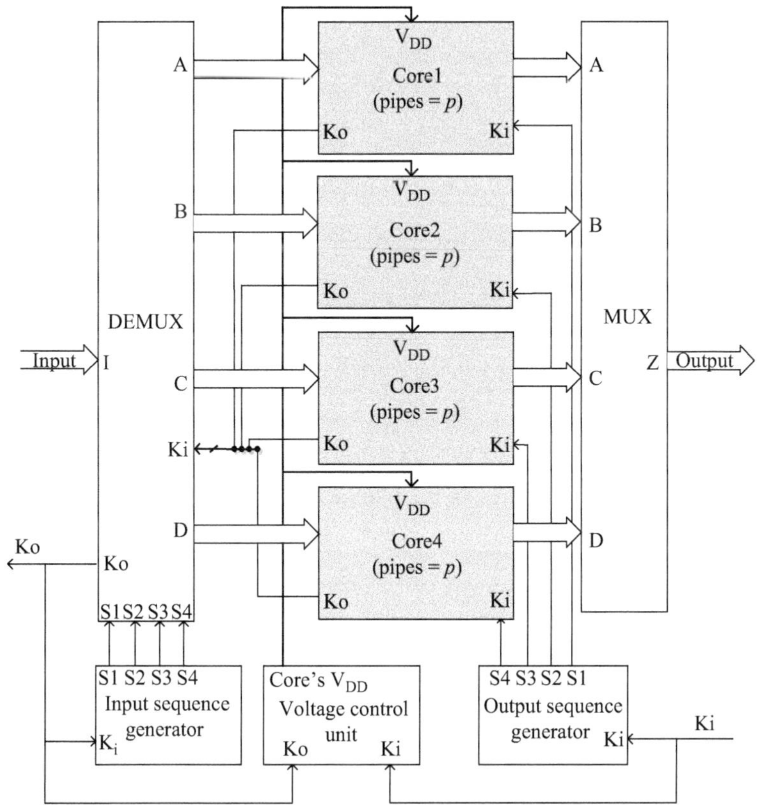

Figure 3.5 Homogeneous platform with four cores and voltage control unit

at maximum speed. The core voltage is the second domain, which is tuned up and down according to data input rate. When the data input rate is high, the cores work at the maximum voltage supply for best performance. On the other hand, the second domain voltage drops, and performance is traded off for energy efficiency.

The voltage control unit (VCU) in Figure 3.6 implements dynamic voltage scaling on the platform. The basic function of the VCU is detecting the input data rate variation and quantizing the variation into reference in a range of minimum and maximum supply voltage. The latency of the MTNCL pipeline is used to design detection circuit. With various scenarios of input data variation, the prediction circuit is designed to make the VCU efficient in more complex situation.

3.2.2 Pipeline latency and throughput detection

The latency of a pipelined circuit is the delay from the first input data to the first output data. In a Boolean pipelined architecture, the latency of the circuit depends on the clock period and number of pipeline stage. And the clock frequency is related to the register timing, combination delay and clock skew. So, the

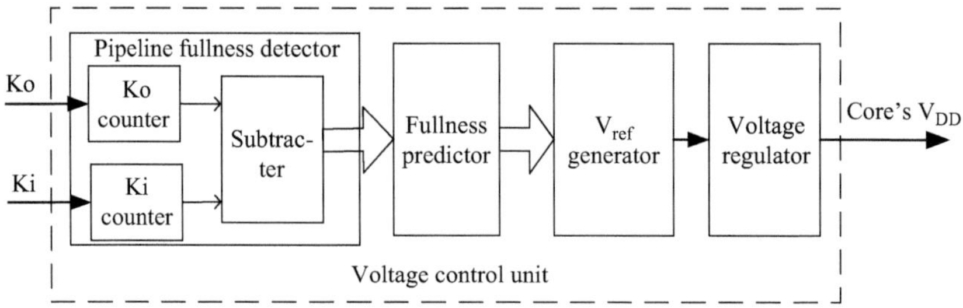

Figure 3.6 Internal structure of the voltage control unit

synchronous logic has the worst-case performance in terms of latency. The latency of the synchronous pipeline cannot be used for data input quantization because they both depend on clock frequency. Extra circuit such as first-in-first-out (FIFO) buffer has to be implemented when applying DVS on synchronous pipeline. In the asynchronous world, each DATA cycle propagates through the register, the combination block and the completion detection block at the initialized NULL stages. The latency of the MTNCL pipeline is the propagation delay from the input port to the output port. It is independent of the input data rate and can serve as a natural quantize for performance.

Inside the voltage controller, the latency of the asynchronous pipeline serves as a timing period to quantize the input data rate. In the latency of the MTNCL pipeline, if the data input rate is high, the DATA/NULL patterns could fill the whole pipeline. If the input data rate is low, each data could propagate through all the NULL cycles to arrive the output port before the next data enters the pipeline. The *Ko* signal at the input side indicates the data entering the pipeline and the *Ki* signal at the output side indicates the data exiting the pipeline. The pipeline fullness detector, as shown in the detection block of Figure 3.6, is used to accumulate the *Ko*'s rising edge and subtract the *Ki*'s rising edge. The value of the fullness detector indicates the number of data inside the pipeline during the latency time. Assuming that there is no delay between the *Ki* signal toggling and the DATA or NULL transition at the output port, the pipeline fullness is used as the quantization of the input data rate.

3.2.3 Pipeline fullness and voltage mapping

The voltage control loop is based on the information from the fullness detector. Voltage will be tuned high if more data present in the pipeline otherwise it lowers down. As an example, the homogenous platform is instantiated with 4 FIR cores, each with 8 taps as the computing units. The fullness of the platform is observed with the core's V_{DD} fixed to various voltage supplies and maximum workload. When the supply voltage is high, the processing core works fast, and pipeline fullness stays low. With maximum workload for the observation, the pipeline accumulates maximum number of data at the minimum operating voltage. Table 3.1 shows the pipeline fullness variation with the supply voltage in an adjustable range with IBM

Table 3.1 Pipeline fullness observation

Core's V_{DD}	**0.6 V**	**0.7 V**	**0.8 V**	**0.9 V**	**1.0 V**	**1.1 V**	**1.2 V**
Fullness	12	10	9	8	7	6	5

130 nm process. A linear characteristic is used to construct a voltage divider network, with maximum fullness in the platform pipeline converted to 1.2 V and minimum fullness mapping to 0.6 V.

3.2.4 Workload prediction

As the MTNCL circuit is delay insensitive, the platform is able to tolerate the delay overhead caused by adjusting V_{DD}, without losing data or malfunctioning. For certain applications where input data bursts are common, the throughput adjusting may lag behind the input variations and degrade the overall performance. Even though a long data buffer could be applied to register all input data, the overhead will be worse in terms of energy consumption. Therefore, a workload predictor is developed to enhance the DVS control mechanism.

As an example, in the homogenous platform implemented with four FIR cores, the pipeline fullness detector has a 4-bit binary output, with an entire state space comprising 16-fold history. However, implementing 16 states in hardware causes high overhead. As the pipeline fullness in the platform is always continuously changing with the handshaking signals, the simplified algorithm is developed to predict the acceleration of the pipeline fullness, as well as tracing the previous history.

In the prediction circuit, the output of pipeline fullness detector, Q, is latched by the external input signal *sleepin*, which is used to generate NULL cycles in MTNCL. The fullness acceleration is reduced to 3 states, which are *Riseup*, *DonotChange*, and *Lowdown*, in one-hot encoding. The acceleration state is predicted in a finite state machine (FSM) and applies to the registered Q for generating the predicted fullness, *PreQ*. In the following DATA cycle, *PreQ* will be evaluated to produce a miss or hit signal, depending on weather *PreQ* and Q is equal or not. The miss or hit signal will update the FSM and predict the subsequent fullness acceleration.

The state switch mechanism imitates the two-way branch predictor [5] utilized to improve the flow in the instruction pipeline. Five states, *SR* (strongly rise-up), *WR* (weakly rise-up), *SL* (strongly low-down), *WL* (weakly low-down), and *DC* (don't care) are encoded in the FSM. In the states of *SR* and *WR*, the prediction result of q' is *Riseup*. In the states of *SL* and *WL*, the prediction result of q' is Lowdown. In the state of DC, the prediction result of q' is *DonotChange*. Between *WR*, *DC* and *WL*, the state's transition also depends on the value of q besides "miss" and "hit", while in other states, previous acceleration is used besides this signal, as illustrated in Figure 3.7.

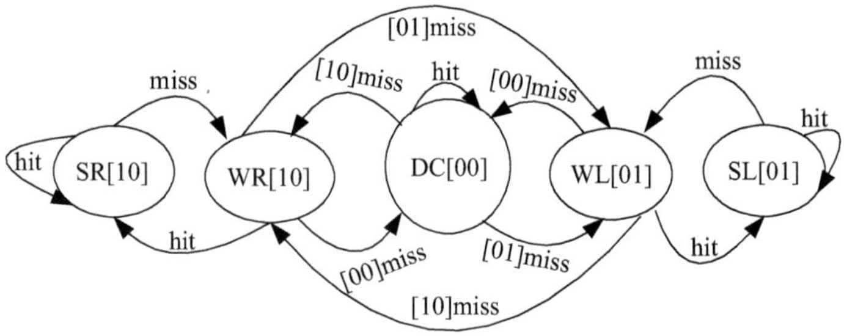

Figure 3.7 State machine for workload prediction

3.2.5 Circuit fabrication and measurement

An 8-tap Boolean and MTNCL FIR filters, as well as the homogeneous platform are taped out in the MITLL 90 nm CMOS FDSOI [6] process run. All the circuit designs are optimized for sub-threshold operation with 300 mV supply voltage. Same I/O logic is used to reduce the number of input/output pads for the physical implementation of the Boolean and MTNCL FIRs. For throughout testing, V_{DD} is fixed at 300 mV and a body-biasing voltage ranging from −1 V to −2 V is applied. The temperature of the test environment is maintained at 25 °C.

The testing result of the Boolean FIR filter and its MTNCL counterpart is shown in Figure 3.8, regarding the energy per data and the performance. The power and energy measurements are taken over a range of operating speed. The results indicate that the Boolean FIR filter operates at a range of speed from 260.5 Hz to 1.303 Hz and the energy per data is from 10.37 nJ to 2,640.8 nJ, at 300 mV V_{DD} and −1.7 V body-biasing voltage. The T_{dd} of the asynchronous FIR filter is ranged from 366.7 Hz to 1.83 Hz with energy per data from 6.3 nJ to 1,352.34 nJ, at 300 mV V_{DD} and −1.55 V body-biasing voltage. Comparing to the results of the Boolean FIR, the MTNCL design has 1.4× higher operating speed and 1.5× lower energy per data on average.

A more complex design based on the homogeneous platform, which consists of four asynchronous FIR filters processing data in parallel, is tested as fully functional with 0.3 V power supply and −1.9 V body-biasing voltage. The energy and performance data is shown in Figure 3.9. Since I/O logic is eliminated from the design, the result is close to the maximum as the input data rate increases. The best result on the FPGA testbench is 49.364 pJ per data with the T_{dd} at 6.02 μs. The energy consumption of the platform rises to 2,784.9 pJ per data when the T_{dd} is 320.1 μs.

3.3 Advanced topics on power-performance balancing

The bottleneck of the parallel processing is the fixed input/output data sequence. If all the internal cores work at the same speed, the platform reaches its maximum speed up; otherwise, the platform performance is bounded up by the slowest core.

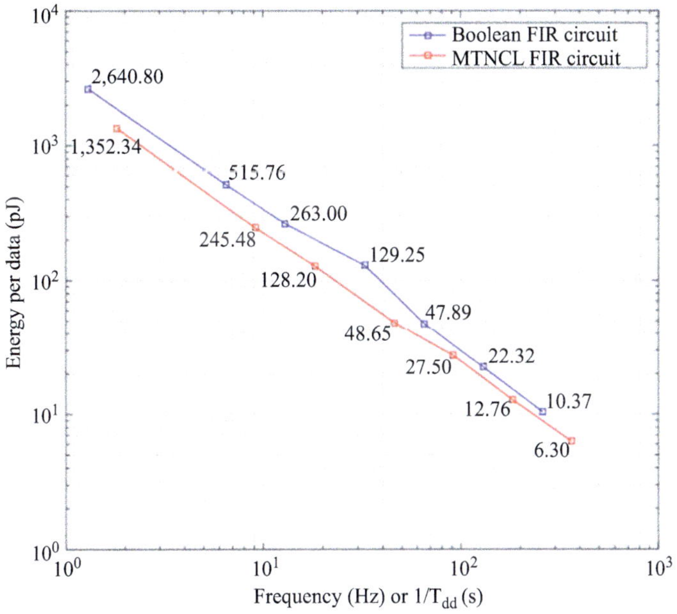

Figure 3.8 Testing result of the Boolean and MTNCL FIR chip

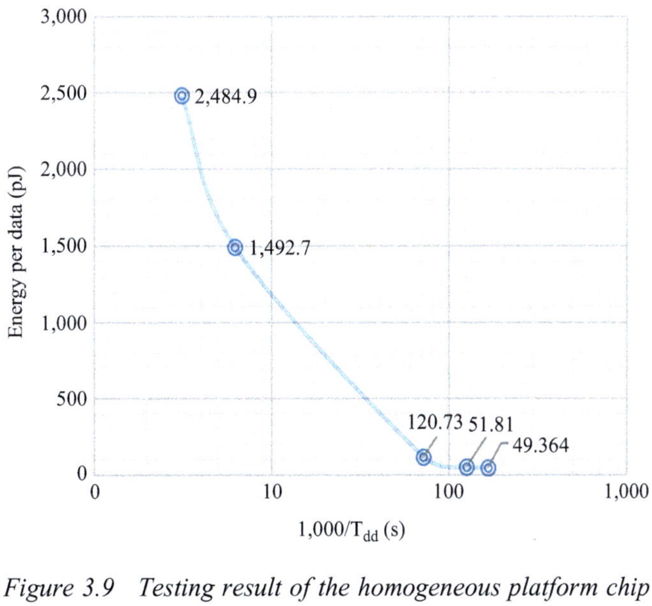

Figure 3.9 Testing result of the homogeneous platform chip

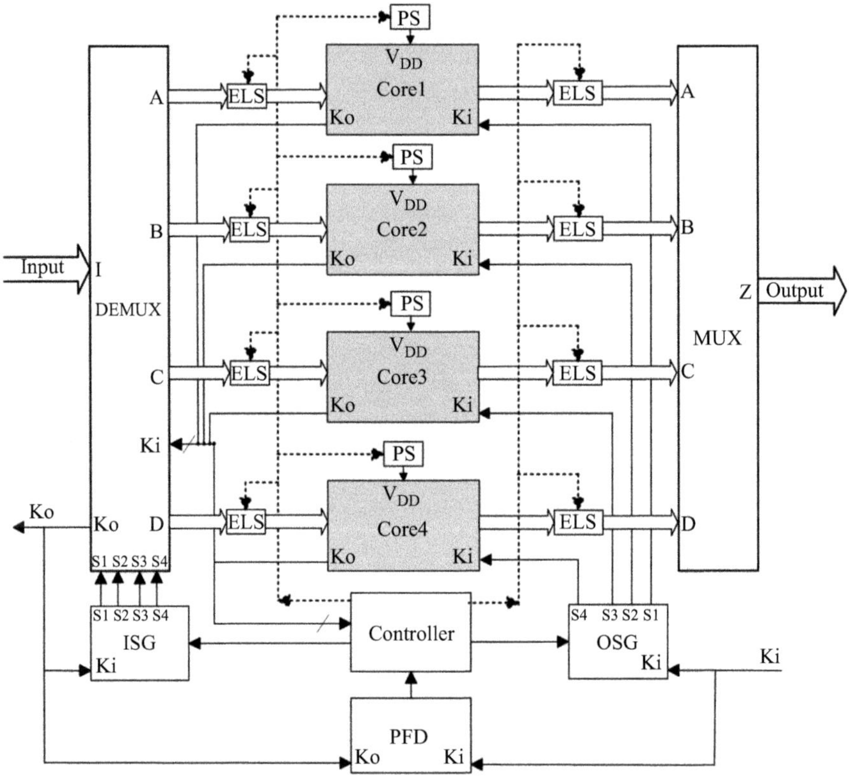

Figure 3.10 Homogeneous platform architecture enabling core disability

In the worst case, the malfunction of one core may cause stall to all. Two improvements are introduced in this section; one is enabling core disability for fine-grain control of the power-performance balancing, the other is redesigning the peripheral circuit to tolerate heterogeneous execution.

3.3.1 Homogeneous platform with core disability

The improved architecture provides fine-grain power and throughput balancing through introducing processing core disabling as an addition to DVS, as shown in Figure 3.10. High threshold voltage transistors are used as power switch (PS) to enable/disable voltage supply to the core. Enable level shifters (ELS) are inserted between the parallel cores and the peripheral components where two voltage levels are different when the core is powered off. Figure 3.11 shows the structure of the ELS. When EN signal is high, signal propagates from IN to OUT and the voltage level is being shifted. While EN signal is low, OUT is held at a constant value through peripheral components power supply V_{DD2}.

Figure 3.12 presents the platform controller with DVS and core disability. It serves as the decision-making unit to adjust supply voltage and/or disable cores

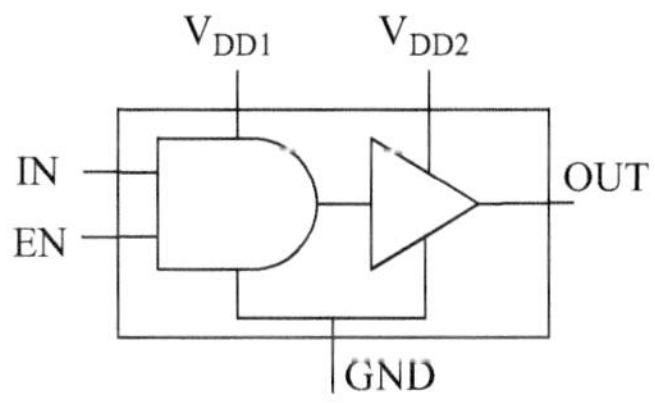

Figure 3.11 Enable Level Shifter (ELS) block

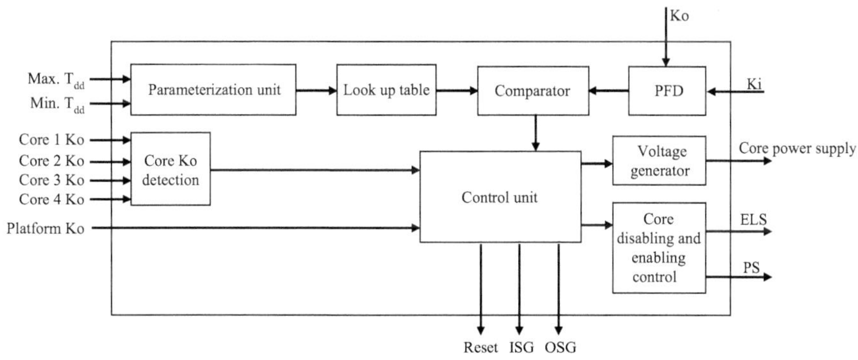

Figure 3.12 Control block for core-voltage core-on pairing

according to platform throughput. Pipeline fullness detector (PFD) realized in Figure 3.6 is implemented to continuously provide real-time platform throughput to the controller. Once the controller receives the throughput information from the PFD, a comparator compares it with throughput already populated in the look up table (LUT) when user configured the controller with max and min core T_{dd}. The comparison results will then yield a core-voltage core-on pairing (CVCOP) so the control unit knows what voltage it needs to set for supply voltage and how many cores it needs to disable. Then the controller starts the disabling/enabling sequence for fine-grained core state control, which are illustrated in the Sections 3.3.1.1 and 3.3.1.2.

3.3.1.1 Core disabling and enabling sequence

The main aspects of the architecture using DVS and core disabling are the capability to provide fine-grain control of power usage based on throughput and accommodate platform workload variation. Once the controller knows the core voltage and number of cores it needs to provide to the platform, it initializes different sequences to either adjust core supply voltage or disable cores. PFD is redesigned for core enable/disable logic, as shown in the flowchart in Figure 3.13. Take the platform with four cores as an example. Initially, all four parallel cores are active, and the first four data are sent to each core for processing. The processed data go through the multiplexer and appear at the outputs in the original order. Once T_{dd} observed by the PFD has moved to the core disable zone, the controller

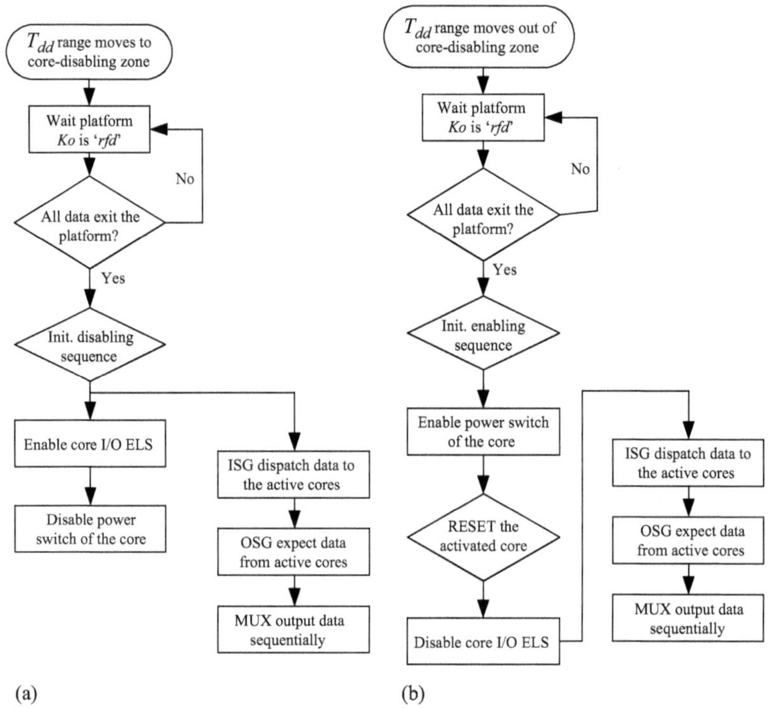

Figure 3.13 Core enabling (a) and core disabling (b) sequence

will initialize the core disabling sequence as shown in Figure 3.13(a). Core disabling starts in the order of Core1, Core2 and Core3. To disable Core1, controller waits for the current four data to exit the platform and platform *Ko* becomes *rfd* (request for DATA). After that, the controller first enables the ELS at Core1's inputs and outputs and then switches off the power switches of Core1. At the same time, the controller instructs the Input Sequence Generator to only dispatch three data to Core2, Core3 and Core4. Output Sequence Generator is also instructed to only take data from those cores.

The core enabling sequence is shown in Figure 3.13(b). When the detected T_{dd} moves to the core enable zone, controller waits for the current data to exit the platform and platform *Ko* becomes *rfd*. Once platform *Ko* becomes *rfd*, controller turns on the power switches of Core1 and resets it. When Core1 pipeline goes into NULL state and all the *Ko* signals become *rfd*, controller disables the ELS for Core1. At the same time, the controller instructs the Input Sequence Generator and Output Sequence Generator to dispatch and receive data from all the active cores.

3.3.1.2 Fine-grained core state control

The homogenous platform incorporating DVS scales the core supply voltage based on a one-to-one mapping relationship in Section 1.2.3, which has seven different core supply voltage to seven platform throughputs. The addition of core-disabling mechanism to DVS extends the mapping range to two dimensions. Combing DVS and core-disabling therefore enables fine-grain control of power usage based on the platform throughput. By enabling the user to configure the maximum and minimum T_{dd} of the parallel cores to the controller, platform throughput and CVCOP mapping are always closely correlated.

To develop the mapping methodology between platform throughput and CVCOP, a case study is conducted on MTNCL homogeneous platform with four FIR filter cores realized at IBM 130 nm technology node. The DVS range is set from 0.8 V to 1.2 V with five steps, and totally 20 CVCOP is applied to the platform with 40 input data patterns so platform total energy and platform average throughput can be measured. Ideally, the platform throughput should have a unidirectional relationship with the platform total energy, the slower the throughput, the lower the energy. However, the platform total energy consumption indicates a nonunidirectional mapping relationship between platform energy and platform T_{dd}. For example, at the points 1.2V/3C and 1.1V/3C, the energy consumption is higher than neighboring points 0.8V/4C and 1.0V/3C with less than 1 ns T_{dd} difference. This implies that when T_{dd} is in the range of 6 ns to 7 ns, instead of choosing 1.2V/3C and 1.1V/3C, 0.8V/4C or 1.0V/3C should be chosen for lower energy consumption and minimal impact on platform throughput.

Applying the above-mentioned methodology, the mapping relationship between platform energy and platform T_{dd} is inversely proportional as demonstrates in Figure 3.14. Instead of using 20 different CVCOP to map to 20 platform T_{dd}, six CVCOP: 1.2V/3C, 1.1V/3C, 1.2V/2C, 1.1V/2C, 1.2V/1C, and 1.1V/1C are removed since these combinations have similar throughput but higher energy consumption compared to other CVCOP. The rest 14 pairs provide improved coherency between platform energy and platform T_{dd}. Better energy and T_{dd} coherency implies that energy is used accordingly to throughput. The relationship between throughput and CVCOP observed in Figure 3.14 can be used to develop the throughput parameterization and subsequently creates a mapping between throughput and CVCOP. Controller is designed to allow the circuit designer of the FIR filter cores to configure the T_{dd} range of the parallel cores between 16 ns and 3 ns. Controller performs a calculation of subtracting 3 from 16 and dividing the

Table 3.2 Fine-grained mapping between platform T_{dd} and CVCOP

	Core voltage				
Core states	**0.8 V**	**0.9 V**	**1.0 V**	**1.1 V**	**1.2 V**
1-Core ON	16 ns	15 ns	14 ns	N.S.[1]	N.S.
2-Core ON	13 ns	12 ns	11 ns	N.S.	N.S.
3-Core ON	10 ns	9 ns	8 ns	N.S.	N.S.
4-Core ON	7 ns	6 ns	5 ns	4 ns	3 ns

[1]Not selected for core-voltage core-on pairing.

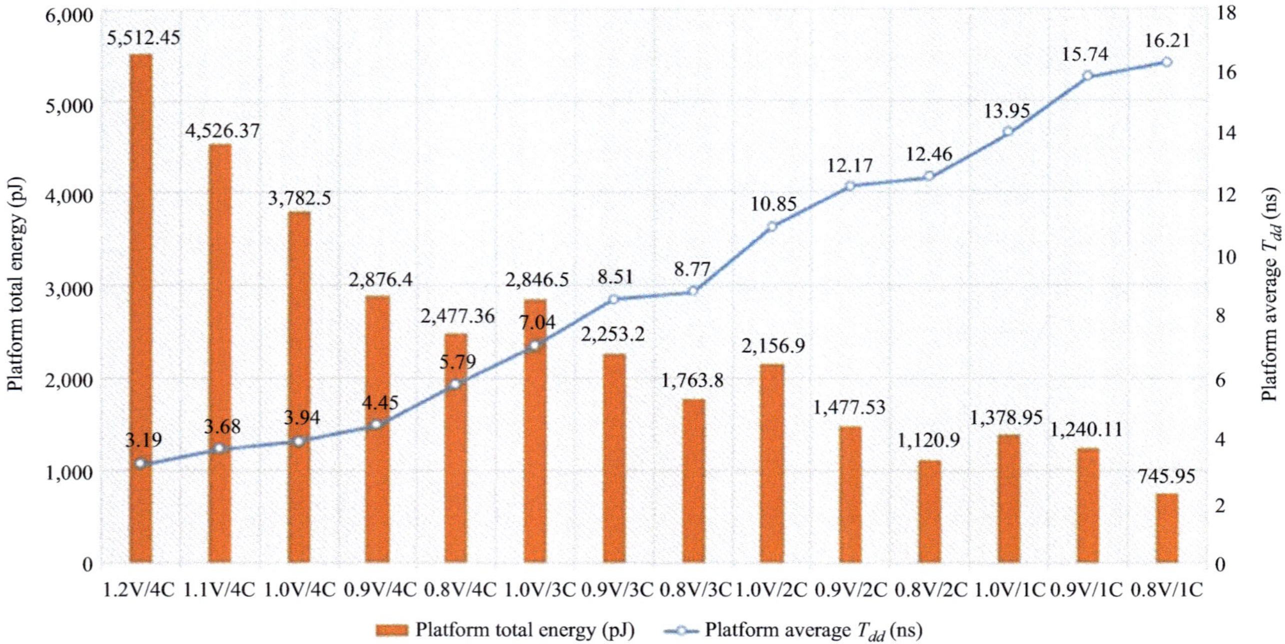

Figure 3.14 Controller throughput—CVCOP mapping strategy

value by 14. Each divided value will be rounded-up before assigned to the 14 CVCOP, creating the mapping table between platform T_{dd} and CVCOP as shown in Table 3.2. After the mapping table is created, the controller continuously monitors the platform T_{dd} and chooses CVCOP for the platform based on the already derived table.

3.3.2 Architecture of the heterogeneous platform

A different approach to break the data input sequence is to enable the platform to dispatch data to a core as soon as it requests for data. However, there could be collisions if more than one autonomous operating core is requesting for data within a short period of time. To prevent collision, an arbitration mechanism is necessary to grant mutually exclusive access to the common data bus of the platform. The worst case of the system throughput could be avoided by assigning the highest priority to the slowest core in the platform when collision happens.

A generic heterogeneous platform incorporating n cores is designed in Figure 3.15. To make the *rfd* of each core mutually exclusive, a generic asynchronous arbiter is in cooperated. After reset, all the internal cores are requesting for DATA and the *Ko* goes to *rfd*, while only one core will be granted by the arbiter to access the external data bus and others will hold their states. The *Ko* signal of the granted core will be deasserted to *rfn* (request for NULL) after the demultiplexer successfully dispatching data. After this initial round, the arbitration network grants another core's request for DATA through the common input data bus. The average waiting time of the cores is minimized by assigning the slowest core to top priority if two or more *rfd*s arrive simultaneously. In other cases, the arbitration network serves in a first-arrive first-grant mode. So, the handshaking signals are guaranteed to be mutually exclusive in *rfd* state.

3.3.2.1 Multiplexer and demultiplexer design with NULL cycle reduction

NULL cycle reduction (NCR) is used to increase the throughput of NCL systems by reducing the NULL cycle on the I/O port in the multicore architectures. In the heterogeneous platform, the external ports for all the handshaking signals of the internal cores facilitate the implementation of the NCR technique in the demultiplexer and multiplexer.

The demultiplexer partitions the common input data bus to n output data paths connecting to the internal cores. The data dispatching operation is controlled by the exclusive *sleepin* signals. Figure 3.16 shows the structure design of the demultiplexer. The *bufm* is an MTNCL buffer with sleep mechanism. When the sleep signal is active, the output is forced to be "0"; otherwise it follows its input. By inserting the *bufm* gate into all the rails of the input data path, the demultiplexer outputs a NULL wave after reset, when all the *sleepin* signals are active. In the heterogeneous platform, the *rfd* states of the cores are mutually exclusive, which means no more than one *sleepin* signals can be deactivated per arbitration; so only the *rfd* granted core's datapath will connect to the common input data bus during the DATA wave. The demultiplexer will automatically generate a NULL wave

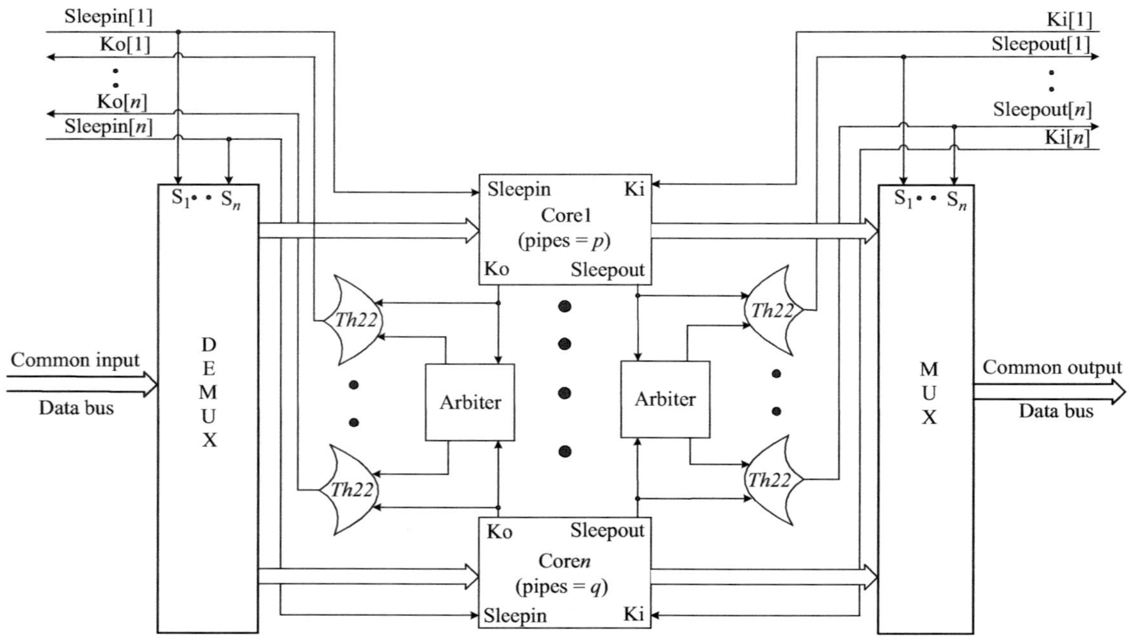

Figure 3.15 Architecture of heterogeneous platform

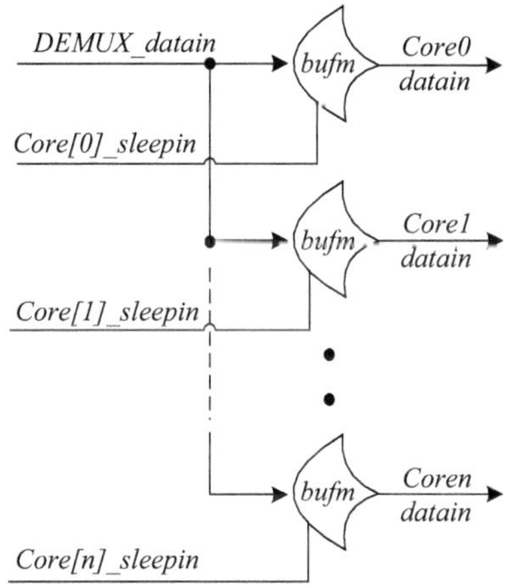

Figure 3.16 Demultiplexer in the heterogeneous platform

onto the datapath of the asynchronous core if its *rfd* is not granted. This simplifies the common input data bus interface, as it does not need to incorporate a NULL spacer when switching among different input data.

The multiplexer is designed in a similar fashion. It multiplexes all the outputs of the internal cores onto one single output data bus for the platform. Again, MTNCL buffer gates—this time with exclusive sleepout signals per core—are employed on all the rails of the core's output data paths to ensure only one core produces DATA states. To eliminate the NULL spacer on the common output bus, the DATA state of the core with output data bus access is held by the OR tree and the C-element gate (TH22) until the next core's data output request is granted. Figure 3.17 shows the structure of the NCR multiplexer with single-bit output from multiple cores. The output from the multiplexer switches between the DATA states of the internal cores following a pattern similar to that of the common input data bus. The output order may be different with the input order. This configuration produces a scalable heterogeneous platform.

3.3.2.2 Asynchronous arbiter design

The handshaking components require that the communication along several input channels is mutually exclusive. The basic circuit needed to deal with such situations is a mutual exclusion element (MUTEX) [7], shown in Figure 3.18. The circuit contains a latch with NAND gates and a metastable filter. The input signals R1 and R2 are two requests that originate from two independent sources, and the task of the MUTEX is to pass these inputs to the corresponding outputs G1 and G2 in such a way that at most one output is active at any given time.

The MUTEX circuit is used to construct the generic arbiter network with N-way inputs. Several architectures, such as mesh, tree, and token ring arbiters, are studied in [8], with the conclusion that the first-arrive first-grant feature is not guaranteed. Without first-arrive first-grant arbitration in the heterogeneous platform, the

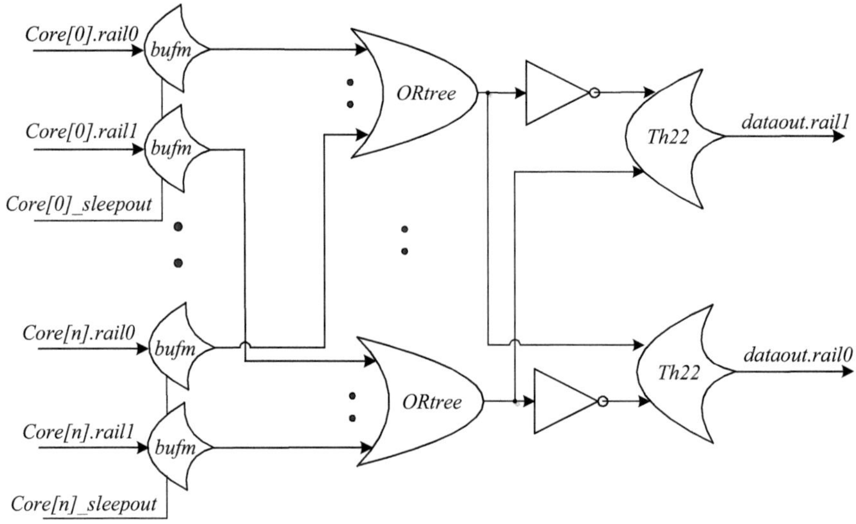

Figure 3.17 NCR multiplexer in the heterogeneous platform

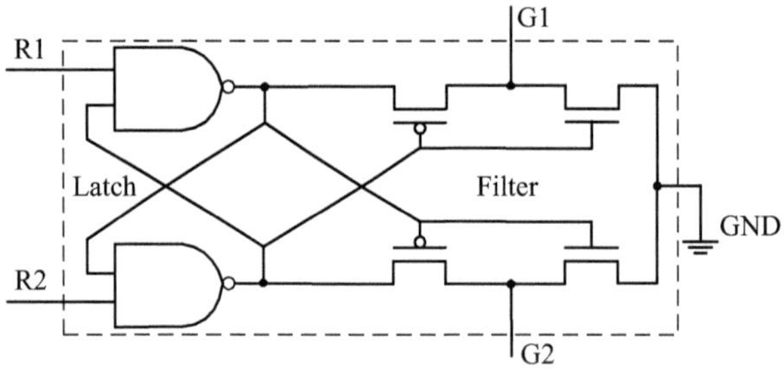

Figure 3.18 Mutual exclusion element (MUTEX) at the transistor-level implementation

rfd competition between two cores could put the third core into starvation even though its rfd has activated. A new architecture is also developed in [8], which needs C_n^2 MUTEXes to prevent the starvation of the N-way requests.

3.3.2.3 Platform cascading

One benefit of the heterogeneous architecture is more flexible for platform cascading without redesigning the data sequence circuit. Connecting the common data bus of the multiplexers and demultiplexers and the handshaking signals will cascade the platform. As shown in Figure 3.19, two generic platforms are scaled horizontally with the same internal cores. In the first platform, two arbiters are implemented to make the *Ko* and sleepout signals from different cores exclusive; while the subsequent platforms just need one arbiter for the sleepout signals since

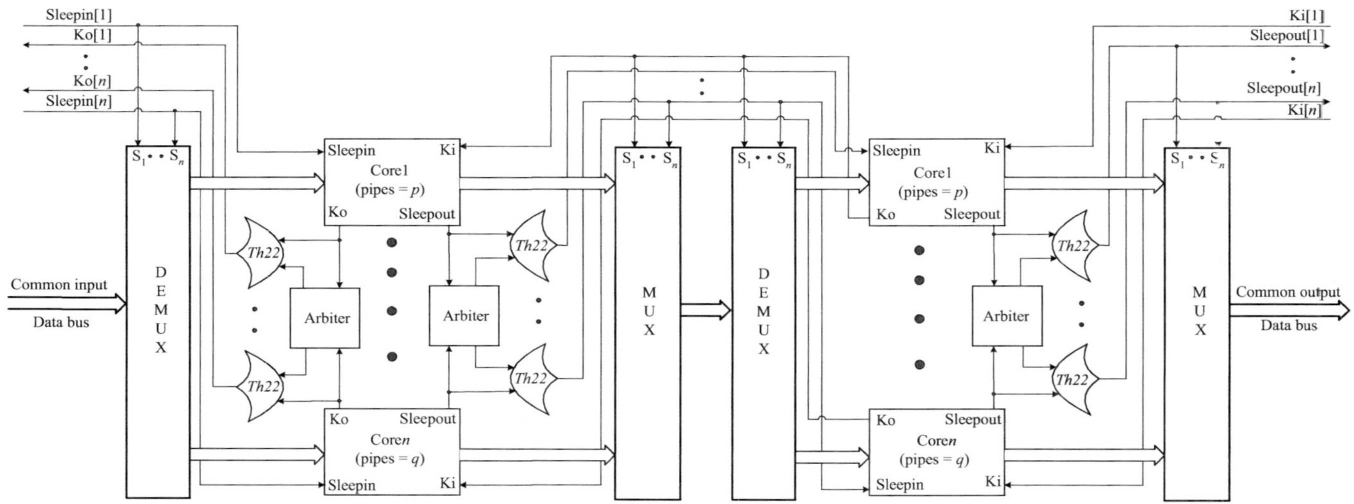

Figure 3.19 Cascading of the heterogeneous platform

the *rfd*s have already become exclusive in the previous platform. The inputs to the first platform are from the common input data bus, and the output data of the first platform is the input data of the subsequent platforms. Cores in the platforms arbitrate for input and output, but compute in parallel. The self-timed nature of delay-insensitive circuit avoids any timing issues between the platform modules. The highly modular interface enables the platform composition with the desired scalability for larger systems.

3.4 Conclusion

Asynchronous circuit designed on delay-insensitive NCL and multithreshold CMOS techniques inherits the benefit on power reduction but degrades the speed. Circuit pipeline and parallel architecture are applied to migrate the performance drawback. In the first part of the chapter, the throughput and latency of the NCL micropipeline are derived for the digital signal processing circuit optimization, including an example on generic FIR design with same performance as its synchronous count part.

Scalable parallel computing architecture that incorporates homogeneous units is designed in Section 3.2 for performance escalation. Besides that, DVS achieves balanced control of performance and power consumption. An effective fullness variance predicting algorithm is implemented to employ the DVS more aggressively in a wider range of system workloads. The platform fabricated using the MITLL 90 nm process consumes 49.364 pJ per data with the best performance when the DATA to DATA cycle time is 6.02 μs.

The schemes on fine-grain core states control and heterogeneous architecture are presented as research topics on power-performance balancing. Core enable and disabling sequence and fine-grain state control earns the maximum benefit of DVS. Common data I/O ports with NULL cycle reduction and asynchronous arbitration network are incorporated in the heterogeneous platform to make a highly modular interface for both horizontal and vertical scaling. Those methodologies demonstrate the advantage of asynchronous circuit in large scale, multithreads and scalable computing applications.

References

[1] Fant, Karl M. and Scott A. Brandt. "NULL convention logic™: a complete and consistent logic for asynchronous digital circuit synthesis." *Proceedings of International Conference on Application Specific Systems, Architectures and Processors, 1996, ASAP 96*, IEEE, 1996.

[2] Smith, Scott C. "Completion-completeness for NULL convention digital circuits utilizing the bit-wise completion strategy." *Proceedings of the International Conference on VLSI, VLSI'03*, Las Vegas, Nevada, USA, June 23–26, 2003, pp. 143–149.

[3] Smith, Scott C. "Speedup of self-timed digital systems using early completion." *Proceedings of IEEE Computer Society Annual Symposium on VLSI. New Paradigms for VLSI Systems Design. ISVLSI*, IEEE, 2002.
[4] Bailey, Andrew, Ahmad Al Zahrani, Guoyuan Fu, Jia Di, and Scott C. Smith. "Multi-threshold asynchronous circuit design for ultra-low power." *Journal of Low Power Electronics* 4, no. 1–12 (2008): 337–348.
[5] Yeh, Tse-Yu and Yale N. Patt, "Two-level adaptive training branch prediction." *Proceedings of the 24th Annual International Symposium on Microarchitecture, ACM (1991)*, pp. 51–61.
[6] Vitale, Steven, Peter W. Wyatt, Nisha Checka, Jakub Kedzierski, and Craig L. Keast. "FDSOI process technology for subthreshold-operation ultralow power electronics." *Proceedings of the IEEE* 98, no. 2 (2010): 333–342.
[7] Seitz, Charles L. "Ideas about arbiters." *Lambda* 1, no. 1 (1980): 10–14.
[8] Liu, Yu, Xuguang Guan, Yang, and Yintang Yang. "An asynchronous low latency ordered arbiter for network on chips." In *2010 Sixth International Conference on Natural Computation (ICNC)*, Vol. 2, 2010, pp. 962–966.

Chapter 4

Asynchronous circuits for ultra-low supply voltages

Chien-Wei Lo[1]

Modern digital systems based on complementary metal-oxide-semiconductor (CMOS) integrated circuits (IC) are increasingly sensitive to power consumption and heat generation which has direct impact on the system's performance and reliability. Power consumption of a system can be effectively reduced by techniques such as supply voltage scaling, downsizing transistors, or limiting switching activity. Supply voltage scaling is amidst one of the most effective way to reduce power dissipation. The continuance reduction of supply voltage will require transistors to operate in subthreshold region. Process technology developed with transistors optimized for subthreshold operation offers the essential building blocks to construct digital systems that are capable of operating at ultra-low supply voltage and consuming significantly less power.

Asynchronous circuits have demonstrated unique capability in near-threshold operating voltage regions [1] since it's not subject to timing constraints and less susceptible to signal integrity issues such as noise and cross-talk effect compared to synchronous circuit. Therefore, asynchronous circuit is suitable to operate at ultra-low supply voltage and takes advantage of the inherent low power consumption. To demonstrate asynchronous circuits are capable of functioning at ultra-low supply voltage with transistors operating in subthreshold region, a series of digital asynchronous circuits are designed and fabricated with MIT Lincoln Lab's (MITLL) 90 nm ultra low power (XLP) fully depleted silicone on insulator (FDSOI) CMOS process utilizing transistors operating at 300 mV supply voltage.

The digital circuits designed using the process are synchronous finite impulse response (FIR) filter, asynchronous FIR filter, and asynchronous homogeneous parallel data processing platform with four FIR filter cores. Physical testing is conducted to measure the circuits' operating speed and power consumption to substantiate asynchronous circuits' capability of real-world ultra-low supply voltage operation and to assess the benefit of reduced power dissipation. Asynchronous FIR

[1]Computer Science and Computer Engineering Department, University of Arkansas, Fayetteville, AR, USA

filter is compared with synchronous FIR filter to demonstrate that asynchronous designs can operate at higher speed with lower power consumption in the subthreshold operating voltage region. Synchronous ring oscillators and asynchronous ring oscillators are designed and placed at difference locations on the die and the oscillating frequencies are observed to characterize inter-die process variation.

4.1 Introduction

4.1.1 Subthreshold operation and FDSOI process technology

The most efficient way to reduce power dissipation of digital system is to reduce the supply voltage [2]. The dynamic power equation, $P_{dyn} = C_L V_{DD}^2 f$ [3] indicates that supply voltage scaling can result in 2× power reduction. With the continuing demand of low power applications such as mobile devices [4], supply voltages have been scaled lower than transistor threshold voltages, thereby signifying the importance of subthreshold optimized transistor and the corresponding process technology. For transistors operating in the subthreshold regime, the gate voltage is lower than the threshold voltage. As a result, the surface potential is controlled by the depletion region beneath the gate and is nearly constant from the source to the drain, resulting in near zero drift current. Therefore, the transistor on-state current is dictated by the diffusion current instead of drift current [5]. Transistors operating in the subthreshold region are much more power efficient than those operating in strong inversion where drift current is dominant.

In addition to diffusion current, subthreshold swing is another important characteristic of the subthreshold transistor. Subthreshold swing can be defined as $\ln(10)\frac{kT}{q}(1+\frac{C_d}{C_{ox}})$ [6]. The ideal subthreshold swing is calculated by fixing the subthreshold slope factor $(1+\frac{C_d}{C_{ox}})$ to 1 (oxide thickness ~0) hence $S = \ln(10)\frac{kT}{q}$. Ideal subthreshold swing is 60 mV/dec at 300 K room temperature [7]. For bulk CMOS technology, the oxide thickness will not be zero, therefore the subthreshold swing will always be higher than ideal. The smaller the subthreshold swing the faster the transistor can switch between on and off states. To take advantage of the power efficiency of the subthreshold transistor, lowering the supply voltage seems to be the most straightforward way. However, lowering the supply voltage results in a significant increase of subthreshold leakage current [8]. Compared to bulk CMOS technology, fully depleted silicon-on-insulator (FDSOI) technology exhibits inherently low leakage, and have superior subthreshold swing [9], making it suitable for low power CMOS applications. Combining the advantages of FDSOI with transistors optimized for subthreshold operation, the dynamic power and leakage power of a digital system may be reduced [10].

The MITLL 90 nm XLP FDSOI CMOS process provides a novel transistor technology based on metal-gate FDSOI devices [10] that is optimized to operate at 300 mV. A 90% reduction of switching energy can be achieved by the novel

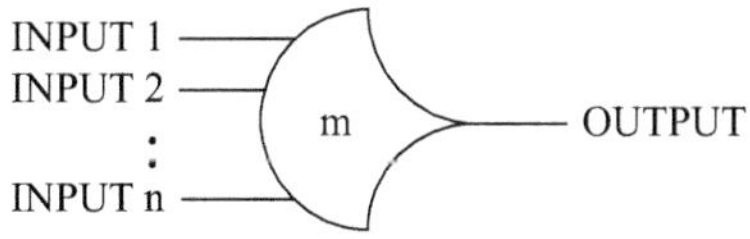

Figure 4.1 Thmn threshold gate in NCL

transistor technology compared to regular 1.2 V bulk-silicon CMOS [10]. The FDSOI CMOS process provides five metal layers and four devices, with a sub-threshold NMOS and PMOS set used for core logic and a superthreshold NMOS and PMOS set used for I/O.

4.1.2 NULL conventional logic and multithreshold NULL conventional logic

The asynchronous circuits were designed using the NULL conventional logic (NCL) and multithreshold NULL conventional logic (MTNCL). NCL is a delay-insensitive (DI) asynchronous paradigm. NCL circuits function correctly as long as the transistors switch accordingly. Therefore, NCL circuits are not accountable to timing constraints like synchronous circuits. The building blocks for NCL circuits are the 27 fundamental gates [11]. A NCL gate is shown in Figure 4.1 in generic representation. NCL gate's unique property is the capability to hold state through hysteresis [11], once *m* of the *n* inputs are asserted, the output will be asserted. For the output to be deasserted after it's asserted, all *n* inputs must be deasserted. The delay insensitivity characteristic of NCL circuits is achieved through the implementation of dual-rail or quad-rail signals. Dual-rail signals utilized two wires D^0 and D^1. Both wires are utilized to describe four different states of NCL logic: $D^0 = 1$ and $D^1 = 0$ represent DATA0 state in NCL. $D^0 = 0$ and $D^1 = 1$ represent DATA1 state in NCL. $D^0 = 0$ and $D^1 = 0$ represent NULL state in NCL, implying the data are not available at the moment. $D^0 = 1$ and $D^1 = 1$ describe an invalid state in NCL. Request and acknowledge signals, represented as *Ki* and *Ko*, are used to control the DATA/NULL cycle.

Sleep transistors can be added to NCL gates for the purpose of power gating and NULL state production. NCL gates with sleep transistor are called MTNCL gates. The output of the MTNCL gate can be pulled low by setting the gate's sleep signal to high, thus creating a NULL state. For a set of current and preceding DATA state, a NULL state is required to be inserted in between the two DATA states to prevent overlapping. The sleep transistors are also utilized to limit leakage power by gating the current flowing from voltage source when the gate is in NULL state. Request and acknowledge signals, *Ki* and *Ko*, are used to control the DATA/NULL cycle of MTNCL circuit. MTNCL has the advantage of reduced design complexity and improved energy efficiency compared to NCL.

4.2 Asynchronous and synchronous design

Five synchronous and asynchronous circuits are designed to operate at ultra-low supply voltage of 300 mV. The circuits include: synchronous ring oscillator, synchronous FIR filter, asynchronous ring oscillator, asynchronous FIR filter, and asynchronous homogeneous parallel data processing platform.

4.2.1 Synchronous and asynchronous (NCL) ring oscillator

The synchronous ring oscillator consists of 11 inverters and an AND gate. The AND gate was used to disable the oscillator. The asynchronous ring oscillator used eleven NCL registers, consisting of ten registers that reset to NULL and one register that resets to DATA0. The enable signal for this circuit was used as the reset for all registers. When enabled, ten of the eleven registers reset NULL. The last register resets to DATA0. Once the reset is disabled, and the circuit is allowed to oscillate at its own speed, the registers begin propagating the one DATA wave within the pipeline, which was originally output by the register that resets to DATA0.

4.2.2 Synchronous FIR filter

The Boolean FIR had a feed-forward structure built with three basic components: the adder, the multiplier, and the shift register. The adders were implemented with generic ripple carry adders (RCA), and the multiplier was built with carry-save adders (CSA). In the synchronous design, the shift register was a series of D flip-flops. An 8-tap FIR with hard-coded coefficients, as shown in Figure 4.2, was taped out as the test vehicle. The input to the FIR circuit was an 8-bit integer, and the output was 22 bits after convolution. For the physical implementation, simple I/O logic was added to reduce the number of input/output ports. The input logic was a shift register with 8 D flip-flops. Only one input pad was used to shift the data in serially, and then the data were loaded to the input ports of the FIR every 8 input clock cycles. The output logic was the reverse of the input logic, with the function of parallel in and serial out as shown in Figure 4.3. It had 22 shift registers, and the input of each register was connected to the output of a 2-to-1 MUX. The MUX was controlled by an external signal called "load_shift" (L/S) to decide if it was going to load the output of the FIR circuit to the output logic or shift the loaded data out of the chip.

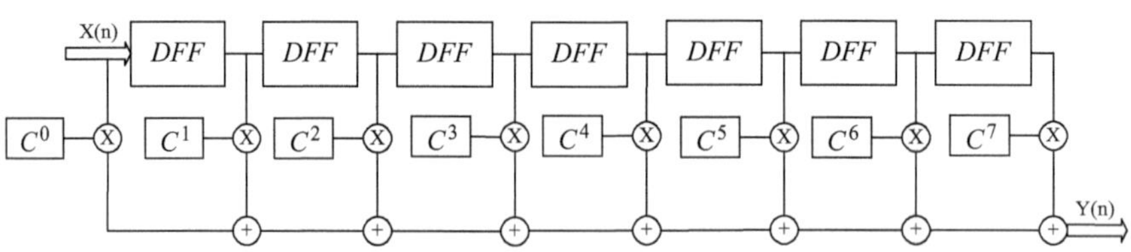

Figure 4.2 Structure of 8-tap synchronous FIR filter

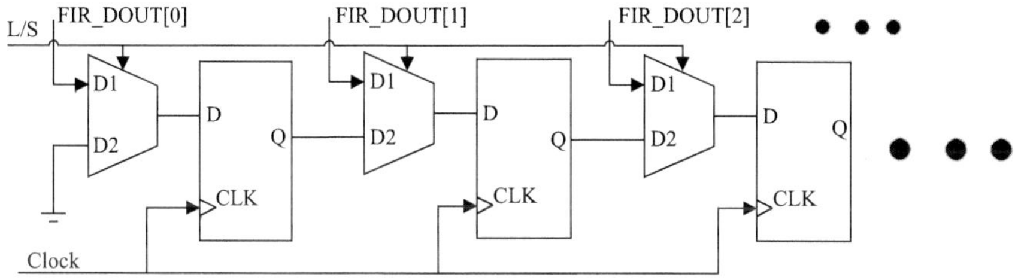

Figure 4.3 Output logic in the synchronous FIR design

4.2.3 Asynchronous (MTNCL) FIR filter

Asynchronous paradigm has been demonstrated to be beneficial for digital signal processing (DSP) applications [12]. A classic DSP application consists of three elements: A/D converters, digital processing circuits, and D/A converters. Asynchronous ADC with intermittent level-corssing sampling scheme is found to have improved energy efficiency through benefiting from signals with statistical properties as indictaed in [13,14]. Unlike traditional ADC which processed intermittent front-end stream into a clock-sampled sample stream, asynchronous ADC is best used when processing intermittent sampling with asynchronous digital processors. FIR filter is mainly responsible for performing convolution operation which is the basic building block of DSP applications. Therefore, FIR is chosen to be implemented in asynchronous circuit and compared with synchronous FIR for operating speed and energy efficiecny.

The asynchronous FIR followed the same structure of its synchronous counterpart, which was based on the unsigned 8×8 multiplier, the generic adder, and shift register with reset to NULL (regnm) and reset to DATA (regdm) functionality, as shown in Figure 4.4. The main difference between the synchronous and asynchronous designs was the shift registers. For the synchronous one, the shift register was just a series of DFFs. Whereas, in the asynchronous design, a data cycle was required for data to be inserted into the shift register after reset to keep the delay. As the generic design scales up, more reset to NULL registers needed to be inserted into the delay path to maintain the throughput [15]. For the physical implementation in the tape-out, the 8-tap asynchronous FIR was launched with the same hard-coded coefficients values as the synchronous one. I/O logic with similar functionality was also implemented in the asynchronous FIR design for input/output port reduction.

4.2.4 MTNCL homogeneous parallel asynchronous platform

The computing performance of a single processing core is limited by voltage and frequency scaling, which can only be scaled as high before the power consumption and heat generation of the core starts having negative impact on the core's performance and reliability. Parallelism has been introduced in large-scale, high-performance computing to improve energy efficiency without sacrificing throughput. One form of parallel computing is multicore processing system [16]. The system employed a fixed number of low-frequency low-supply-voltage processing cores in place of a single high-frequency

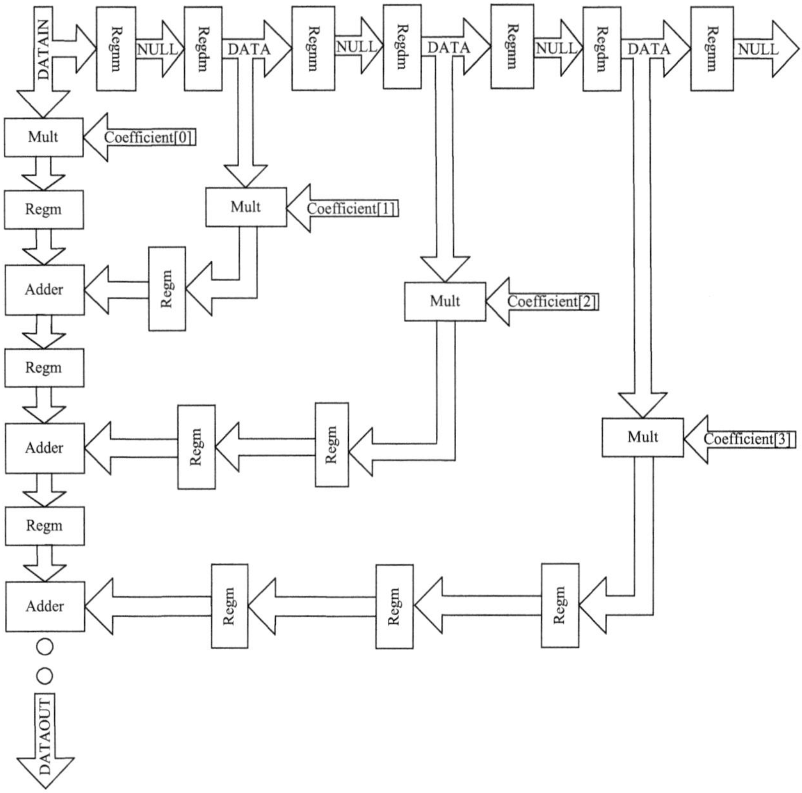

Figure 4.4 Structure of the asynchronous FIR design

high-supply-voltage core. The maximum speed allowed by Amdahl's law can be achieved by dispatching input data to the multiple processing cores and merging the outputs. Parallel architecture can be adopted by MTNCL systems for improved performance and energy efficiency as indicated by previous research [17].

To further improve the throughput and power consumption of MTNCL system and demonstrate the advantage of parallelism, a homogeneous platform was designed for parallel data processing. The platform could incorporate multiple cores with the same functionality such as with four FIR cores; the throughput could then become four times better than the single FIR core. It was a tradeoff between area and performance. The homogeneous platform architecture is shown in Figure 4.5 with top-level components, besides the computing cores, demultiplexer, and input sequence generator were designed to dispatch input data while the multiplexer and output sequence generator guarantee the proper data exit the platform [18]. A voltage control unit (VCU) was implemented to realize dynamic voltage scaling on the platform. The VCU sense the input data rate change and map the change into a range of minimum and maximum supply voltage. When the VCU detects increased input data

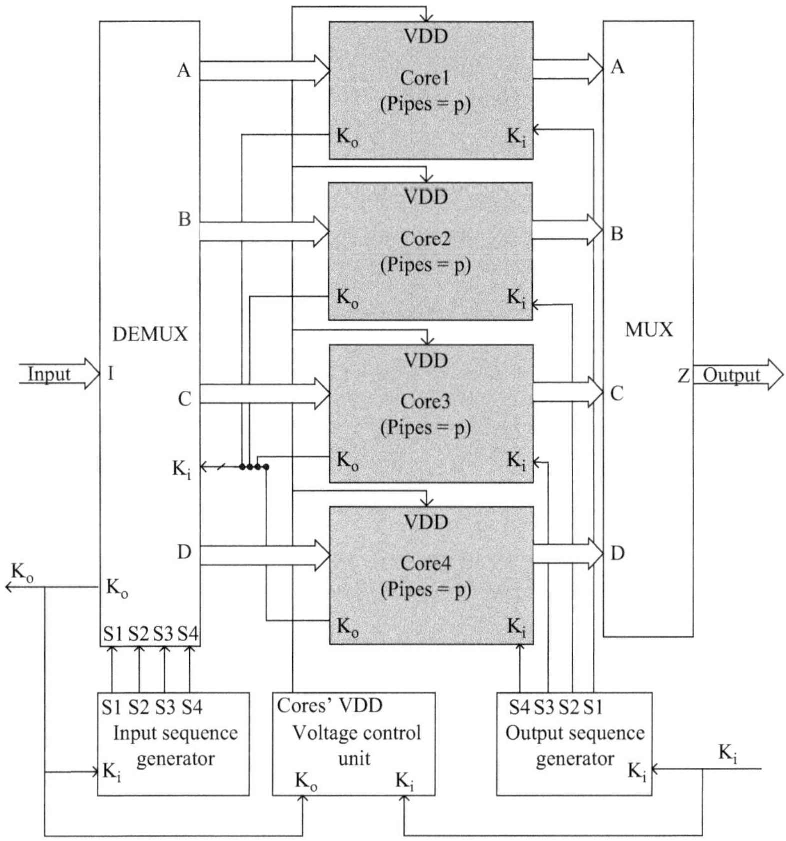

Figure 4.5 Architecture of the homogeneous platform

rate, it raises the core's supply voltage to the maximum value for best performance. The supply voltage is lowered to the minimum value when input data rate drops, where performance is no longer required and is traded off for improved energy efficiency.

The operating principle of the platform is to send the first four data sequentially to the four active FIR cores for processing. Once processing is finished for the four data, the multiplexer will then pass the processed data to the outputs in the same order as inputs. The platform will then request a new set of four data for processing. It is necessary for the input sequence generator, output sequence generator, and the multiplexers to function at maximum speed to guarantee the platform is receiving and dispatching data correctly. Therefore, even though the four parallel cores are operating at 300 mV V_{DD}, the sequence generators and multiplexers are operating at 1.2 V V_{DD}, which is the maximum supply voltage specified by the process technology.

The platform fabricated consists of four 8-tap asynchronous FIR filters incorporated as the processing units. Since the circuit was large enough to place all the input/output pads, I/O logic was not used in the design.

4.3 Physical testing methodologies

Due to the ultra-low supply voltage of the synchronous and asynchronous circuits fabricated with the MITLL 90 nm XLP FDSOI CMOS process, physical testing methodologies are developed specifically to measure the current drawn from the circuits. However, the ultra-low supply voltage of the circuits presents unique challenge when performing functionality testing and measuring power consumption. The noise from the measurement system such as oscilloscope, probes, and connection methods are easily superimposed on to the 300 mV signals. A series of measures has been taken to substantially reduce the measurement system noise and improve the accuracy of functionality testing and current measurement. The measures taken are using the 50 ohm input instead of the 1 Mohm input for oscilloscope, limiting the oscilloscope bandwidth to 20 MHz, and used the high-resolution mode of the oscilloscope which averages the signal to further reduce random noise.

The measurement system includes the utilization of a FPGA to send and receive data to and from the design under test (DUT). A custom designed PCB to upshift and downshift the signals voltage from DUT outputs and FPGA outputs. A testing PCB to host the DUT, and a mixed signal oscilloscope to verify circuit functionalities and obtain power measurement results.

A Xilinx FPGA was used to send test signals to the DUT and also read DUT outputs. The DUT is operating at 300 mV V_{DD}, which is significantly lower than the FPGA operating voltage of 1.8 V. A level shifter PCB is designed to downshift the FPGA output signals voltage from 1.8 V to 300 mV to be recognized by the DUT. The DUT output signals voltage is upshifted by the PCB from 300 mV to 1.8 V to be recognized by the FGPA. Throughout testing, DUT V_{DD} is fixed at 300 mV since the transistors are optimized for 300 mV subthreshold operation and a body-biasing voltage ranging from -1 V to -2 V is applied to each design as recommended by the process technology specification. The temperature of the test environment is maintained at room temperature around 25°C.

A Tektronix mixed signal oscilloscope and a current adapter [19] are used to acquire power measurement results. The current adapter is selected due to its enhanced current range resolution that can measure digital circuit current drawn down to micro amp (±0–1,000 μA) and nano amp range (±0−1,000 nA). Given these capabilities, the current adapter is deemed suitable for power measurement of low power designs [19]. The current adapter can be set in one of the three settings: 1 mV/mA, 1 mV/μA, and 1 mV/nA. It is designed such that the measured current has a 1-to-1 relationship with the voltage output [19]. For example, with a setting of 1 mV/μA scale and the voltage output is 5 mV, the current acquired will be 5 μA. The current drawn from the DUT over a period of time is measured using the mixed signal oscilloscope connected to the current adapter.

4.4 Physical testing results

Totally 12 die are received from the foundry with 5 circuits on each die. Each die is diced to separate the five circuits. Each circuit is attached to a dual-in-line package (DIP) using a silver-based high conductivity epoxy to ensure proper back biasing required by FDSOI process. Probe station is first used to verify the functionality of the circuits before moving on to performance and power measurement using mixed signal oscilloscope and current adapter. The Xilinx FPGA is used to provide input signals and the level shifter PCB is incorporated to downshift the input signal voltage from 1.8 V to 300 mV. The 300 mV input signals are then presented to input probes connected to input pads on the circuits. The outputs from the circuit are acquired through the oscilloscope connected to the output probes. Once the circuit is verified functional by the probe station, it will be moved to the test setup as outlined in Section 4.2.3 for operating speed and power measurement.

The oscillating frequency is measured for both synchronous and asynchronous ring oscillators. Total power consumption and energy consumption per data are measured for synchronous and asynchronous FIR filters, and asynchronous homogeneous parallel data processing platform with four FIR filter cores.

4.4.1 Synchronous designs

Two synchronous circuits are being designed, fabricated, and tested to be operable at 300 mV V_{DD}. The two circuits are ring oscillator and FIR filter.

The synchronous ring oscillator is the first being tested due to its simple construction and basic functionality. The ring oscillator functionality testing is with 300 mV V_{DD} and the die is backside biased at a range of −2.5 V to −3.5 V per process technology recommendation. The enable signal of the ring oscillator is first set to 0 and then set to 1 to enable the ring oscillator for output. When the ring oscillator is enabled, the oscillating frequency is documented. To observe inter-die process variation, two ring oscillators are being placed at different locations on each die. Since there are 12 die manufactured, a total of 24 ring oscillators are measured for oscillating frequency as shown in Figure 4.6.

The frequency measurement results indicate that the ring oscillators placed at location #1 has oscillating frequency ranging from 0.8 MHz to 2.5 MHz with an average frequency of 1.78 MHz. The ring oscillators placed at location #2 has oscillating frequency ranging from 0.725 MHz to 2.5 MHz with an average frequency of 1.57 MHz. Therefore, the oscillating frequency of synchronous ring oscillators at location #1 are 13% faster than those placed at location #2. A higher oscillating frequency for ring oscillators placed at location #1 suggests that the ring oscillator power rails may be fabricated with slightly greater width for that particular location due to random process variation. Another possibility is that interconnects between the inverters inside the ring oscillators are shorter for the particular location. Combining the effects of slightly wider power rails and shorter interconnects, the oscillating frequency is increased.

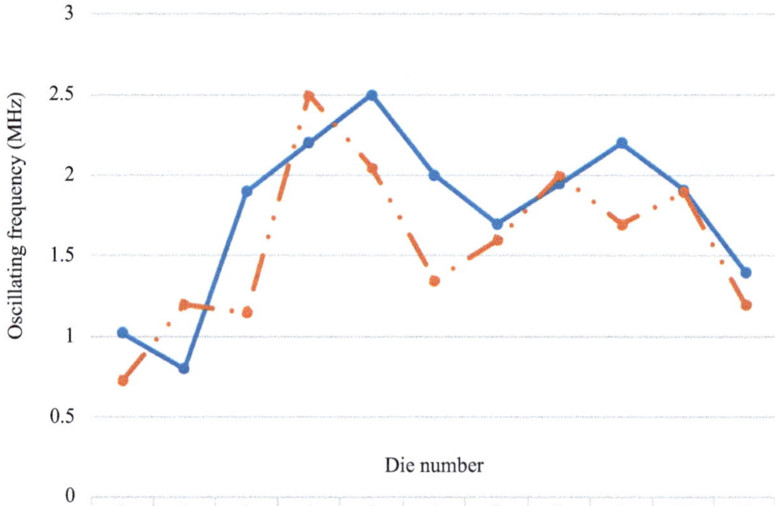

	0	1	2	4	5	6	7	8	9	10	11
Location #1	1.02	0.8	1.9	2.2	2.5	2	1.7	1.95	2.2	1.91	1.4
Location #2	0.725	1.2	1.15	2.5	2.05	1.35	1.6	2	1.7	1.9	1.2

Figure 4.6 Frequency measurements of ring oscillators

The synchronous FIR filter is the second in line being tested. The functionality testing of the FIR is performed with 300 mV V_{DD} and the die is backside biased at a range of −1.94 V to −3.04 V per process technology recommendation. The reset signal of the FIR filter is first set to 1, after the output becomes 0, the reset will be set to 0 and input data will start presenting to the FIR filter by the FPGA. After the reset goes low, current drawn from the power supply to the filter is measured by the current adapter. Based on the current drawn from the power supply by the FIR filter, total power consumption and energy consumption per data are measured over a range of operating speeds as shown in Figure 4.7 and Figure 4.8, respectively.

Based on the energy consumption per data results, the synchronous FIR filter could function at 300 mV V_{DD} and −1.7 V body-biasing voltage with a frequency range from 260.5 Hz to 1.303 Hz. The corresponding energy consumption per data is from 10.37 nJ to 2,640.8 nJ. The slower speed measured from the FIR filter is due to the fact that the space for placing I/O pads around the FIR filter core is limited. The height and the width of the core dictates the number of pads that can be placed surrounding it. After placing the power and ground pads to supply power to the core, the remaining spaces are not enough for primary input and output pads. Therefore, input data from FPGA can't be presented simultaneously to FIR filter inputs. To resolve the issue, I/O logic is designed and implemented in the core to shift data into and out of the FIR filter. I/O logic contributes majorly to the low operating frequency measured. The slower operating speed has negative impact on the energy consumption per data of the FIR filter. The FIR filter total power consumption is ranging from 2.7 μW to 4.21 μW.

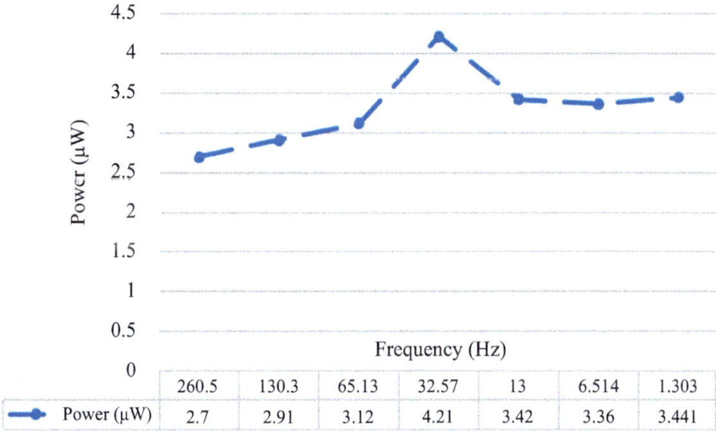

Figure 4.7 Synchronous FIR filter total power consumption

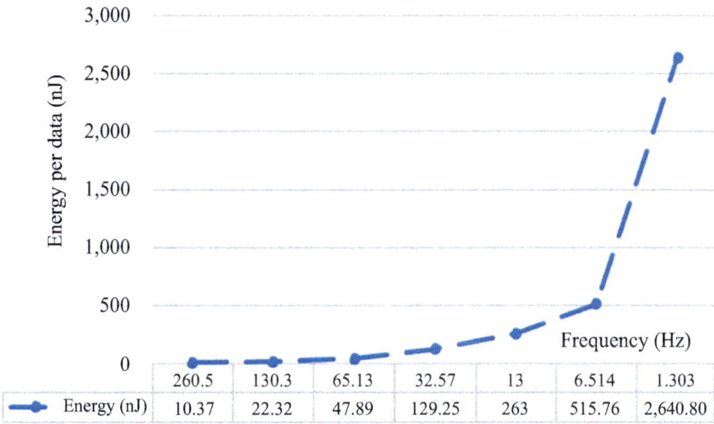

Figure 4.8 Synchronous FIR filter energy consumption per data

4.4.2 Asynchronous designs

Three asynchronous circuits are being designed, fabricated, and tested to be operable at 300 mV V_{DD}. The three circuits are NCL ring oscillator, MTNCL FIR filter, and MTNCL homogeneous parallel data processing platform. The asynchronous designs all have a repeating DATA and NULL cycle, and the handshaking signals *Ki* and *Ko* represents the flow of DATA and NULL cycle in the design. In contrast to synchronous circuits which the speed can be measured by clock frequency, the performance of asynchronous circuits are quantified by measuring DATA-to-DATA cycle time, represented as T_{dd}. Clock period in synchronous circuits is directly equivalent to T_{dd} in asynchronous circuits.

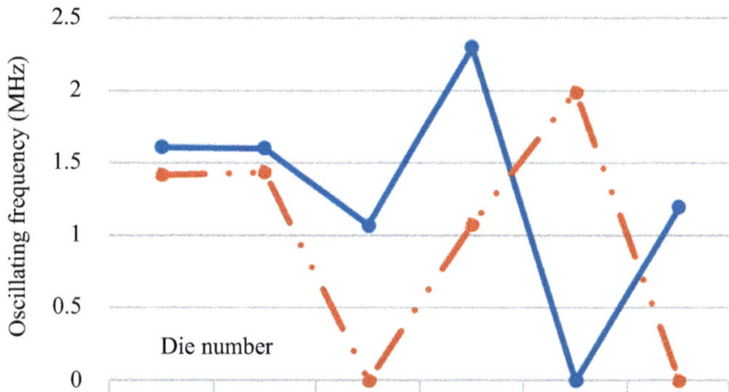

	4	5	6	7	10	11
Location #1	1.61	1.6	1.074	2.3	0	1.196
Location #2	1.42	1.44	0	1.08	1.994	0

Figure 4.9 Frequency measurements of ring oscillators

The first asynchronous design tested is NCL ring oscillator. The ring oscillator functionality testing is with 300 mV V_{DD} and the die is backside biased at a range of −2.5 V to −3.5 V. Similar to synchronous ring oscillator, the NCL ring oscillator is placed at two different locations on the same die with a total of 12 NCL ring oscillators measured for oscillating frequency as shown in Figure 4.9. The frequency measurement results indicate that the NCL ring oscillator placed at location #1 has an oscillating frequency range from 1.196 MHz to 1.61 MHz with an average frequency of 1.3 MHz. NCL ring oscillators placed at location #2 has an oscillating frequency range from 1.08 MHz to 1.994 MHz with an average frequency of 0.99 MHz. The 0 MHz oscillating frequency for location #1 die #10 and location #2 die #6 represents nonfunctioning ring oscillators.

The oscillating frequency of NCL ring oscillators at location #1 is faster by about 31.3% than the ones placed at location #2. The NCL ring oscillator frequency difference is similar to synchronous ring oscillator where those placed at location #1 is faster than the ones placed at location #2. A higher oscillating frequency for ring oscillators placed at location #1 suggests that the ring oscillator power rails may be wider for that particular location due to random process variation. Another possibility is that interconnects between the inverters inside the ring oscillators are shorter for location #1. The combined effects of wider power rails and shorter interconnects increased the oscillating frequency.

When comparing synchronous and asynchronous ring oscillator together, the synchronous ring oscillator is found to have higher oscillating frequency for both location #1 and location #2, which is 1.78 MHz and 1.57 MHz. The NCL ring oscillator frequency is at 1.3 MHz and 0.99 MHz. NCL ring oscillator having slower frequency than its synchronous counterpart is expected due to the implementation of NCL pipelining architecture which increased the time required for data propagation.

The second asynchronous design tested is MTNCL FIR filter. The FIR filter functionality testing is with 300 mV V_{DD} and the die is backside biased at a range of

−0.84 V to −2.6 V. The MTNCL FIR filter has three outputs *Ko*, *sleepout*, and *IOout*. After reset is pulled 0, the output signal *Ko* should be toggling if the circuit is functioning correctly. Once the handshaking mechanism is established in the FIR filter, the current drawn from the power supply by the MTNCL FIR filter is measured by the current adapter. Based on the current drawn, the total power consumption and energy consumption per data are quantified and shown in Figures 4.10 and 4.11. The T_{dd} captured has a range from 2.727 ms to 545.3 ms with corresponding energy per data from 6.3 nJ to 1,352.34 nJ. To make the energy consumption per data and total power consumption comparable between synchronous FIR filter and MTNCL FIR filter, the MTNCL FIR filter is not pipelined and the functioning speed is also limited by the I/O logic implemented to shift data in and out of the filter. Similar to synchronous FIR filter, the nonpipelined design and the I/O logic has negative impact on the energy consumption per data for the MTNCL FIR filters. The total power consumption measured has a range from 2.23 μW to 2.52 μW.

The MTNCL FIR filter is compared with the synchronous FIR filter for operating speed and energy efficiency as shown in Figure 4.12. Based on the measured clock period of synchronous FIR filter and the T_{dd} captured of MTNCL FIR filter, the MTNCL FIR filter has 1.4× higher operating speed compared to synchronous FIR filter at 300 mV V_{DD}. For energy consumption per data, the MTNCL FIR filter is 1.5× lower than its synchronous counterpart. The operating speed and energy consumption comparison indicates that FIR filter based on MTNCL has higher speed and lower energy consumption when operating within the subthreshold region at 300 mV V_{DD}. For FIR filter ultra-low supply voltage operation, the filter implemented with MTNCL is able to reduce more power consumption without sacrificing throughout compared to the filter implemented with conventional Boolean logic. Therefore, the MTNCL FIR filter is proven to be more energy efficient than asynchronous FIR filter when transistors are operating

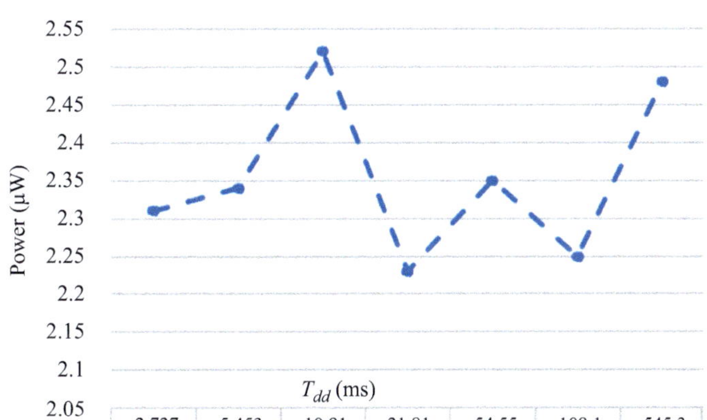

	2.727	5.453	10.91	21.81	54.55	109.1	545.3
Power (μW)	2.31	2.34	2.52	2.23	2.35	2.25	2.48

Figure 4.10 MTNCL FIR filter total power consumption

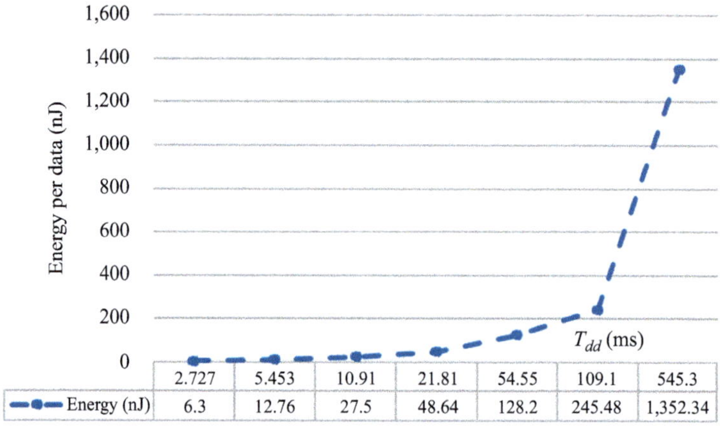

Figure 4.11 MTNCL FIR filter energy consumption per data

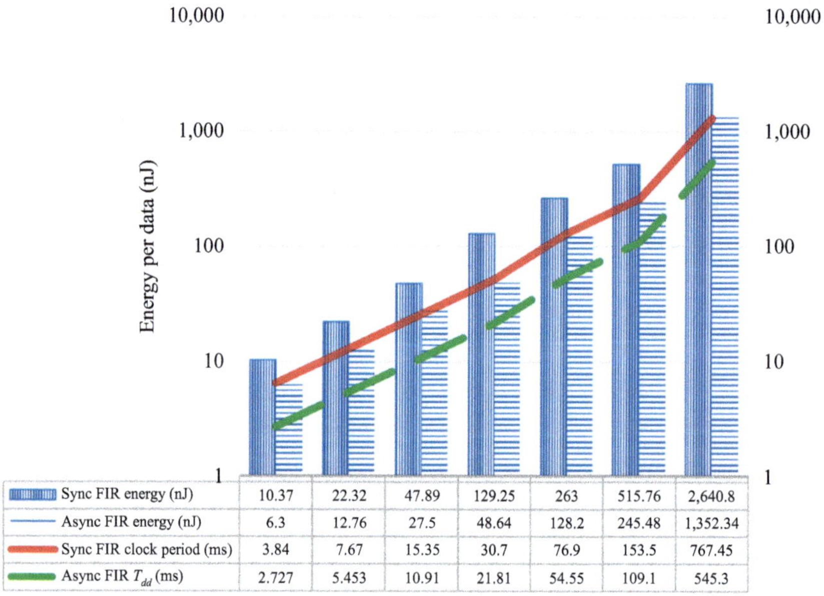

Figure 4.12 Performance and energy comparison for synchronous and MTNCL FIR filter

in the subthreshold regime. The drawback of asynchronous FIR filter is the larger gate count and area required to implement handshaking logics specific to MTNCL.

The last asynchronous design tested is MTNCL homogeneous parallel data processing platform with 4 FIR filter cores. The platform functionality testing is with 300 mV V_{DD} and the die is backside biased at a range of -1.9 V to -2.24 V.

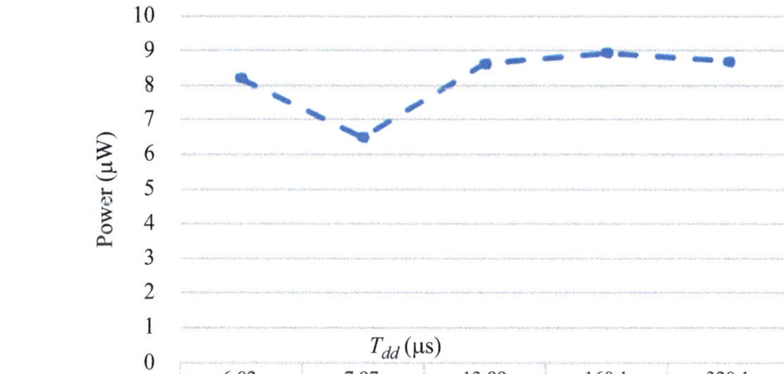

T_{dd} (μs)	6.02	7.97	13.99	160.1	320.1
Power (μW)	8.2	6.5	8.63	8.93	8.7

Figure 4.13 MTNCL platform total power consumption

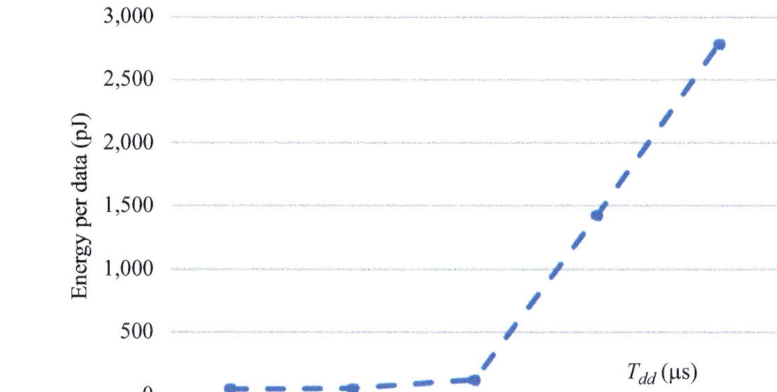

T_{dd} (μs)	6.02	7.97	13.99	160.1	320.1
Energy (pJ)	49.364	51.81	120.73	1,429.7	2,784.9

Figure 4.14 MTNCL platform energy consumption per data

The platform has two handshaking signal outputs, *Ko* and *sleepout*. After reset, *Ko* should be toggling if the circuit is functioning correctly. Once the handshaking mechanism is established in the platform, the current drawn from the power supply by the platform is measured for a period of time by the current adapter and mixed signal oscilloscope. Based on the current drawn, the total power consumption and energy consumption per data are quantified and shown in Figures 4.13 and 4.14. The T_{dd} captured has a range from 6.02 μs to 320.1 μs with corresponding energy per data from 49.364 pJ to 2784.9 pJ. The operating speed of the platform with 4 MTNCL FIR filter cores is higher than previously tested single MTNCL FIR filter due to the platform's implementation of parallelism. The parallel architecture realized by incorporating 4 MTNCL FIR filter cores has significantly improved throughout compared to a single MTNCL FIR filter. The MTNCL homogeneous parallel data processing platform is a complex design and therefore having larger core size. The increased height and width of the core allowed the placement of power and ground pads with enough space for all the primary input and output pads. There is no I/O logic required in the platform since enough I/O pads are implemented for data inputs and outputs. Due to the platform's much higher operating speed, the energy consumption per data is much lower than the MTNCL FIR filter. The total power consumption of the platform ranges from 6.5 μW to 8.93 μW.

4.5 Conclusion

Complex digital system's performance and long term reliability is directly related to the system's ability to use energy efficiently. Digital systems designed with energy efficiency in mind will enjoy the benefits of reduced power consumption and operating cost as well as lowered heat generation and improved system robustness. Reduce the supply voltage of digital system is one of the most effective methods of limiting dynamic power consumption. With the continuing demand to improve energy efficiency, the system supply voltage will surely be reduced further and further. Once supply voltage is scaled lower than the transistor's threshold voltage, process technology with transistor optimized for subthreshold operation becomes critical to ensure the ultra-low power operation of the digital system without sacrificing performance and reliability. Lowering the supply voltage and operate transistors in subthreshold region has the advantage of reduced power consumption. However, subthreshold leakage current will increase with the lowering of supply voltage. To prevent excessive power consumption induced by subthreshold leakage current, FDSOI technology is being adopted over bulk CMOS technology due to its inherently low leakage characteristic. The MITLL 90 nm XLP FDSOI CMOS process has provided the capability of ultra-low supply voltage operation while preventing excessive subthreshold leakage power consumption.

Asynchronous logic like MTNCL presents itself as a suitable choice for designing circuits intended for ultra-low supply voltage operation. MTNCL designs are delay-insensitive, not subject to timing constraints, and won't be affected by signal integrity issues like noise and cross-talk. Designs operating at ultra-low supply voltage has the advantage of significantly reduced switching power consumption. However, the ultra-low voltage level will induce signal integrity issues for signals traveling in wires throughout the design. The lower the voltage level, the easier a signal will be corrupted by noise of nearby components or becomes a cross-talk victim of nearby wires. For synchronous designs where clock tree timing and data path timing are critical to the design's correct operation, extra shielding are required for clock routing. For data path with long wires, extra buffers are required to segment the wire to reduce noise and prevent cross-talk effect from negatively affecting the path timing. Asynchronous designs are insensitive to delay and doesn't require clock signals to propagate data, which makes the design much more robust against noise and cross-talk effect even at ultra-low supply voltage level. Complex system designed in asynchronous style will be able to benefit from the significantly reduced switching power consumption of ultra-low supply voltage operation without sacrificing system throughput and long-term resiliency to degradation introduced by process, voltage, and temperature variation.

Five synchronous and asynchronous circuits are designed and implemented with MITLL 90 nm XLP FDSOI CMOS process. The process provides transistors optimized for subthreshold operation at 300 mV V_{DD}. The ultra-low supply voltage and FDOSI technology have enabled significant reduction in switching and leakage power consumption of both synchronous and asynchronous circuits. All of the five synchronous and asynchronous circuits fabricated are physically tested to be functioning correctly in the subthreshold region of 300 mV V_{DD} with body biasing between −1 V and −2 V. The synchronous and NCL ring oscillators are observed to have oscillating frequency from 0.99 MHz to 1.78 MHz. Both synchronous and NCL ring oscillators placed at location #1 on the die has higher oscillating frequency than those placed at location #2, reflecting inter-die process variation. The synchronous FIR filter has total power consumption ranging from 2.7 μW to 4.21 μW while the MTNCL FIR filter has total power consumption ranging from 2.23 μW to 2.52 μW. The synchronous FIR filter is inferior in terms of operating speed and energy efficiency compared to MTNCL FIR filter. The MTNCL FIR filter has an approximately 1.4× higher functioning speed and 1.5× lower energy consumption per data, indicating that MTNCL designs are advantageous for ultra-low supply voltage operation. Furthermore, the successful functional, performance, and power testing of MTNCL homogeneous parallel data processing platform with four FIR filter cores suggests that complex multicore systems can take advantage of the ultra-low supply voltage offered by the process with transistors operating in subthreshold region to achieve improved energy efficiency and system throughput, which in the long run, is beneficial to the system's operating cost and reliability.

References

[1] Beerel PA, Roncken ME. Low power and energy efficient asynchronous design. Journal of Low Power Electronics. 2007;3(3):234–53.

[2] Pelloie JL. SOI for low-power low-voltage-bulk versus SOI. Microelectronic Engineering. 1997;39(1–4):155–66.

[3] Rabaey J. *Low Power Design Essentials*. Springer Science & Business Media; 2009.

[4] Ellis CS. Controlling energy demand in mobile computing systems. Synthesis Lectures on Mobile and Pervasive Computing. 2007;2(1):1–89.

[5] Arora ND. *MOSFET Models for VLSI Circuit Simulation: Theory and Practice*. Springer Science & Business Media; 2012.

[6] Sze SM, Ng KK. *Physics of Semiconductor Devices*. John Wiley & Sons; 2006.

[7] Cheung KP. On the 60 mV/dec@ 300 K limit for MOSFET subthreshold swing. In *2010 International Symposium on VLSI Technology Systems and Applications (VLSI-TSA)*, April 26, 2010 (pp. 72–73). IEEE.

[8] Yeo KS, Roy K. *Low Voltage, Low Power VLSI Subsystems*. New York: McGraw-Hill; 2005.

[9] Colinge JP. *Silicon-on-Insulator Technology: Materials to VLSI*. Springer Science & Business Media; 2004.

[10] Vitale SA, Wyatt PW, Checka N, Kedzierski J, Keast CL. FDSOI process technology for subthreshold-operation ultralow-power electronics. Proceedings of the IEEE. 2010;98(2):333–42.

[11] Smith SC, Di J. Designing asynchronous circuits using NULL convention logic (NCL). Synthesis Lectures on Digital Circuits and Systems. 2009;4(1):1–96.

[12] Thian R, Caley L, Arthurs A, Hollosi B, Di J. An automated design flow framework for delay-insensitive asynchronous circuits. In *2012 Proceedings of IEEE Southeastcon*, March 15, 2012 (pp. 1–5). IEEE.

[13] Allier E, Sicard G, Fesquet L, Renaudin M. A new class of asynchronous A/D converters based on time quantization. In *2003 Proceedings of the Ninth International Symposium on Asynchronous Circuits and Systems*, May 12, 2003 (pp. 196–205). IEEE.

[14] Alacoque L, Renaudin M, Nicolle S. Irregular sampling and local quantification scheme AD converter. Electronics Letters. 2003;39(3):1.

[15] Men L, Di J. An asynchronous finite impulse response filter design for digital signal processing circuit. In *2014 IEEE 57th International Midwest Symposium on Circuits and Systems (MWSCAS)*, August 3, 2014 (pp. 25–28). IEEE.

[16] Herbert S, Marculescu D. Variation-aware dynamic voltage/frequency scaling. In *IEEE 15th International Symposium on High Performance Computer Architecture, 2009. HPCA 2009*, February 14, 2009 (pp. 301–12). IEEE.

[17] Men L, Di J. Asynchronous parallel platforms with balanced performance and energy. Journal of Low Power Electronics. 2014;10(4):566–79.

[18] Men L, Hollosi B, Di J. Framework of an adaptive delay-insensitive asynchronous platform for energy efficiency. In *2014 IEEE Computer Society Annual Symposium on VLSI (ISVLSI)*, July 9, 2014 (pp. 7–12). IEEE.

[19] Jones D. Projects and circuits: the μCurrent—use this precision multimeter current adaptor to make truly accurate readings. Everyday Practical Electronics. 2011:40(5):10.

Chapter 5
Asynchronous circuits for interfacing with analog electronics

Paul Shepherd[1] *and Anthony Matthew Francis*[2]

Very few integrated circuits, whether they are microprocessors or ASICs, are designed with purely digital or analog components. Mixed-signal circuits and systems straddle this divide, and this chapter examines how analog components may be interfaced with asynchronous logic. This chapter presents two example systems, which cover three methods of closing the feedback loop to maintain asynchronous operation. In the first example, an asynchronous serializer/deserializer (SerDes), the analog components have known end-states and are physically included in the loop of the asynchronous logic stage. Since the completion detection occurs directly, this system maintains its quasi delay-insensitive (DI) operation. For the second example, a successive approximation analog-to-digital converter (SAR ADC), the circuit cannot be fully included in the loop, and two options for maintaining asynchronous operation are described.

Due to the nature of the authors' interests and respective circuit applications, the example circuits also work in spatially distributed implementations. In these cases, different parts of the circuit may have significantly different operating environments, and may even be designed in different IC fabrication processes.

Some readers of this chapter will be analog or mixed-signal designers who are curious about how delay insensitive logic can be used in their work. For that audience, a brief overview of the ring oscillator metaphor is presented here and will clarify the behaviors described later in the examples.

5.1 The ring oscillator metaphor

DI logic can be divided up into registration stages, much like standard or "clocked" register transfer logic (RTL). Rather than being completed by a register, DI logic registration stages are "completed" by completion logic represented as a DI register. Figure 5.1 shows two such DI registers with a cloud of combinational logic between them. A cycle of events proceeds around the ring formed by the

[1]Benchmark Space Systems, South Burlington, VT, USA
[2]Ozark Integrated Circuits, Fayetteville, AR, USA

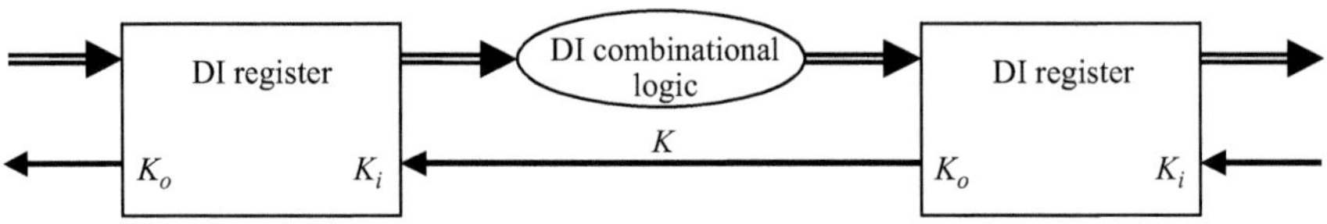

Figure 5.1 A pair of DI registers creates a feedback loop

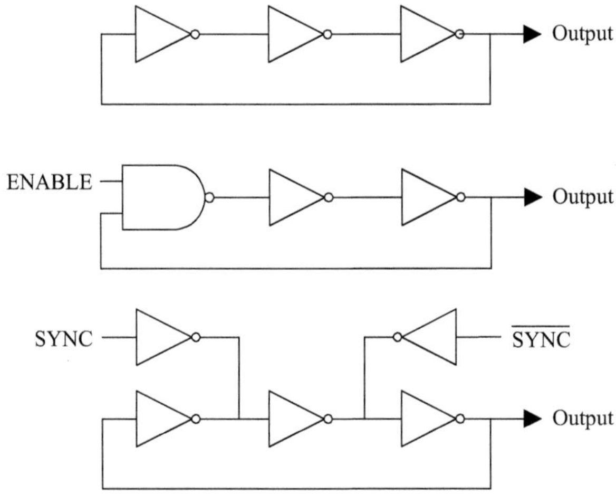

Figure 5.2 A simple ring oscillator (top) can be modified to become a gated ring oscillator (middle) or a synchronized ring oscillator (bottom)

registers and internal logic, and the events inside one loop interact with the events in the surrounding loops. Events that are moving on the forward side of the ring are called wavefronts (NULL or DATA), and events which propagate on the backward side of the ring are called completion signals (request-for-data or request-for-null). The speed at which these events propagate around a single ring is dependent upon many factors such as how large the ring is physically and logically, the environmental (temperature, supply voltage) impacts on transistor performance, and the power supply voltage and device parasitics. This event propagation is very similar to the behavior of a Boolean ring oscillator.

A ring of N inverters is essentially a simple state machine. When N is even, the state machine has two stable states, but when N is odd, the state machine has $N + 1$ unstable states, and is a ring oscillator. Replacing one inverter with a NAND gate turns a simple ring oscillator into a gated ring oscillator. Adding parallel inverters provides a synchronization input. In both cases, the ring oscillator's internal state is made dependent on the behavior of surrounding circuits. These variations are shown in Figure 5.2 below.

A single DI stage has two gating inputs: the logic input on the left side, and the completion input (*Ki*) on the right side. It also has two outputs: the logic output on the right side, and the completion output (*Ko*) on the left. The internal state of the stage is defined by the signal (either DATA or NULL), and the internal completion line (either request-for-data (RfD) or request-for-null (RfN)). The state machine transitions when one of the inputs changes state, and the outputs change along with the internal state variables. Figure 5.3 shows the DI stage and its state diagram.

The *Ki* and signal inputs act as gating inputs, but a more complex stage is necessary to illustrate synchronization of the stage with adjacent stages. Consider the stages shown in Figure 5.4: the DI register on the lower-left feeds both registers

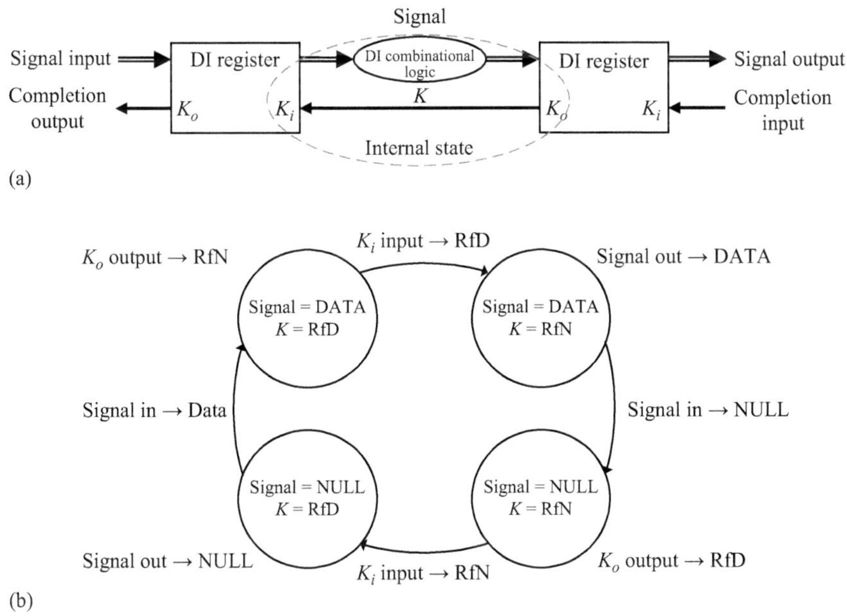

Figure 5.3 (a) A single DI stage with the internal state indicated and (b) the state-transition inputs and outputs

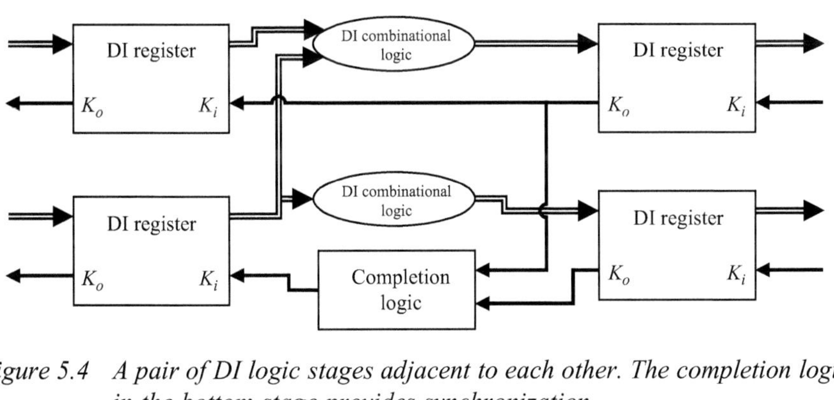

Figure 5.4 A pair of DI logic stages adjacent to each other. The completion logic in the bottom stage provides synchronization

on the right. The completion logic on the bottom acts to synchronize the state transitions along the top and bottom paths. Like the synchronized ring oscillator, internal nodes of the simple state machine become inputs in this case.

The asynchronous SerDes shown below demonstrates the ring oscillator behavior clearly and is a good first case study for those new to DI mixed-signal systems.

5.2 Example applications

5.2.1 An asynchronous serializer/deserializer utilizing a full-duplex RS-485 link

One of the most common intersections of analog and digital circuitry is at a communication physical layer. RS-485 is an incredibly robust and common link that is commonly used in environments requiring noice-immune communications and vary in complexity from point-to-point, single-direction transmission to multipoint bidirectional communication.

As will be presented, RS-485 communication can be accomplished by combining NCL and analog differential signaling (RS-485 physical layer) to seamlessly transmit data over short or long distances and wide operating temperatures [1]. Synchronous logic requires universal asynchronous receiver/transmitters (UARTs) to interface with the RS-485 link, and the UART includes, at a minimum, a serializer/deserializer (SerDes), transmit-clock generation and recovery circuits, and transmit/receive control logic. The clock recovery circuit is often implemented as a phase-locked loop frequency multiplier. Figure 5.5 shows the trade-in complexity between a system with a UART and the DI system described in this chapter. The DI system requires a full-duplex RS-485 link but does away with the clock generation and recovery circuits (naturally).

The idea of using an RS-485 transceiver pair to send NCL logic signals over long distances came from the recognition that an RS-485 bus has three states: TRUE, FALSE, and IDLE. These bus states map easily to the TRUE, FALSE, and NULL states of NCL logic. Transceivers contain both a driver and receiver and most can be used in full-duplex mode, where the send and receive pairs are unconnected, which mimics the forward and completion path of NCL. The most significant difference between RS-485 and NCL is that RS-485 transceivers are designed to interface with Boolean logic and incorporate failsafe biasing to report an IDLE bus as a logic 0 [2]. Failsafe biasing causes the RO output to be a FALSE when the bus is in its IDLE state.

An additional IDLE detection was added to a standard RS-485 transceiver to make it compatible with NCL (Figure 5.6). The bus idle output (BIO) signal can be combined with the receive output (RO) signal to generate the valid NCL rail states. The circuit does not do away with failsafe biasing; it actually mirrors it on a second comparator. When the output of the two comparators agrees, the bus state is a DATA signal, but when they disagree, the bus is IDLE, corresponding to an NCL NULL signal. The internal circuitry of the receiver is shown in Figure 5.7.

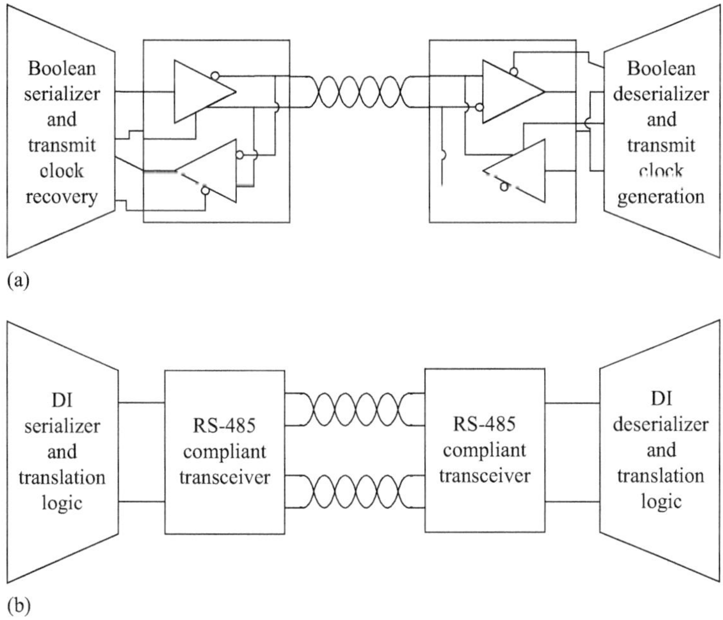

Figure 5.5 Comparison of a point-to-point RS-485 link in a synchronous logic system (a) versus the DI RS-485 link described below (b)

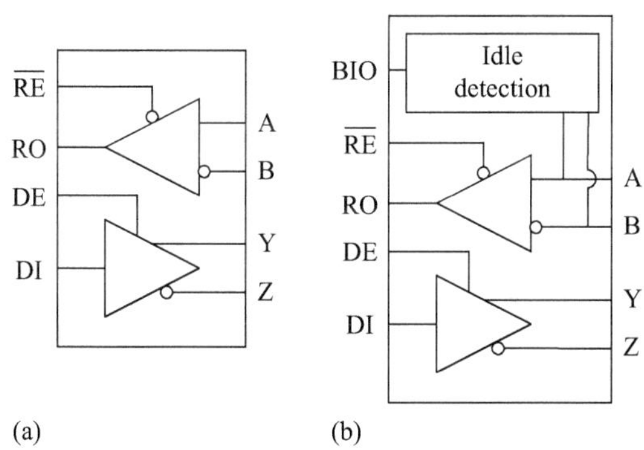

Figure 5.6 A common full-duplex RS-485 transceiver (a) compared with one modified to recognize a bus IDLE state (b)

With the addition of the bus idle detection, the complete 8-bit SerDes is realized with slave-side and master-side finite state machines and a small amount of glue logic. The complete system is shown in Figure 5.8. A key feature of this implementation is that the glue logic, RS-485 transceivers, and twisted pair could be removed from the system, and the slave and master finite state machines would

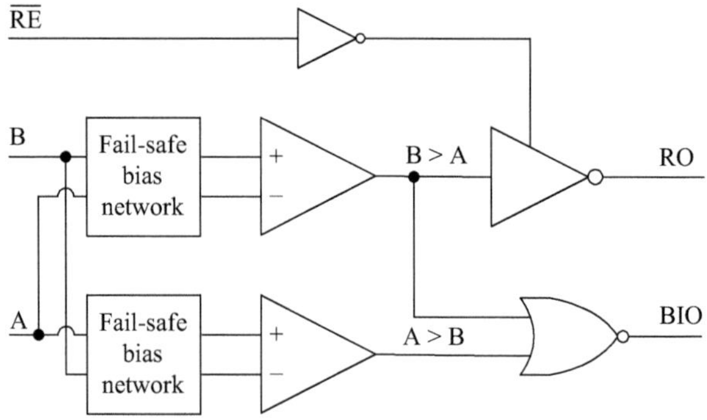

Figure 5.7 Generating the bus idle output using two comparators with symmetric fail-safe biasing

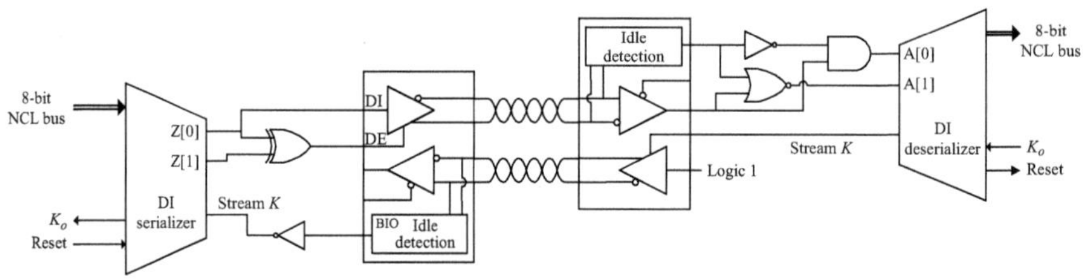

Figure 5.8 Complete asynchronous SerDes with RS-485 link. From left to right are the DI serializer (slave FSM), slave glue logic, slave RS-485 transceiver, full-duplex twisted pair, master RS-485 transceiver, master glue logic, and DI deserializer (master FSM)

still function perfectly. They implement a more complex FSM similar to the one presented in Figure 5.3, as shown in Figure 5.9. The block diagram of the complete asynchronous serializer/deserializer is shown in Figure 5.10.

5.2.2 Fully asynchronous successive approximation analog to digital converters

Even more than communication, the most obvious intersection of digital logic and analog circuitry is at the data converter, in both digital-to-analog (DAC) and analog-to-digital (ADC) operations. While "clocked" converters are essentially synchronous in nature, as will be shown the DI paradigm can be incorporated (with power consumption advantages) in such converters through judicious integration of "completion" concepts inherent in most ADC designs.

The successive approximation analog-to-digital converter (SAR ADC) is a common circuit used in mixed-signal systems. Any N-bit SAR ADC can be thought of as a finite state machine (FSM) with $N + 3$ states as shown in Figure 5.11. The WAIT state is the trivial state between active conversions, and the other states can

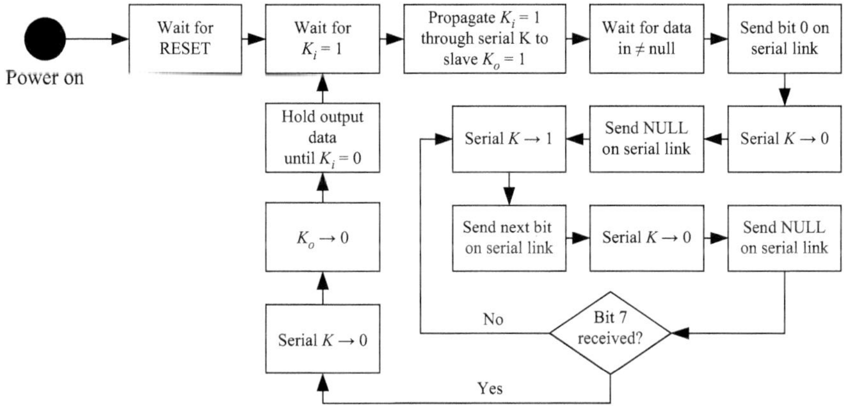

Figure 5.9 DI serializer/deserializer finite state machine

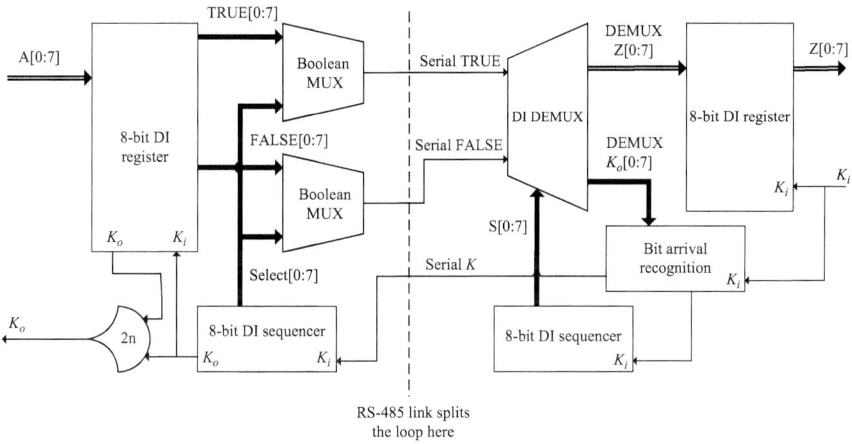

Figure 5.10 The complete asynchronous serializer/deserializer without the RS-485 link in the loop. The serializer is on the left side of the dashed line, and the deserializer is on the right side of the dashed line

be further subdivided into two sets: the SAMPLE and HOLD states, and the COMPARE states.

The concept of a DI SAR ADC is not a new one [3], but the term asynchronous SAR ADC has been commonly applied to a class of circuits which aren't fully asynchronous. These circuits transition through the COMPARE states asynchronously but are generally designed for bounded-delay operation in order to work with external sampling clocks. In order to differentiate the circuits described here, the names *Fully Asynchronous SAR ADC* or *Delay-Insensitive (DI) Asynchronous SAR ADC* are used.

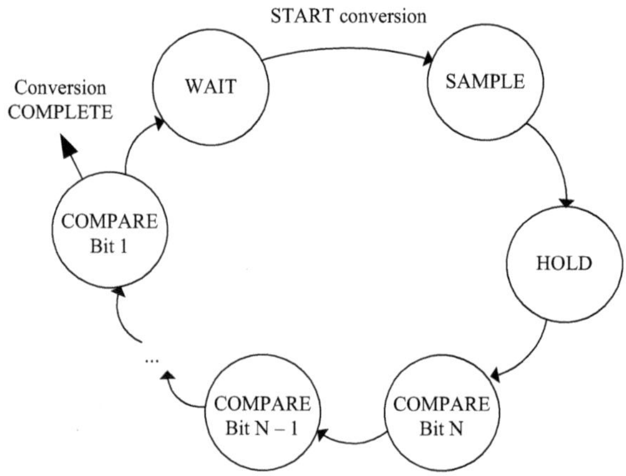

Figure 5.11 *FSM diagram showing the N + 3 states of an N-bit SAR ADC. For a synchronous system, the clock drives the transitions between states*

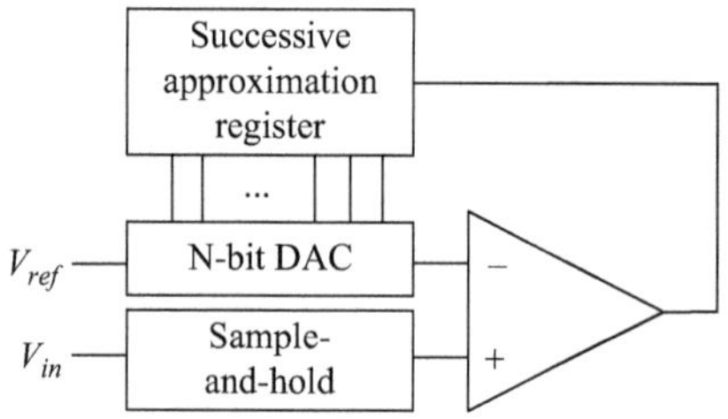

Figure 5.12 *The most basic blocks of a successive-approximation ADC. The N-bit DAC is the switched capacitor array which generates the successive approximation voltages from V_{ref}*

5.2.2.1 Basic operation of the successive approximation analog-to-digital converter

There are many excellent references on the operation of the traditional SAR ADC [4,5]. An array of switched capacitors is used to create the successive-approximation voltage. This combination of switches and capacitors is often referred to as the internal DAC since the switching is digitally controlled and leads to an analog voltage for comparison to the input voltage. A block diagram of the fundamental SAR ADC is shown in Figure 5.12. A subtle variation is the charge-redistribution SAR ADC, shown in Figure 5.13, where the switched capacitor network stores the comparison voltage minus the input voltage and compares this value to ground. In this chapter a singled ended and a differential implementation of the charge-redistribution SAR ADC are considered.

5.2.2.2 Asynchronous input voltage sampling

As previously mentioned, most SAR ADCs described as asynchronous still require a sampling clock for correct operation. Overcoming this requirement to achieve a truly clockless system is an area of active research, but two promising design patterns are described here. For a fully asynchronous SAR ADC, the key challenge is determining the duration of the SAMPLE state. During this state, the input voltage is stored on a capacitor as shown in Figure 5.13. The minimum charge time required during the SAMPLE state is a function of the series resistance from the input and the required precision. (A SAR ADC with a certain number of bits may have a lower effective number of bits (ENOB) due to noise and reduced sampling time, among other causes.) The duration of the HOLD state is often trivial in comparison to the duration of SAMPLE state. During this state the input voltage is disconnected, and any other preparations are made to start the COMPARE states. The equivalent circuit during a COMPARE state is shown in Figure 5.14.

There are two approaches to setting the duration of the SAMPLE state. The first is to specify what the minimum duration should be (the same way it would be done for a synchronous SAR ADC) and introduce that delay into the completion detection of the asynchronous FSM controlling the system. This approach is simple and requires few additional components, but inserting a fixed delay turns an otherwise DI system into a bounded-delay system. The second option is to model the charging process with another circuit that has known start- and end-states,

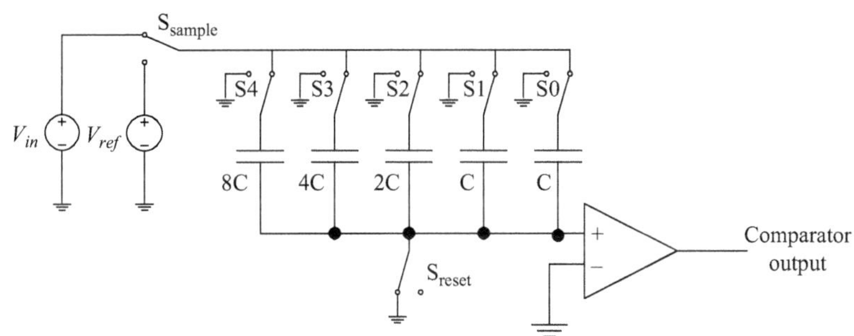

Figure 5.13 Charge redistribution circuit of a four-bit single-ended SAR ADC. Switch positions are shown for the SAMPLE state

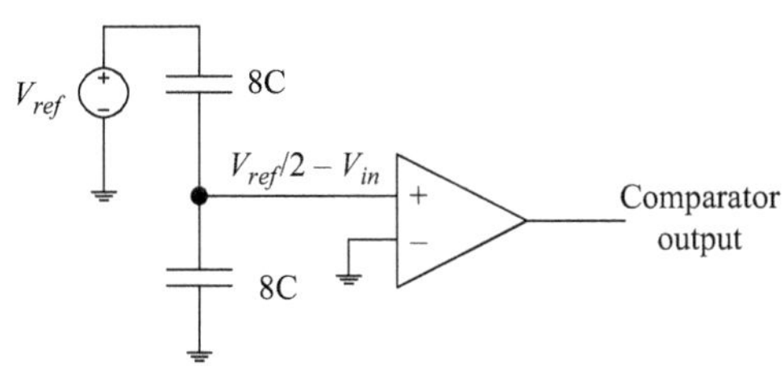

Figure 5.14 Equivalent circuit while testing Bit 4 (MSB)

allowing an indirect method to observe completion of the SAMPLE state. This has the benefit of maintaining the DI nature of the system, but at the cost of significantly higher complexity. This approach requires two input buffers with shared biasing to match their performance, and a second sampling capacitor and comparator.

In both cases, an estimate of the minimum duration of the SAMPLE phase is necessary to design the completion circuit. Any charging error will be unrecoverable later during the COMPARE states. The charging time is a function of the current capacity of the input buffer, the sampling capacitance, and the difference between the input voltage and the starting voltage (ΔV). The worst-case scenario is when ΔV is equal to the full-scale voltage.

$$\Delta V_{\text{sample}}(t) = \frac{1}{C}\int_0^{t_{\text{sample}}} I(\tau)d\tau$$

$$\%\ \text{Error}(t) = \frac{\Delta V_{\text{sample}}(t)}{V_{\text{fullscale}}}$$

$$\%\ \text{Error}(t) = \frac{V_{fs}}{C}\int_0^{t_{\text{sample}}} I(\tau)d\tau$$

For an N-bit converter, the sampling error should be less than ½ to ¼ LSB, or $1/2^{N+2}$. The minimum sampling time can be computed for a given value of reference voltage and an output resistance of the input buffer. If the output impedance of the voltage buffer is approximately resistive, then the time required is directly related to the RC time constant of the circuit (I_{sc} is the short-circuit output current):

$$\frac{1}{2^{N+1}} \geq \%\ \text{Error}(t) = e^{-t\frac{I_{sc}}{V_{fs}C}} \rightarrow \log\left(\frac{1}{2^{N+2}}\right) \geq -\frac{tI_{sc}}{V_{fs}C}$$

$$t \geq \frac{V_{fs}C}{I_{sc}}\log\left(2^{N+2}\right) = \tau\log\left(2^{N+2}\right),\ \frac{V_{fs}C}{I_{sc}} = \tau$$

This value is a useful practical sampling time. For example, 8-bit SAR ADCs should sample for at least 7τ, 12-bit SAR ADCs for at least 10τ, and 16-bit SAR ADCs for at least 12.5τ (Table 5.1). For the case where the input buffer acts as a perfect current source/sink into the capacitors, the charging time could be much shorter.

For the bounded-delay approach shown in Figure 5.15, the delay is set by the RC time constant and threshold voltage of the Schmitt trigger. Determination of the charging time must take into account circuit parasitics and process corners, or even better, be measured experimentally. It is possible to use this method without an active input buffer, but the source impedance of the input voltage must also be considered. In the simplest implementation, this approach can be extremely space-efficient, especially if the capacitor and resistor were realized as discrete components.

Table 5.1 Minimum time to sample input voltage for a converter with a given number of bits

N bits	**Minimum sampling time**
8	6.93τ
12	9.70τ
16	12.48τ

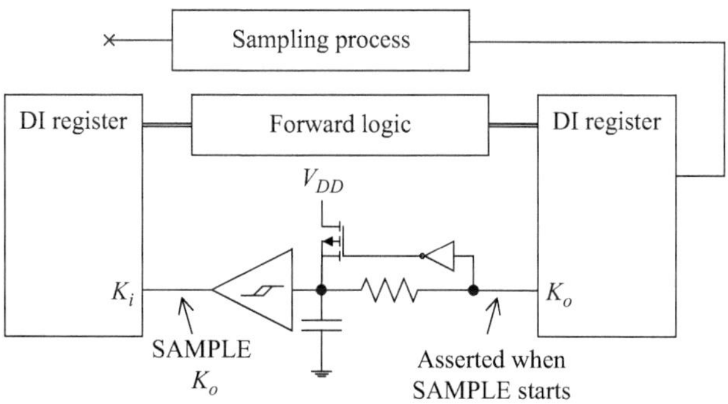

Figure 5.15 Inserting a timing delay in the completion path of the FSM. Since the sampling process speed isn't measured, the local circuit becomes bounded-delay

The second approach, shown in Figure 5.16, is more robust at the cost of higher complexity. In this approach, two input buffers with shared biasing are used. The primary input buffer charges the sampling capacitor to V_{in}, while the secondary buffer charges a timing capacitor. The secondary buffer's output stage and the timing capacitor can be scaled down but require a shared biasing circuit with the primary buffer in order to maintain consistent a ratio. The SAMPLE completion detection signal will be generated once the sampling capacitor crosses the Schmitt trigger threshold.

The HOLD state can be much shorter in duration than the SAMPLE state. The HOLD state provides an opportunity to set all switches to the appropriate states before starting the COMPARE state, and timing can be handled with the bounded-delay approach of Figure 5.15. This approach can also be used for each of the following COMPARE states, although there is also a DI solution for completion detection during these states.

5.2.2.3 Asynchronous voltage comparisons

The techniques developed for Asynchronous SAR ADCs are equally applicable to the fully Asynchronous SAR ADC during the COMPARE states. A differential

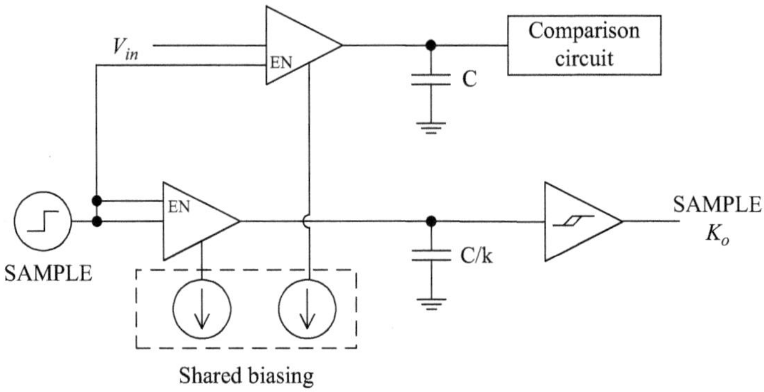

Figure 5.16 Mirroring the charging process to maintain the DI completion path. The biasing of the charging amplifier, capacitor value, and Schmitt trigger threshold must be sized to match the worst-case charging time for the comparison capacitor

sampling/comparison network combined with a differential-in/differential-out comparator allows completion detection in addition to determining the bit value [6]. In the context of a sampling clock, these circuits assume bounded-delay asynchronous operation, and the benefit that is touted is that most of the COMPARE states are completed quickly, leaving extra time for comparisons where the input voltage difference is small and the comparator response time is long. In fact, an internal timeout circuit similar to the bounded-delay completion detection in Figure 5.15 is recommended for these circuits to avoid incomplete conversions [7]. The sample and comparison process of the clocked, asynchronous, and fully asynchronous SAR ADC are shown in Figure 5.17.

The risk of an incorrect conversion in a bounded-delay asynchronous SAR ADC is called metastability and is a function of the comparator's regeneration time and the time available for the conversion. Comparators have a characteristic regeneration time, τ, and theoretically for a 0 V differential input, they will never converge to a logic-level output state. Fortunately, in the presence of noise, this cannot be the case. In [8], the author shows that the probability of an incomplete comparison (P) decreases as a function of the time available for the comparison (T_{avl}):

$$P = \frac{2V_{\text{logic}}e^{-\frac{T_{avl}}{\tau}}}{V_{fs}}$$

In a system which is designed to be DI instead of bounded-delay, the probability will go to zero, and all conversions can be completed. Although incomplete conversion is no longer an error source, other sources such as offset and noise still exist. τ is a function of the device transconductance, parasitic capacitance, and load capacitance, and can be reduced with wider devices up to a point [9].

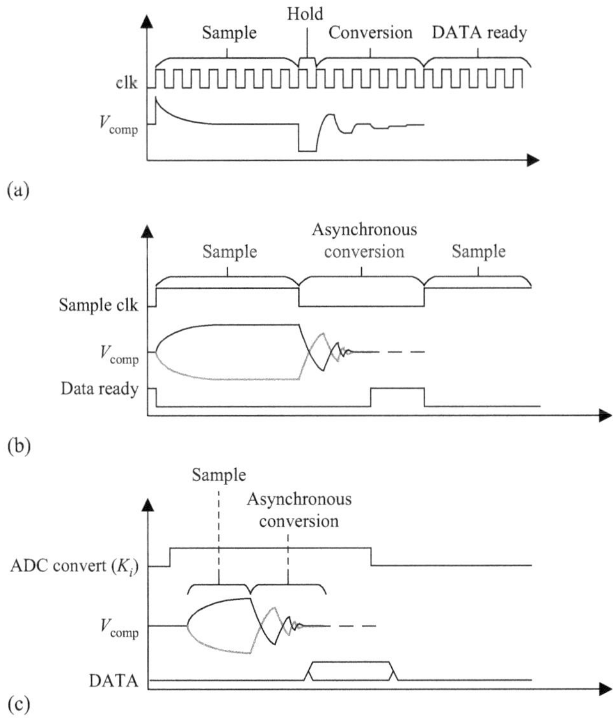

Figure 5.17 Clocked SAR ADC (a), asynchronous SAR ADC (b), and fully asynchronous SAR ADC (c). Note that the asynchronous and fully asynchronous SAR ADCs must be differential, and the positive and negative comparator inputs are shown in black and gray

5.3 Conclusion

In many ways the interface of asynchronous logic to analog circuitry has the same considerations of interface between clocked logic and analog circuitry. A strategy for transitioning between the analog world and digital must be considered, where analog design must consider many process, temperature, and other environmental concerns that digital circuitry can more easily abstract. As has been illustrated, especially with respect to communication, asynchronous DI interfaces can lend themselves to elegant, simplified interfaces. Fully embracing the clockless paradigm simplifies the analog designer's job in comparison to designing clocked communication interfaces, where the challenges of clock domain coherency are commonly left as a challenge to the analog designer.

However, there are also challenges, in the case of data conversion, where the interface of the clockless paradigm to a fundamentally clocked universe requires outside the box thinking, for example, what does a sample mean if not captured with a coherent time-base? As illustrated, the solution requires embracing ownership of the clockless concept to the analog design and embracing methods to map the equivalent DI logic concept to the analog domain (delay insensitivity, logic completion, power supply insensitivity, etc.). While challenging, the rewards in terms of circuit performance, power consumption, and process variation insensitivity can be immense and well worth the effort.

References

[1] P. Shepherd, S. C. Smith, J. Holmes, A. M. Franics, N. Chiolino and H. A. Mantooth, "A robust, wide-temperature data transmission system for space environments," in *2013 Aerospace Conference*, Big Sky, MT, 2013.

[2] T. Kugelstadt, "RS-485 failsafe biasing: old versus new transcievers," Texas Instruments Incorporated, Dallas, TX, 2013.

[3] T. Kocak, G. R. Harris and R. F. Demara, "Self-timed architecture for masked successive approximation analog-to-digital conversion," *Journal of Circuits, Systems and Computers*, vol. 16, no. 1, pp. 1–14, 2007.

[4] R. J. Baker, *CMOS Circuit Design, Layout, and Simulation*, 3rd edition, Wiley-IEEE Press, 2010.

[5] T. Kugelstadt, "www.ti.com," February 2000. [Online]. Available: http://www.ti.com.cn/cn/lit/an/slyt176/slyt176.pdf. [Accessed January 2019].

[6] S.-W. M. Chen and R. W. Brodersen, "A 6b 600MS/s 5.3mW asynchronous ADC in 0.13/spl mu/m CMOS," in *2006 IEEE International Solid State Circuits Conference—Digest of Technical Papers*, San Francisco, CA, 2006.

[7] Y. Zhu, C. Chan, S.-P. U and R. P. Martins, "A 10.4-ENOB 120MS/s SAR ADC with DAC linearity calibration in 90 nm CMOS," in *2013 IEEE Asian Solid-State Circuits Conference (A-SSCC)*, Singapore, 2013.

[8] P. M. Figueiredo, "Comparator metastability in the presence of noise," *IEEE Transactions on Circuits and Systems I: Regular Papers*, vol. 60, no. 5, pp. 1286–1299, 2013.

[9] C.-H. Chan, Y. Zhu, S.-W. Sin, B. Murmann, S.-P. U and R. P. Martins, "Metastability in SAR ADCs," *IEEE Transactions on Circuits and Systems—II: Express Briefs*, vol. 64, no. 2, pp. 111–115, 2017.

Chapter 6
Asynchronous sensing

Montek Singh[1]

A high-level view of a typical sensing system is shown in Figure 6.1. The underlying physical phenomenon is first converted to an electrical signal using a transducer, which may then be amplified before analog-to-digital conversion (ADC). The digital value is then subjected to digital signal processing (DSP) either in hardware or by software running on a processor. In some applications, for example, networked sensors, the processor may optionally also communicate the value to other sensor nodes.

There is a large universe of designs for each step of the sensing pipeline. In this chapter we focus on a few examples that best illustrate the promise of asynchronous techniques in the design of such sensing systems. We first focus on image sensors since they are complex enough to highlight several challenges posed by globally rigid timing approaches, and allow us to highlight the benefits of asynchronous pixels. Next, we discuss sensor network processors, and focus on the key twin demands of highly energy-efficient operation and low-area-cost implementation. Finally, we focus on DSP approaches and make the case for asynchronous, that is, continuous-time, signal processing for both increased energy efficiency, as well as lower spectral noise and lower aliasing by eliminating clock-based sampling.

6.1 Image sensors

While CCD image sensors were historically very successful, they have been all but replaced by CMOS sensors over the past decade or so. The reason for the proliferation of CMOS sensors is threefold: CMOS photodetectors are easier to integrate with peripheral processing logic, consume lower power, and offer higher capture rates. Furthermore, CMOS sensors also allow small amounts of processing logic to be embedded inside each pixel—for example, an amplification stage that significantly reduces noise—giving rise to what are called "active pixels sensors" (APS). Since CMOS APS sensors are now ubiquitous, we will focus our discussion in this section to such sensors only.

[1]Department of Computer Science, University of North Carolina, Chapel Hill, NC, USA

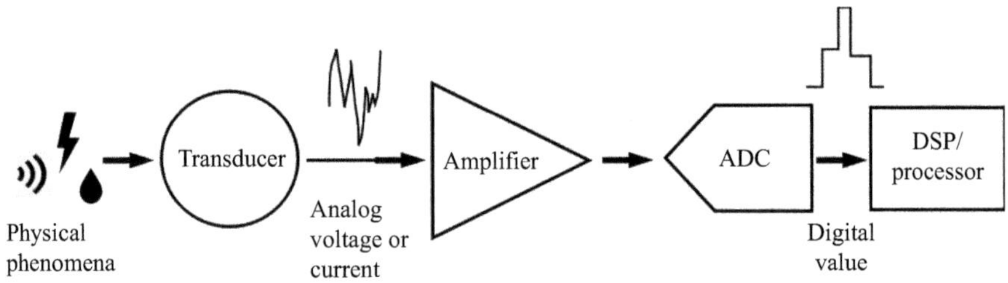

Figure 6.1 A high-level view of a sensing system

6.1.1 Frames versus frameless sensing

Traditional camera sensors are rooted in the notion of timed frames in which every pixel reports a new value proportional to the light gathered over a fixed period in a synchronized manner. This unnecessary synchronizing constraint imposes significant costs in terms of dynamic range, precision, and sensitivity. We contend that, given recent advances in hardware and software, a paradigm shift toward asynchronous sensing is both possible and tremendously advantageous.

Consider the key challenge of increasing the dynamic range that can be imaged. This range is the ratio of maximum measurable intensity before the sensor is saturated to the minimum intensity that can be distinguished from the noise floor. Saturation occurs when the photocurrent has completely discharged the voltage across the photodetector. A number of *high dynamic range* (HDR) approaches have been introduced to address the saturation problem. Common to many HDR approaches is the idea that pixels receiving a higher light intensity should be activated for a shorter time interval to avoid overflow. Similarly, darker pixels should be allowed to collect light for a sufficiently longer interval. The notion of using variable exposure times for different pixels is not a good match with the fixed time frame paradigm. Consequently, this problem is a good candidate for an asynchronous (i.e., frameless) solution.

We will describe two classes of asynchronous sensing approaches that abandon the notion of timed frames, and instead allow each pixel to independently measure light intensity free from the constraints of global timing. Let us emphasize at the outset that these approaches do not necessarily have to be implemented using clockless circuits. Instead, asynchrony is exploited at a higher level to eliminate the need for a fixed global shutter time. However, a clockless circuit implementation is certainly a natural fit for these sensing approaches, thereby allowing asynchrony to percolate from the system level down to the circuit level.

We first give background on traditional image sensing architectures. Then we present the two classes of asynchronous imaging systems: (i) spiking pixel sensors that convert light intensity into a pulse frequency; and (ii) sensors used in silicon retinas that logarithmically convert light intensity into voltage.

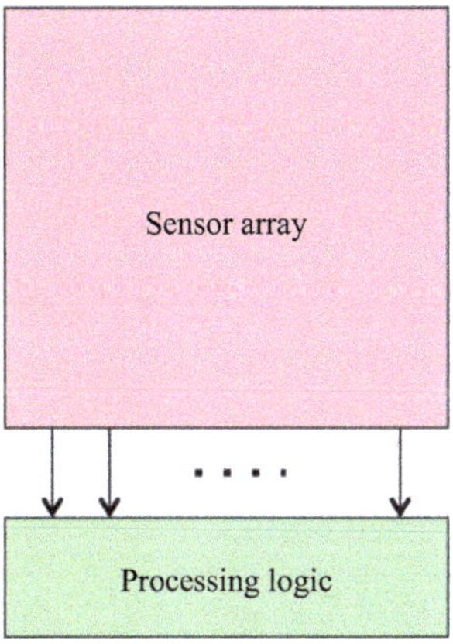

Figure 6.2 An image sensor

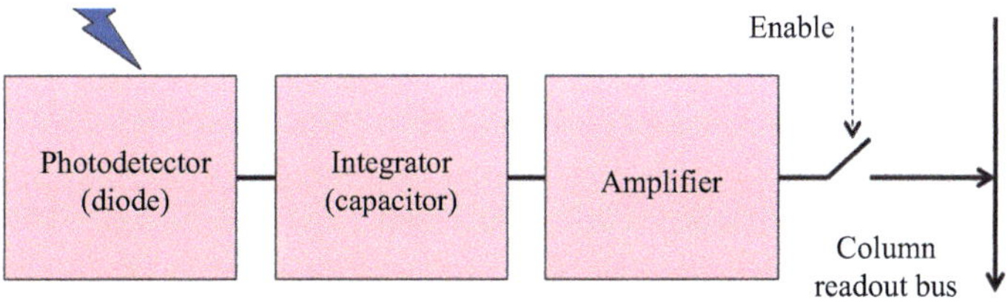

Figure 6.3 Inside a pixel: photodiode, integrator, and readout circuitry

6.1.2 Traditional (synchronous) image sensors

A typical camera sensor consists of an array of sensing elements ("sensels" or "pixels") and processing logic on its periphery. The sensor array is organized as multiple columns. All pixels within a column share the communication or readout circuitry that connects them to the processing logic. Most modern sensors use one analog-to-digital converter (ADC) per column. All modern sensors use what are called "active pixels", which include an in-pixel amplifier for amplifying the signal before transmitting it on the column readout bus.

Figure 6.2 shows the overall sensor array, and Figure 6.3 shows the main circuit components inside a single active pixel. The photodetector generates current proportional to the intensity of light incident on it. For good quality sensing, the photodetector must be physically sized as large as possible to collect the most number of photons. The integrator is a capacitor that integrates the photodetector current over the measuring interval. High intensities can cause the integrator to saturate. At low intensities, various sources of noise overwhelm the signal. The ratio of the maximum to minimum measurable intensities is the dynamic range of the sensor. The in-pixel amplifier (present in all "active-pixel sensors") and column readout vary, but typically consist of a small number of transistors (as few as 2 to 6).

The simplest implementation of an active pixel is shown in Figure 6.4. A reverse-biased diode is used as the photodetector. As photons are incident on the diode, photocurrent flows through the diode proportional to the light intensity. The diode's junction capacitance, C_{diode}, acts as the integrator of this photocurrent. In particular, before sensing, a reset pulse is applied to the reset transistor (M_{rst}), which recharges the voltage across the diode to a reset value (actually, $V_{rst}-V_{th}$, where V_{th} is the threshold drop across the reset transistor). When light is incident on the diode, the photocurrent I_{ph} discharges the diode voltage at a rate equal to I_{ph}/C_{diode}. The diode voltage is amplified by a source follower (M_{sf}), which provides isolation between the diode output and the column readout bus. At the end of the measurement interval, the switch (M_{sel}) connects the pixel's output to the column bus when the row containing this pixel is selected. Thus, central to the operation of a pixel is the notion of a measurement interval (i.e., frame time), over which the photocurrent is integrated to yield a measure of the light intensity.

The pixel's analog value is transferred to the readout bus that is shared by all the pixels in a column. Typically one analog-to-digital converter is used per column (per-pixel ADC is costly and hence only used in experimental or niche applications). The quantization step of this ADC determines the measurement granularity. All pixels in a row share the readout enable. The rows are read out sequentially, and typically result in a rolling shutter artifact since each pixel row is effectively imaged at a slightly different time instant. But there are global shutter designs where all pixel values are sampled at the same instant and buffered internally for sequential readout.

Though there are several slight variations on the theme, most commercial camera sensors today implement this architecture. Some implementations add an extra transistor to enable a sample-and-hold-operation, which allows global shutter, that is, all pixel outputs are sampled at the same instant even though the readout is performed sequentially by rows. In all of these approaches, obtaining a high dynamic range (HDR) remains a key objective.

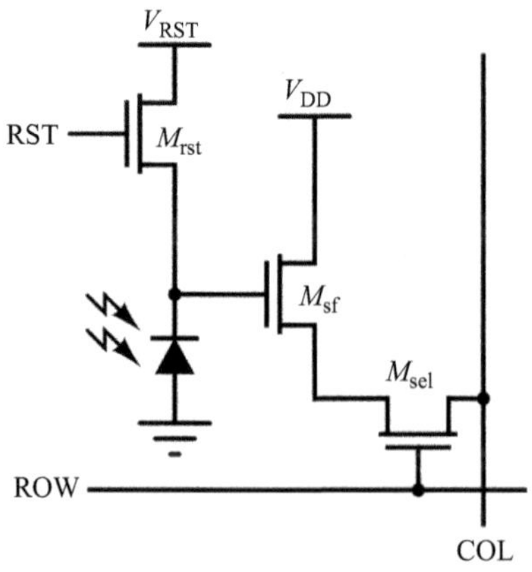

Figure 6.4 A three-transistor active-pixel sensor

Although a number of modifications have been proposed to obtain HDR operation, this is an area where asynchronous solutions outshine others. In the next two sections, we present two classes of asynchronous imaging sensors: spiking pixel sensors and logarithmic sensors.

6.1.3 Asynchronous spiking pixel sensors

Spiking pixel sensors are asynchronous image sensors that convert photocurrent into a pulse train whose frequency is proportional to the light intensity [1–9]. The key idea is to reset the integrator once a threshold is reached, record this event by emitting a "tick", and start over, thereby avoiding saturation. In effect, this is 1-bit ADC or delta-sigma modulation.

Figure 6.5 shows the basic circuit of a spiking pixel. Once the diode voltage has been discharged below a reference voltage (V_{ref}), the comparator kicks off a reset operation which quickly recharges the diode voltage. This in turn causes the comparator to reset, thereby generating a pulse on the output, and restarting the integration. The frequency of the pulse train generated is proportional to the incident light intensity.

There are two different ways in which the pixel value can be computed from this pulse train: either by (i) counting the number of ticks emitted per frame time; or (ii) measuring the time between ticks. The drawback of counting ticks is coarse-grain integer-only readout: no partial ticks are generated, resulting in large errors (banding) in the dark regions. In contrast, measuring time between ticks can have significant benefits, but care must be taken to avoid both complex in-pixel timing circuitry, and contention over global timing resources.

One recent approach that calculates intensity by *measuring the time between pulses* is the recent design of Singh *et al.* [9]. This approach, in effect, measures the *reciprocal* of the incident light intensity. We devote the remainder of this section to discussing this approach in detail.

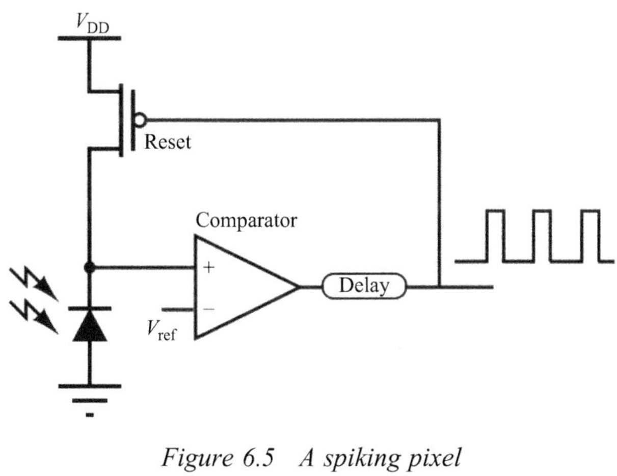

Figure 6.5 A spiking pixel

Pixel architecture. The first novel feature introduced by this design is that the train of events goes through a "prescaler" or "decimator" that divides the frequency of the event train by an appropriate factor (typically a power of two) before passing through to the column readout (see Figure 6.6). The role of the decimator is to reduce the frequency of events joining the column readout stream at a given pixel so as not to overwhelm the traffic along a column. Decimation is key to obtaining a high dynamic range because the high pulse frequency emanating from a very bright pixel is decimated before readout, thereby conserving the bandwidth of the column circuitry.

Using a power of 2 for this *decimation factor* keeps the implementation simple with low area cost. In effect, this is a form of *rate control.* For the best results, the decimation factor should be settable on a per-pixel basis. One approach is to have the processing logic assign the decimation factor for each pixel. Another approach is to allow each pixel to determine its own decimation factor autonomously. In particular, a global relatively low-speed timing reference signal (e.g., 1 KHz) is provided, which each pixel uses to determine its own optimal decimation factor. Note, however, that this reference interval is not an actual "exposure time"; the sensor operates in a frameless manner. The goal is generally to produce 2–3 events during this interval, so enough information is generated to calculate the intensity, and no more.

Communication architecture. The second novel feature is a completely different architecture of the column readout (see Figure 6.7). Instead of a bus, each column is a fully registered pipeline (FIFO), which allows each pixel to insert event tokens without the need to exclusively acquire the entire bus. Each pixel inserts its events into the stream through its associated merge node. The events travel downwards in this pipeline, where they are finally received and processed by the processing logic.

Event encoding. Each event carries two pieces of information: (1) *Row*: an integer specifying the pixel that generated this event, that is, its row number; and (2) *Decimation Factor*: the prescaling factor being used by this pixel. The decimation factor will be conveniently set to be a power of two (2^D), thus only D needs

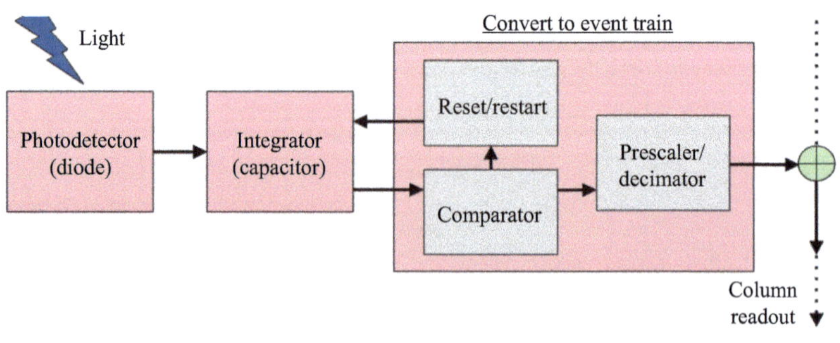

Figure 6.6 Spiking pixel architecture (from [9])

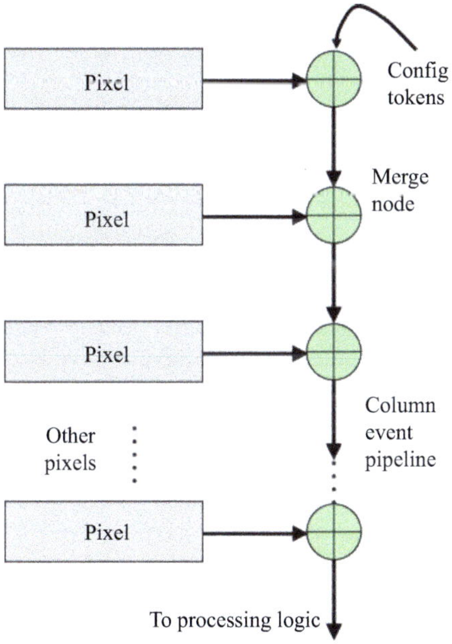

Figure 6.7 Column architecture of spiking pixel sensor (from [9])

to be sent. This information is needed to properly rescale the computation of the light intensity value in the processing logic. Thus: *Event* := *<Row, D>*.

Time stamping. The third novel feature is that *events are not time-stamped* at the pixel where they are generated. Instead, events are time-stamped at the final destination in the processing logic. This is a key design feature deliberately chosen to keep the in-pixel circuitry small, and also to avoid the wiring needed for transmitting timing information throughout the sensor array. If the traffic along the column is uncongested, there will not be any appreciable transport jitter, and therefore the timing relationship between pixel events will be preserved. Note that while low jitter is a requirement, it is not crucial, however, for the pipeline to have a low end-to-end propagation delay because only the time difference between events is relevant. Low jitter is achieved through the technique of frequency decimation, which ensures that bright pixels do not overwhelm the pipeline.

Pipeline implementation. While the overall design is frameless and, therefore, a natural candidate for implementation using clockless circuits, it can also be implemented using synchronous circuits. However, a clockless implementation of the column pipeline has some benefits: significantly less energy consumption, and eliminating the need to distribute a high-speed clock throughout a multimegapixel sensor array (thereby also reducing noise due to clock spikes).

Direct floating-point readout $<\Delta t, D>$**.** The value of the incident light intensity sensed by a pixel is computed from the time difference Δt between events received by the processing logic from that pixel. If 2^D is the decimation factor currently used by the pixel, the light intensity, *I*, is given by:

$$I \propto \frac{2^D}{\Delta t}$$

This formula lends itself to a direct floating-point representation, where: $1/\Delta t$ is the mantissa, and D is the exponent. As a result, simply reporting $(\Delta t, -D)$ without performing any computation on it suffices as the reciprocal of the computed intensity in floating-point representation.

Results. While this approach has not yet been implemented in silicon, initial validation has been performed via simulation [9]. For comparison, an existing experimental intensity-to-frequency sensor [4] was also modeled as a base case. Overall, the proposed sensor was able to capture a dynamic range of 22 bits at the equivalent speed of 1,000 "frames"/sec, and delivered power SNR $>$120 dB. Since the scene's dynamic range was far greater than is possible to show on paper, we include two intensity-windowed versions of the results, allowing us to focus on the darkest and the brightest regions of the intensity range. Figure 6.8(a) shows that the proposed approach can provide significant detail and smooth gradations in the dark regions, whereas the existing sensor offers much less information and shows significant banding. Figure 6.8(b) shows that in the brightest regions, the proposed approach can provide significant detail, whereas the existing sensor has blown highlights.

6.1.4 Asynchronous logarithmic sensors

The second type of asynchronous sensors presented here are logarithmic sensors, which have been used successfully in silicon retinas. Unlike linear sensors, which integrate photocurrent over time to generate a voltage, logarithmic sensors directly convert photocurrent into voltage without integration.

The key idea is to use a MOS transistor operating in subthreshold mode to do instantaneous conversion [10,11] (Figure 6.9). For a given photocurrent, I_{ph}, the gate-to-source voltage across the load transistor M_{load}, operating in subthreshold regime, is related logarithmically to it:

$$V_{\mathrm{DD}} - V_{\mathrm{diode}} = m \cdot \log\left(\frac{I_{\mathrm{ph}}}{I_0}\right)$$

Thus, the photocurrent is directly converted to voltage by the subthreshold transistor as well as logarithmically compressed.

This log-compression obtains 15–20 bits of dynamic range squeezed into about 0.5 V of voltage. A further benefit is that the response time of pixels can be quite fast (μs to even sub-μs). As a result, such sensors (e.g., ATIS [12] and DAVIS [13] sensors) have been used in special-purpose robot vision applications where very high speeds are desired for fast tracking, 3D pose estimation, optical flow, and gesture recognition [14].

Logarithmic sensors used in silicon retinas mimic the behavior of neurons in the human retina by performing spatial and temporal contrast detection [15] in the processing circuitry that follows the photodetector. The architecture uses spikes to

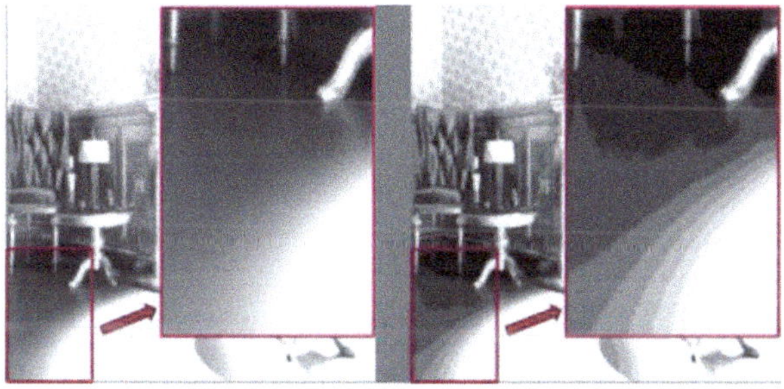

(a) Comparison of dark intensities

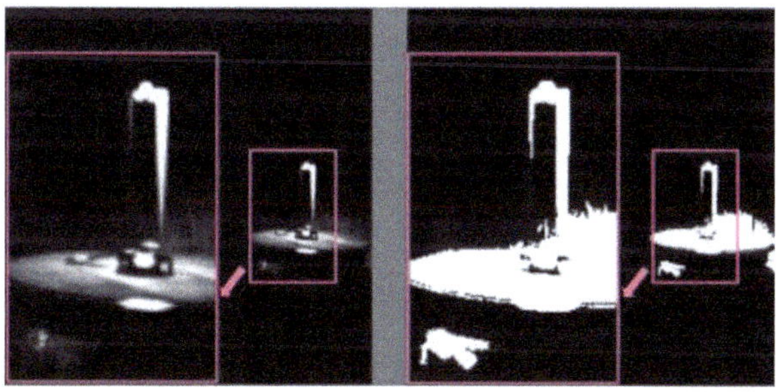

(b) Comparison of bright intensities

Figure 6.8 Comparison of our approach (left) and base approach (right). The HDR image captured by each is shown by "windowing" into the dark and bright ends of the intensity range

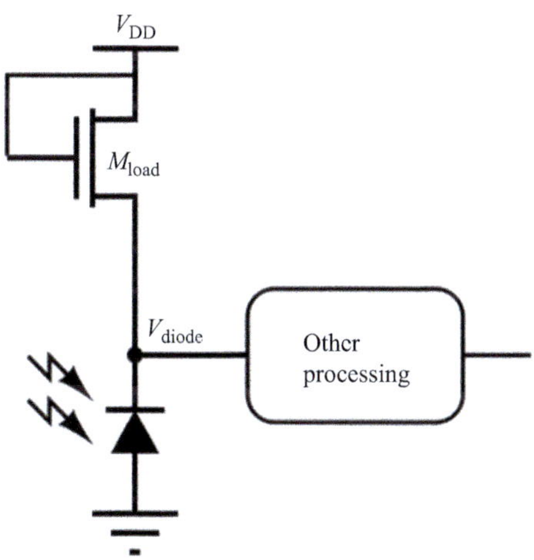

Figure 6.9 Subthreshold load transistor provides logarithmic conversion

represent changes, which are communicated to the peripheral processing logic using address-event representation (AER [15]). Therefore, the bandwidth and power requirements are greatly reduced [14].

But logarithmic sensors have significant challenges as well. There is a large variation in subthreshold MOS characteristics which cause high fixed-pattern noise [16] that cannot simply be corrected by black-frame subtraction or correlated double sampling (CDS) because of the nonlinear nature of the sensing. Furthermore, the SNR achieved is significantly lower: the log-compression means the darker pixels generate very low signal values, and the lack of time integration worsens noise. These challenges are exacerbated in newer CMOS technologies. Furthermore, the communication architecture of these sensors has been based on global arbitration: a pixel must obtain a lock from both a row arbiter and a column arbiter in order to transmit its spike [11]. While such a global arbitration approach has worked in the reported designs, their size has been limited to 240 $\times$ 180 pixels or so. Successfully scaling such a globally arbitrated approach to handle multi-megapixel sensor arrays will be a challenge.

In summary, logarithmic sensors have significant advantages for use in silicon retinas for special applications such as fast tracking and gesture recognition where the main task is feature extraction via rapid contrast detection. These applications do not need high-fidelity intensity values that are required in high-end photography and video, and therefore the relatively low SNR and low pixel count of logarithmic sensors is not a hindrance.

6.2 Sensor processors

Sensor networks are made up of a collection of low-cost nodes that gather, process, and communicate information about the environment within which they are embedded. Each such node, therefore, must include not only a sensing element, but also circuitry to implement the communication.

These networks pose some unique design challenges because of the strict energy budgets they impose, and also because of the dynamic nature of their network topologies. As a result, typically a microprocessor is used to implement the networking, including dynamic routing, message queueing, timestamping, etc. Further, all processing must be highly energy efficient, and the implementation must have low area cost.

6.2.1 SNAP: a sensor-network asynchronous processor

SNAP is an asynchronous processor introduced by Kelly *et al.* [17], which was custom designed to deliver the sensor networking capability at a low energy and area cost (see Figure 6.10). An interesting feature of SNAP is that it was designed to serve dual roles: (i) to serve as the main processor in an actual sensor node; and (ii) to be a component of a network-on-chip (NoC) that was custom designed to simulate sensor networks. Thus, the NoC was to be used as a miniature replica on a single chip of the actual wider area sensor network, but since both the simulation

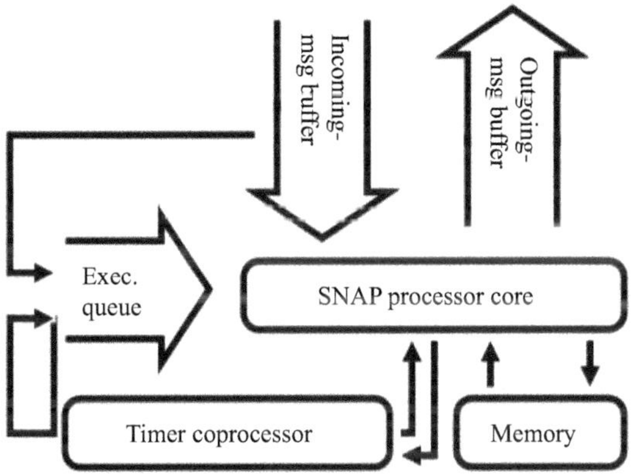

Figure 6.10 High-level block diagram of the SNAP processor

and the actual deployment use the same processor, simulation could occur using the same software. As a result, the simulation faithfully captures all of the capabilities and limitations of the physical hardware, instead of higher level software abstractions.

A key design requirement was to minimize the chip area of the processor. A smaller area allows a smaller size for the sensor node, and also allows a larger number of nodes to be simulated onto the NoC. A number of careful design decisions were made to achieve this goal: no cache, no virtual memory or exceptions, no multiply/divide/floating-point unit, datapath limited to 16 bits (instruction can be 16-bit or 32-bit), and all memory is DRAM.

The processor's instruction set architecture as well as the implementation were highly tuned to optimize the main task of event queue management, including scheduling and canceling events, as well as sending and receiving them. The circuit-level implementation of the processor used quasi-delay-insensitive (QDI) asynchronous circuits. To achieve good throughput without a high area and energy cost, a modest amount of pipelining was implemented. To keep the NoC simulation faithful to the real time, a special timer coprocessor was implemented, dedicated to the task of maintaining a scaled version of the real time, and to timestamping and scheduling events.

Detailed simulations of the layout of the processor (in 0.18 micron) indicate promising performance, both in terms of energy efficiency and speed. In particular, the processor delivered 240 MIPS performance at the nominal 1.8 V supply voltage, but was functionally correct down to 0.6 V with a throughput of 28 MIPS. The energy consumed per instruction was 218 pJ at 1.8 V and 24 pJ at 0.6 V. For extremely low activity levels where fewer than 10 events are being processed per second, the active power drops to 150–550 nW at 1.8 V, and only 16–58 nW at 0.6 V. This power consumption is several orders of magnitude less than commercial microcontrollers with similar capabilities.

6.2.2 BitSNAP

In subsequent work, Ekanayake *et al.* [18] redesigned a follow-on processor using bit-serial datapaths to yield even lower energy consumption. A key feature of the bit-serial datapath was dynamic significance compression: by dropping leading 0s and 1s from integers, and transmitting only the required bits, 30%–80% of the switching energy of the integer datapath can be saved. As a result, BitSNAP can reduce overall energy consumption by about 50% over a comparable parallel-word processor, while delivering a throughput that is only 20%–25% lower, but still more than adequate (6–54 MIPS) for low-power sensor node applications.

6.3 Signal processing

Traditional digital signal processors operate on signals that are both discrete in time and in amplitude. Since real-world phenomena typically produce signals that are continuous in both dimensions, they must first be sampled to discretize them in time and quantized to discretize them in amplitude. Sampling allows the signal to be represented by a finite number of digital words, enabling digital computation. Quantization, also called analog-to-digital conversion (ADC), allows processing to mitigate the adverse effects of analog imperfections, noise, and parameter variations.

A key drawback of the discrete-time approaches is that sampling imposes limitations on the digital signal processor due to aliasing. If the signal changes faster than the limit of what the sampling rate can faithfully capture, then information is lost. Typically, the sampling frequency is rigidly set for the worst-case operating regime (to meet the Nyquist criterion), regardless of the actual signal activity. But in that case, opportunities for savings in power dissipation are squandered. In particular, even during periods of idling or low activity, sampling continues at the high rate, and each new sample triggers new activity in the DSP. This excessive power consumption is a significant impediment to using such signal processors in sensor applications.

Another drawback of discrete-time approaches is that they introduce significantly greater spectral noise within the band of interest. Consider, for example, an input signal that is a pure sinusoid with frequency f_{in}. If it is quantized without sampling, higher harmonics at integral multiples of the fundamental frequency (mf_{in}) are introduced because the shape is no longer a pure sinusoid. But if the signal is sampled in time as well, at the rate f_s, then the harmonic components are aliased to frequencies $\pm nf_s \pm mf_{in}$. As a result, infinitely many frequency components (corresponding to different pairs of values for n and m) now lie within the frequency band of interest, leading to significantly higher in-band quantization noise, which in turn makes subsequent signal processing more challenging.

6.3.1 Continuous-time DSP

A novel approach has been proposed that eliminates the sampling step: continuous-time discrete-amplitude signal processing [19,20]. Figure 6.11 shows the central

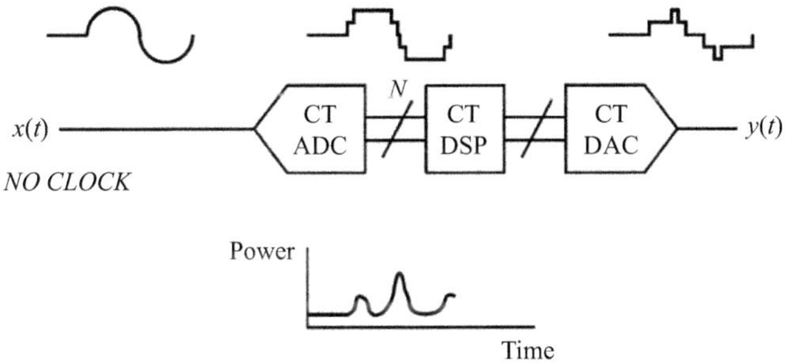

Figure 6.11 Block diagram of a continuous-time digital signal processing (CT DSP) system (from [20])

idea of this approach. The input signal is continuously quantized, and the digital word is then sent to the DSP, whose output is then converted back to analog. In practice, the ADC generates a new quantized value only when the input has changed by a quantum. This type of operation can be achieved by, for example, using a level-crossing ADC as opposed to a clocked ADC. The DSP is only stimulated to process when a new word value is generated by the ADC and communicated to the DSP using a handshake. The entire signal processing pipeline is, therefore, asynchronous.

The complete design of a continuous-time digital FIR filter was reported in [21]. The design was an 8-bit, 16-tap FIR filter chip, implemented in IBM 0.13 micron CMOS technology. The filter was used as part of an ADC/DSP/DAC system. Since there is no sampling clock, the filter pipeline responds quickly to input changes, without waiting for a clock tick. The filter can automatically handle inputs of different or varying sample rates, without requiring any internal adjustment or recalibration. Finally, during periods of input inactivity, the filter exhibits little switching activity. This experiment confirms that the asynchronous approach of continuous-time DSP is a good match for sensing applications.

6.3.2 Asynchronous analog-to-digital converters

Unlike traditional clocked ADCs, which sample the input signal at regular intervals for conversion to digital, asynchronous ADCs employ irregular sampling. Conversion is done when the signal crosses a quantization level, that is, only when the output must change. This idea is illustrated in Figure 6.12 [22].

In [20,23], the input signal is continually compared with the two levels immediately above and below the input value, and when the input crosses one of those levels, an "up" or "down" output event is generated, which is then post-processed into a digital word output. These designs demonstrate a significant reduction in power consumption compared to synchronous designs. While the actual amount of power saved will depend on input signal activity, these designs consume very low power when idling (i.e., input signal is steady), whereas a clock-based ADC will consume significant energy at every sampling instant.

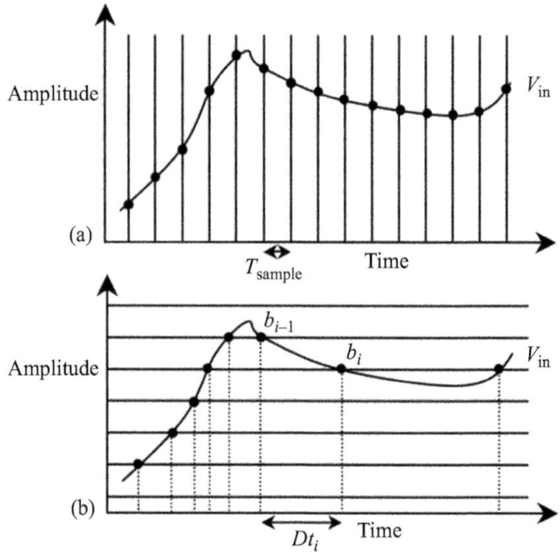

Figure 6.12 (a) Synchronous ADC with regular sampling, versus (b) asynchronous level-crossing ADC (from [22])

6.3.3 A hybrid synchronous–asynchronous FIR filter

At IBM Research, a project was undertaken jointly with Columbia University to develop a mixed synchronous–asynchronous implementation of a finite impulse response (FIR) filter for use in the read channels of modern disk drives [24]. The goal was to reduce the filter's latency over its wide range of operating frequencies. In a synchronous implementation that is deeply pipelined for speed, the latency becomes poorer when the data rate, and hence the clock recovered from it, slows down. The hybrid synchronous/asynchronous implementation replaced the core of the filter with an asynchronous pipelined unit featuring a fixed latency, while the remaining circuitry was kept synchronous. The resulting chip exhibited a 50% reduction in worst-case latency, along with a 15% throughput improvement, over IBM's leading commercial clocked implementation in the same technology.

6.4 Conclusion

In this chapter, we have presented a few examples of designs from the domain of sensors that illustrate the promise of asynchronous techniques in the design of sensing systems. Such systems often impose strict performance requirements. For example, in wide area distributed networked sensors, there are constraints of tight energy budgets and low area costs. In the case of image sensors, a large array of tiny sensory pixels are packed into a single chip, and pose performance challenges if rigid global synchronization is used. Since transduction is usually followed by analog-to-digital conversion (ADC) and digital signal processing (DSP), these processing techniques must also be specialized to be highly energy efficient. Furthermore, sensing often involves long idle periods, so all sensing systems must have very low idle power consumption. In this chapter, we saw examples of frameless image sensors, asynchronous sensor processors, and continuous-time ADC and DSP, all of which exemplify the power and promise of asynchronous design in the field of sensing.

References

[1] L. McIlrath. "A low-power low-noise ultrawide-dynamic-range CMOS imager with pixel-parallel A/D conversion." IEEE J. Solid-State Circuits, pp. 846–853, 2001.

[2] A. Kitchen, A. Bermak, and A. Bouzerdoum. "PWM digital pixel sensor based on asynchronous self-resetting scheme." IEEE Trans. Electron Devices, vol. 25, no. 7, 2004.

[3] A. Kitchen, A. Bermak, and A. Bouzerdoum. "A digital pixel sensor array with programmable dynamic range." IEEE Trans. Electron Devices, vol. 52, no. 12, pp. 2591–2601, 2005.

[4] X.Wang, W. Wang, and R. Hornsey. "A high-dynamic-range CMOS image sensor with in-pixel light-to-frequency conversion." IEEE Trans. Electron Devices, vol. 53, no. 12, pp. 2988–2992, 2006.

[5] Y. Chen, F. Yuan, and G. Khan. "A new wide dynamic range CMOS pulse-frequency-modulation digital image sensor with in-pixel variable reference voltage." Proceedings of the Midwest Symposium on Circuits and Systems (MWSCAS), 2008.

[6] J. Doge, G. Schonfelder, G. T. Streil, and A. Konig. "An HDR CMOS image sensor with spiking pixels, pixel-level ADC, and linear characteristics." IEEE Trans. Circuits Syst., vol. 49, no. 2, pp. 155–158, 2002.

[7] A. Bermak. "VLSI implementation of a neuromorphic spiking pixel and investigation of various focal-plane excitation schemes." Int. J. Robot. Autom., vol. 19, no. 4, pp. 197–205, 2004.

[8] B. Fowler, A. Gamal, and D. Yang. A CMOS area image sensor with pixel-level A/D conversion. Stanford University Tech Report, 1995.

[9] M. Singh, P. Zhang, A. Vitkus, K. Mayer-Patel, and L. Vicci. "A frameless imaging sensor with asynchronous pixels: an architectural evaluation." Proceeding of Internatinal Symposyum on Asynchronous Circuits and Systems (ASYNC-17), San Diego, May 2017.

[10] S. Kavadias, B. Dierickx, D. Scheffer, A. Alaerts, D. Uwaerts and J. Bogaerts. "A logarithmic response CMOS image sensor with on-chip calibration." IEEE J. Solid-State Circuits, vol. 35, no. 8, 2000.

[11] P. Lichtsteiner, C. Posch, and T. Delbruck. "A 128X128 120 dB 15 us latency asynchronous temporal contrast vision sensor." IEEE J. Solid-State Circuits, vol. 43, no. 2, pp. 566–576, 2008.

[12] C. Posch, D. Matolin, and R. Wohlgenannt. "A QVGA 143 dB dynamic range frame-free PWM image sensor with lossless pixel-level video compression and time-domain CDS." IEEE J. Solid-State Circuits, vol. 46, pp. 259–275.
[13] R. Berner, C. Brandli, M. Yang, S.-C. Liu, and T. Delbruck. "A 240 × 180 10mW 12us latency sparse-output vision sensor for mobile applications." IEEE Symposium on VLSI Circuits (VLSIC), 2013.
[14] A. Amir, B. Taba, D. Berg, *et al.* "A low power, fully event-based gesture recognition system." IEEE Conference on Computer Vision and Pattern Recognition (CVPR), 2017, pp. 7243–7252.
[15] C. A. Mead and M. A. Mahowald. "A silicon model of early visual processing." Neural Networks, vol. 1, pp. 91–97, 1988.
[16] D. Joseph and S. Collins. "Transient response and fixed pattern noise in logarithmic CMOS image sensors." IEEE Sensors J., vol. 7, no. 8, pp. 1191–1199, 2007.
[17] C. Kelly, V. Ekanayake, and R. Manohar. "SNAP: a sensor-network asynchronous processor." Proceedings of the Ninth International Symposium on Asynchronous Circuits and Systems, 2003, Vancouver, BC, Canada, 2003, pp. 24–33.
[18] V. N. Ekanayake, C. Kelly, and R. Manohar. "BitSNAP: dynamic significance compression for a low-energy sensor network asynchronous processor." 11th IEEE International Symposium on Asynchronous Circuits and Systems, New York City, NY, USA, 2005, pp. 144–154.
[19] Y. W. Li, K. L. Shepard, and Y. P. Tsividis. "Continuous-time digital signal processors." 11th IEEE International Symposium on Asynchronous Circuits and Systems, New York City, NY, USA, 2005, pp. 138–143.
[20] B. Schell and Y. Tsividis. "A continuous-time ADC/DSP/DAC system with no clock and with activity-dependent power Dissipation." IEEE J. Solid-State Circuits, vol. 43, no. 11, pp. 2472–2481, 2008.
[21] C. Vezyrtzis, W. Jiang, S. M. Nowick, and Y. Tsividis. "A flexible, event-driven digital filter with frequency response independent of input sample rate." IEEE J. Solid-State Circuits, vol. 49, no. 10, pp. 2292–2304, 2014.
[22] E. Allier, G. Sicard, L. Fesquet, and M. Renaudin. "A new class of asynchronous A/D converters based on time quantization." Proceedings of the Ninth International Symposium on Asynchronous Circuits and Systems, 2003, Vancouver, BC, Canada, 2003, pp. 196–205.
[23] F. Akopyan, R. Manohar, and A. B. Apsel. ''A level-crossing flash asynchronous analog-to-digital converter." Proceedings of the 12th IEEE International Symposium on Asynchronous Circuits and Systems (Async 06), IEEE Press, 2006, pp. 11–22.
[24] M. Singh, J. A. Tierno, A. Rylyakov, S. Rylov, and S. M. Nowick. "An adaptively pipelined mixed synchronous-asynchronous digital FIR filter chip operating at 1.3 gigahertz." IEEE Transactions on Very Large Scale Integration (VLSI) Systems, vol. 18, no. 7, pp. 1043–1056, 2010.

Chapter 7

Dcsign and test of high-speed asynchronous circuits

Marly Roncken[1] and Ivan Sutherland[1]

This chapter explores the design and test of high-speed complementary metal oxide semiconductor (CMOS) self-timed circuits. Section 7.1 describes how the properties of CMOS technology itself limit how fast a self-timed circuit can run. Section 7.2 presents our Link and Joint model, a unified point of view of self-timed circuits that allows reasoning about them independently of circuit families and handshake protocols. The model separates communication and storage, done in *Links*, from computation and flow control, done in *Joints*. The model also separates *actions* from *states*. Special *go* signals enable or disable Joint actions on an individual basis. The individual *go* signals make it possible to initialize, start, and stop self-timed operations reliably, which is crucial for design as well as for at-speed test, debug, and characterization. Section 7.3 examines design and test aspects of the Weaver, a self-timed nonblocking 8 × 8 crossbar switch designed using the Link and Joint model. We report measured test results from a working Weaver chip in 40 nm CMOS with speeds up to 6 Giga data items per second. With 72 bit wide data items, this amounts to 3.5 Tera bits per second for the full crossbar.

7.1 How fast can a self-timed circuit run?

How fast can a self-timed circuit run? What are its fundamental speed limits? What design considerations are important for digital circuits intended to operate at close to maximum speed? How does the potential speed of self-timed circuits compare with the speed of externally clocked circuits?

Lacking an external clock to drive their actions, all self-timed circuits must act on their own. In place of an external clock, all self-timed circuits and systems use the oscillations of logic rings to drive their actions.

Just as the tick of an external clock can be used as a unit of time, this chapter uses the term *gate delay* as if it were a unit of time. Of course, delay varies from

[1]Asynchronous Research Center, Maseeh College of Engineering and Computer Science, Portland State University, Portland, OR, USA

one logic gate to another, but high-speed circuits tend to size their transistors so that their logic gates have similar delay, giving the notion of gate delay a rational basis.

A gate delay is also a partly topological notion of the time it takes a signal to pass through an inverting logic gate, however long that might be. Except for pass gates, all individual CMOS logic gates invert logic signals, and so gate delays also count logic inversions.

7.1.1 Logic gate delays

A good model for CMOS logic gates operating at full power supply voltage associates delay with output transition time as illustrated in Figure 7.1.

When the input of the logic gate reaches a switching threshold the output of the gate begins an approximately linear change in voltage. The rate at which the output voltage changes depends on the *drive strength*, or simply *strength*, of the gate and how much *load* the gate drives. This makes it possible to formulate the notion of gate delay as the ratio of load driven to strength:

$$delay = load\ driven/drive\ strength.$$

Figure 7.1 illustrates the input and output voltage transitions of a logic gate driving respectively a light, medium, and heavy load. It uses as gate delay how long the gate output takes to reach the switching threshold of the next gate. Let us assume for now that the output voltage ramp starts at a power or ground supply rail. As Figure 7.1 illustrates, the actual delay of a logic gate of given drive strength depends on the load it drives and the switching threshold of the next logic gate. Assuming switching thresholds are about midway between the power and ground supply rails, the delay of each logic gate is approximately half the time its output would take to swing from one rail to the other. Transitions in signal voltage are the

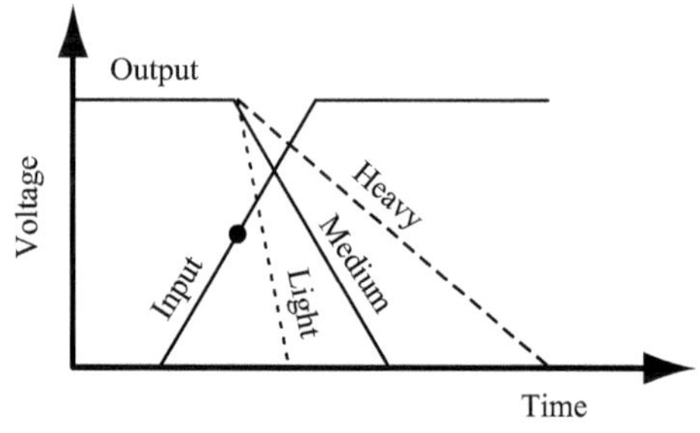

Figure 7.1 CMOS logic gate delay model. The outputs of CMOS logic gates change relatively slowly. The picture shows three possible output transitions for a logic gate driving a light, medium, or heavy load. The output begins to fall when the input voltage exceeds a logic threshold, marked here with a dot at approximately half of the supply voltage. Because the next logic gate acts only when its input reaches its threshold, the delay associated with each gate is roughly half as long as the full output transition time of the gate

major source of delay and energy consumption in CMOS logic. In the usual operating range, the voltage transitions are approximately linear ramps.

If, however, the output ramp starts at a voltage other than a power or ground supply rail, the delay of the gate also depends on that starting voltage. The delay may be very small if the output ramp starts at nearly the switching voltage of the next gate. Moreover, different starting voltages cause different delays. A ring of logic gates with widely varying delay may still oscillate but the output signals of its slower gates may fail to swing rail to rail and may thus create a steady state behavior that is very different from the ring's starting cycles. Reliable operation is best achieved by making sure that the output transitions of all gates start and end near the voltage of a power or ground supply rail.

7.1.2 Rings of logic gates

Ring oscillators internal to a self-timed circuit drive the circuit's actions. It is well known that a CMOS ring oscillator must have an odd number of at least three inverting logic gates. It is less well known that for reliable oscillation the logic gates in a small ring must have very nearly matching delays. One can design self-timed circuits that operate at maximum possible speed, that is, at the speed of a three-gate ring oscillator, like 4-2 GasP [1]. However, such high speed requires very careful design. Practical self-timed circuits, like the 6-4 GasP circuits used in the Weaver and discussed in more detail in Section 7.3, tend to run no faster than the speed of a five-gate ring oscillator. Nevertheless, the Weaver circuits still offer an impressive throughput of one data element every ten fanout-of-four gate transitions, which is about twice the typical speed of a clocked system.

In this section, we examine and compare three-gate rings to five-gate rings. Let us start with a ring oscillator with three logic gates, like the one in Figure 7.2. If all three gates have equal delay, all three signals—*a*, *b*, and *c*—will reach full swing, as illustrated in the upper timing diagram of Figure 7.2. If, however, gate *c* is twice as slow as the others, its output will barely reach full swing, as illustrated in the lower timing diagram. For all three signals of a three-gate ring oscillator to reach full swing, the delays in the three gates must match within about a factor of two. Any greater delay mismatch risks uncertain switching delay and erratic behavior.

Figure 7.3 shows a ring oscillator with five logic gates. If all five gates have equal delay then all signals will reach full swing, as illustrated in the upper timing diagram of Figure 7.3. In the middle timing diagram, signal *e* suffers twice the delay of signals *a*, *b*, *c*, and *d*, and yet *e* still dwells at the power and ground rails. The lower timing diagram shows that *e* has to be about four times slower to barely reach full swing. In comparison, rings with five logic gates offer the following three benefits over maximum-possible-speed rings with three logic gates.

1. **Greater flexibility in accommodating gate delay mismatches:** Rings of five logic gates better tolerate variation in the delays of their individual parts be it from variations in wire capacitance or in manufacturing or otherwise. Rather than matching the logic gate delays within a factor of two as required by a three-gate ring, a ring of five logic gates requires that the delays match within

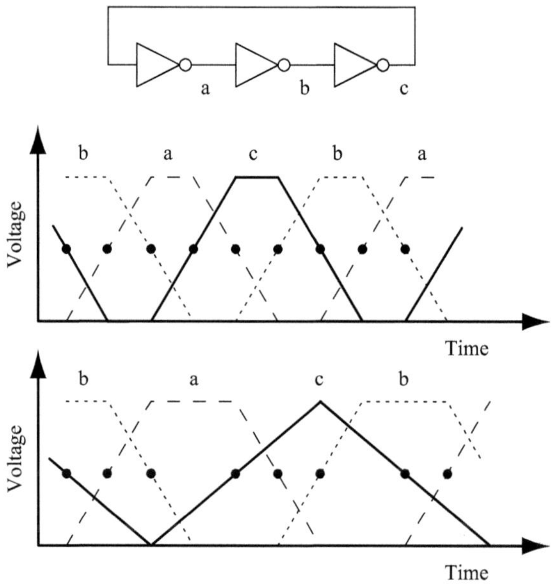

Figure 7.2 Delay in a three-inverter ring. A ring with three inverting gates of similar delay produces full-swing signals, as illustrated in the upper timing diagram. In the lower timing diagram, signal c suffers twice the delay of a and b. Were signal c driven any slower, it would fail to reach full swing. For all three signals to make full transitions, the delays of the gates must match within about a factor of two

about a factor of four. A factor of two requires careful design. A factor of four is easy to achieve.

2. **Greater robustness in signaling:** The signals produced by rings of five logic gates dwell at the power and ground supply rails longer than those of rings of three logic gates. In contrast, the signals in three-gate rings curve and sometimes even form sharp points near the rails as they change direction from rising to falling. The dwell feature in rings of five gates separates cleanly the successive rise and fall ramps of each signal.
3. **Greater flexibility in accommodating logic computations:** The topology of five-gate rings provides more logic gates to do logic. Rings with three gates often fail to have enough stages to invert particular signals and must instead resort to duplicating the entire ring in true and complement form.

Their better tolerance for delay mismatch and greater logic flexibility make ring oscillators with five logic gates not only more robust than those with three gates but also easier to design. In general, the more gates a logic loop has the greater disparity it permits between its slow and fast gates.

As mentioned at the start of Section 7.1, self-timed circuits and systems use the oscillations of logic rings to drive their actions. Going around the ring twice creates

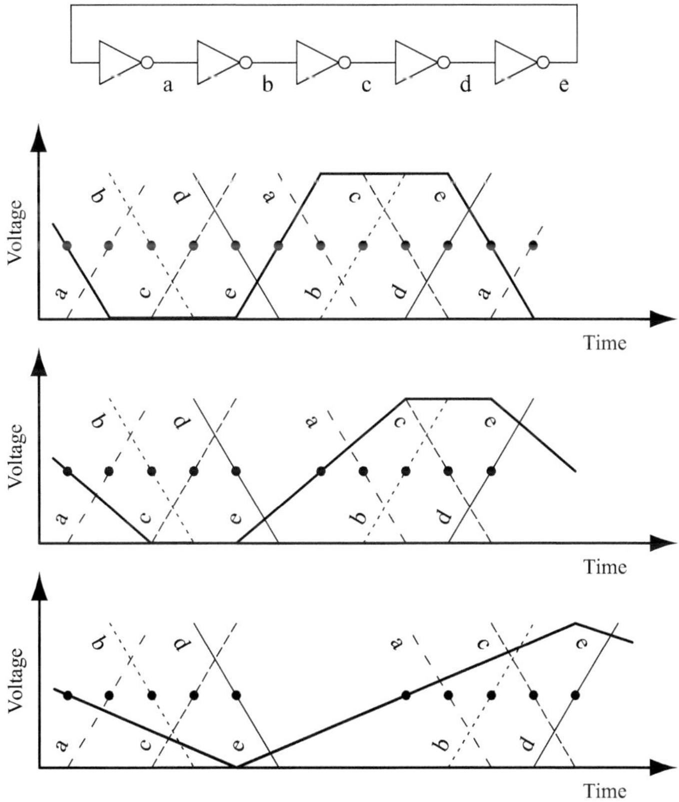

Figure 7.3 Delay in a five-inverter ring. A ring with five inverting gates of similar delay produces full-swing signals, as illustrated in the upper timing diagram. In the middle timing diagram, signal e suffers twice the delay of the other signals and yet still dwells at the power and ground rails. For all five signals to make full transitions, the delays of the gates must match within about a factor of four, as illustrated by the lower timing diagram

[Note: For clarity, the diagrams in Figure 7.3 omit the power-rail and ground-rail voltage levels for signals a through d and show only their rising and falling voltage transitions.]

a high-to-low-to-high or low-to-high-to-low pulse on each signal. The rings in Figures 7.2 and 7.3 generate approximately symmetric pulses with more or less equal high-to-low-to-high and low-to-high-to-low pulse widths. It is also possible to use the signal changes on a logic ring to generate asymmetric pulses, as in Figure 7.4.

When given wider than three gate delay pulse inputs, the two pulse generators in Figure 7.4 create asymmetric pulses with three gate delay high-to-low-to-high pulse widths in (a) and three gate delay low-to-high-to-low pulse widths in (b). One

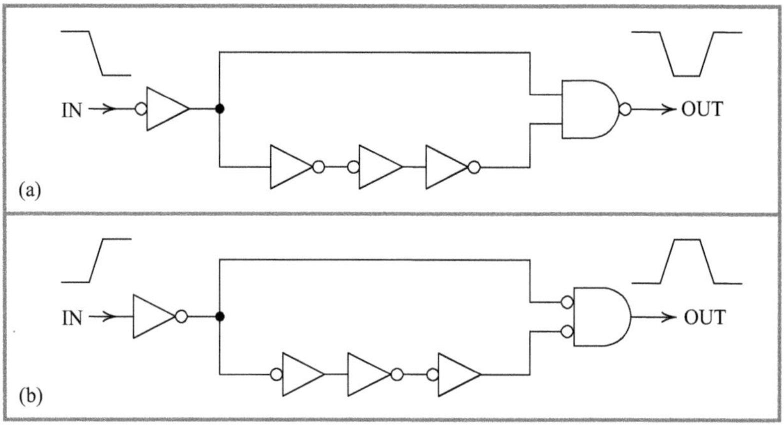

Figure 7.4 Asymmetric pulse generators. A falling transition on IN in (a) makes the output, OUT, of the NAND gate fall and then rise in rapid succession by generating a three gate delay high-to-low-to-high pulse, as suggested by the waveforms shown above signals IN and OUT. Likewise, a rising transition on IN in (b) creates a low-to-high-to-low pulse of three gate delays at the output, OUT, of the input-inverted AND gate—also known as a NOR gate. For full-swing transitions and a three gate delay output pulse, the gates must be sized appropriately and may require artificial load inverters not shown here. All input signals on IN must be high for at least three gate delays and low for at least three gate delays

[Note: Figure 7.4 draws inverting gates with an inversion symbol "○" either at the gate's output or at the gate's input to indicate whether the key transition they drive is a falling or rising one. For instance, the key transition for OUT in (a) is a falling transition, because it starts the pulse on OUT. Therefore, we draw the NAND gate that drives it with the inversion symbol at its output. Likewise, the key transition for OUT in (b) is a rising transition, and so we draw the gate that drives it with the inversion symbol at its inputs. This notation is a matter of taste, but where it matters it enables us to emphasize the meaning of the circuit.]

can style pulse generators similar to those in Figure 7.4 to generate wider pulses of a more or less fixed width.

Locally generated pulses—either symmetric or asymmetric—can be used as "local clock" signals to drive local latches or flipflops or other types of storage elements to update the local state information changed during each action. Local clock pulses that drive many storage elements may need amplification to obtain enough drive strength. Section 7.1.3 discusses how one can amplify pulse signals.

7.1.3 Amplifying pulse signals

To drive a large load from a relatively weak source one can use a series of inverters with exponentially increasing drive strengths, as shown in Figure 7.5.

To support the *strength*, *load*, and *step-up* numbers indicated in the context of Figure 7.5 and the subsequent Figure 7.6, let us define the terms associated with these numbers. All definitions are relative to a unit inverter, which is the smallest inverter allowed in a particular circuit family or manufacturing process. The unit inverter presents a load of 1 at its input and has a drive strength of 1.

- *Drive strength or strength* for short indicates the ability of a logic gate to drive load. Its strength is how many times stronger the logic gate is than a unit inverter in driving load at its output. Typically, one makes the transistors in a logic gate wider or—equivalently—puts them in parallel to increase the drive strength of the gate.
- *Load presented or input load* is how much input charge a logic gate takes to turn its transistors on or off relative to the charge required to switch a unit inverter. In other words, the load that a logic gate presents to its input is how much more difficult it is to turn its transistors on or off than it is to turn on or off the transistors of a unit inverter. Wider transistors are more difficult to drive.
- *Step-up* is the ratio of load driven to strength:

  $$\textit{step-up} = \textit{load driven}/\textit{drive strength}.$$

  Note that we used the same formula for the delay of a logic gate—see page 114. Successive gates that use the same step-up, or delay, are fastest overall.* We chose the strengths in Figures 7.5 and 7.6 to give each gate an equal step-up of four.
- *Amplification or gain* is the ratio of load driven, at the output of a gate or series of gates, to load presented at the input of a gate or series of gates. A logic gate can drive multiple other gates. The input loads presented by the other gates add up to the total load driven by the logic gate. Figures 7.5 and 7.6(a) and (b) show the load presented at the input, *IN*, of each two-stage pulse amplifier and the total load that can be driven at the output, *OUT*, to give each stage a step-up of four. The corresponding amplifications from *IN* to *OUT* are 16, 72, and 60, respectively.

Figure 7.5 shows two series inverters that form a two-stage amplifier with a uniform step-up of four per stage. Together the two stages provide an amplification of 16—from presenting a load of 1 at *IN* to driving a load of 16 at *OUT*.

However, given that an inverter's rise time almost always differs from its fall time, each stage will either retard or advance its output transition relative to its input transition. Each stage in the pulse amplifier of Figure 7.5 has two opportunities to change its output pulse width relative to its input pulse width—once by retarding or advancing the rising transition of the pulse and once again by retarding or advancing the falling transition of the pulse. The accumulated change will inevitably either lengthen or shorten a pulse from the amplifier's input, *IN*, to its output, *OUT*. This is particularly problematic when amplifying short pulses from *IN* to *OUT*.

*An alternative but equivalent guideline for achieving fastest delay follows in Section 7.1.4.

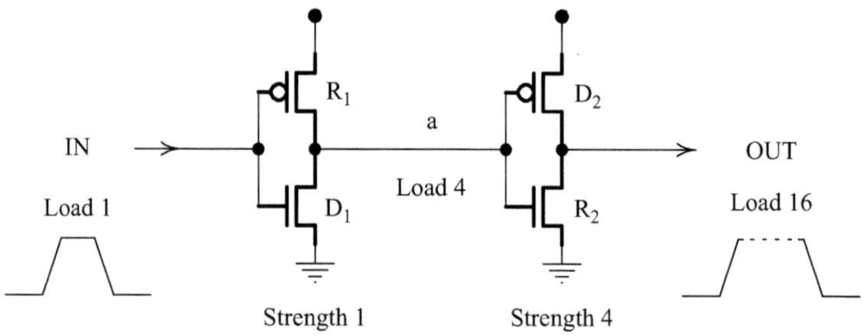

Figure 7.5 Pulse amplifier. A simple two-stage amplifier consisting of two inverters in series can amplify a pulse. The strengths shown give each inverter a step-up of four. The inverter pair provides a 16-fold gain. The broken line in the pulse waveform at signal OUT indicates that the pulse width at OUT can vary and thus be wider or narrower than the pulse width at IN. The variation is due to accumulated differences in rise and fall times at each inverter stage, and makes this simple solution less suitable for amplifying short pulses

If you must amplify a short pulse it is best to do so by avoiding accumulation of delay changes from intermediate stages. To do so, use feedback at the output to control the output pulse width directly, as in Figure 7.6.

The two circuits in Figure 7.6 illustrate a technique called *post-charge logic* described in an expired patent [2] from the late Bob Proebsting that:

> "[· · ·] permits propagation of a pulse through an arbitrary number of stages with the pulse width remaining essentially unchanged."

To understand how each circuit of Figure 7.6 works, first consider only the bold transistors labeled D_1 and D_2. These transistors match the corresponding transistors in Figure 7.5—they turn on the pulse signal at each stage. In the circuits of both Figures 7.5 and 7.6, when the input signal, *IN*, rises, transistor D_1 drives signal *a* low, turning on transistor D_2 to drive the output signal, *OUT*, high. Each stage has more drive strength than its predecessor, as indicated by the strength numbers for each stage in each circuit.

But unlike the two inverter stages in Figure 7.5, the two stages in Figure 7.6(a) and (b) avoid wasting input charge to control the reset transistors, R_1 and R_2. Instead, the *post-charge logic* of Figure 6(a) and (b) drives R_1 and R_2 from the amplified output signal. For each circuit in Figure 7.6, when *OUT* rises, its inverted and amplified signal, *c*, falls to raise its inverted and amplified signal, *d*, turning on transistor R_2 to reset *OUT* to low. Meanwhile, the falling signal *c* turns on transistor R_1 to reset *a* to high, and—for (b)—also turns off transistor D_{1A} to avoid fighting *IN* pulses that might be somewhat wider than three gate delays.

In addition to stabilizing the pulse width, the *post-charge logic* technique offers a much higher amplification. The load numbers for *IN* and *OUT* in Figure 7.6 show an amplification of 72 for (a) and 60 for (b) versus only 16 for Figure 7.5.

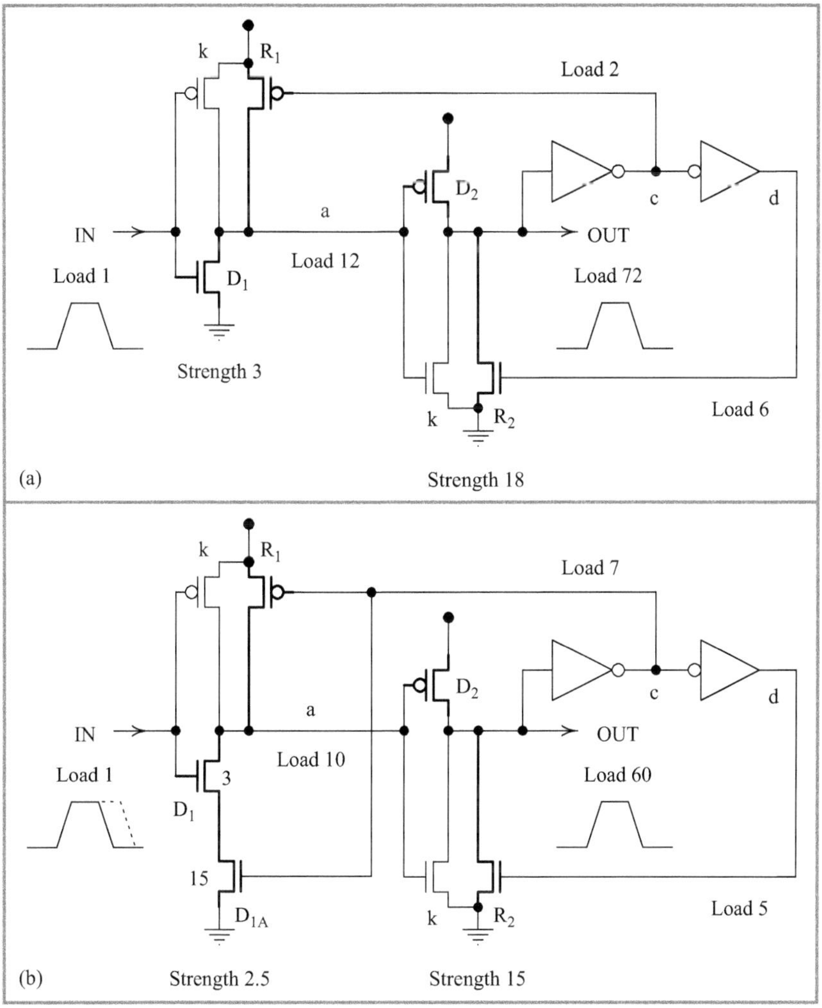

Figure 7.6 Pulse amplifiers with higher gain. More amplification is available by postponing the drive of reset transistors, R_1 and R_2. Two feedback loops from the amplified output signal, OUT, drive R_1 and R_2. Because none of the charge on signals IN and a *is spent on transistor R_1 and R_2 respectively, transistors D_1 and D_2 can be stronger than in Figure 7.5. High gain is extremely useful for driving signals with large fan-out, like clock trees or the "local clock" signals that drive the storage elements in many an asynchronous design. Moreover, the two feedback loops maintain the pulse duration, which is particularly important when amplifying short pulses. Both (a) and (b) output fixed three gate delay wide pulses, but (b) will accept somewhat wider input pulses*

In other words, for the same input loading, Proebsting's two-stage *post-charge logic* can drive about four times as much load as two ordinary inverters, and it responds just as fast.

A rationale for the strength and load numbers in Figures 7.5 and 7.6 follows in the itemized calculation narrative below. The calculations assume that a P-type transistor is about two times harder to drive than an N-type transistor with the same output load. These assumptions are valid for the 40 nm CMOS Weaver chip in Section 7.3.

- An N-type transistor and a P-type transistor each of strength 1 make a unit inverter of strength 1, like the first stage inverter in Figure 7.5. Assuming that P-type transistors are twice as hard to turn on or off as N-type transistors, two thirds of the input load of the first stage inverter in Figure 7.5 is due to R_1. When *IN* goes high this two third serves to turn off R_1, leaving one third to turn on D_1 to pull signal a down. Using a step-up of four per amplification stage, transistor D_1 with its drive strength of 1 allows the falling signal a to drive a load of 4.
- Instead of turning R_1 off at the last possible time, one could turn R_1 off ahead of time, as do the amplifiers in Figure 7.6(a) and (b). Thus when *IN* goes high, the first stage inverter in the two amplifiers in Figure 7.6 serves fully to turn on D_1 to pull signal a down. In other words, these first stage inverters can devote their available input load of 1 for *IN* to drive N-type transistor D_1, that is, to drive $\frac{1}{3} \times x$, with x the strength of D_1. This makes it possible to resize D_1 from a strength 1 in Figure 7.5 to a strength 3 in Figure 7.6(a) and (b). Using a step-up of four per stage, transistor D_1 with its drive strength of 3 allows the falling signal a to drive a load of 12 in Figure 7.6(a). The strength-3 transistor D_1 in series with a strength-15 transistor D_{1A} in Figure 7.6(b) yields a combined drive strength of $1/(1/3 + 1/15)$, or 2.5, for the first-stage amplification. Using a step-up of four, the strength-2.5 series pair allows the falling a to drive a load of 10 in (b).
- Similarly, with R_2 in the second stage inverter in Figure 7.6(a) and (b) turned off ahead of time, the available load on a can be devoted entirely to turn on D_2 to pull *OUT* up. In other words, the second stage inverter in Figure 7.6(a) can devote 12 to drive the load of P-type transistor D_2, that is, to drive $\frac{2}{3} \times y$, with y the strength of D_2. This makes it possible to resize D_2 from a strength 4 in Figure 7.5 to a strength 18 in Figure 7.6(a). Using a step-up of four per stage, the strength-18 transistor D_2 allows the rising *OUT* signal to drive a load of 72 in Figure 7.6(a). Likewise, P-type transistor D_2 in Figure 7.6(b) can be resized to a strength of 15, allowing the falling *OUT* signal to drive a load of 60 in (b).
- These calculations can be extended to choose the strengths of the series inverters on *OUT* that drive signals c and d to reset the output of each amplification stage in Figure 7.6(a) and (b). Their strengths can be small—between 1 and 2—and hardly impact the remaining drive load on *OUT*. By fine-tuning the drive strengths of these feedback inverters one can fine-tune the pulse width on *OUT*.

The circuit in Figure 7.6(a) assumes similar input and output pulse widths. The circuit in Figure 7.6(b) accommodates wider input pulses. Between pulses, weak keepers—marked with the letter *k* maintain the charge and corresponding logic high or low voltage level on the output signal of each stage in Figure 7.6(a) and (b).

Figures 7.5 and 7.6 show circuits of only two stages, amplifying a low-to-high-to-low pulse of three gate delays. Similar circuits amplifying high-to-low-to-high pulses or circuits with more stages and wider pulse widths are also possible. All assume that their signals have enough time to reach full swing—see Figures 7.2 and 7.3 for a reminder on full swing rail to rail signal transitions.

7.1.4 The theory of logical effort, or how to make fast circuits

To do its logical function, a NAND gate must have more transistors than an inverter. To drive as much load as a same-strength inverter, a NAND gate not only uses more transistors but also has transistors connected in series that are extra strong. Because it has more transistors, some of them extra strong, a NAND gate has more input load. It is harder to turn the transistors of a NAND gate on or off than it is to turn on or off the transistors of an inverter of the same strength.

The Theory of *Logical Effort* [3] quantifies the "cost" or *logical effort* of doing logic as how much worse input load a logic gate presents than would an inverter of equal strength. In other words:

$$\textit{logical effort} = \textit{input load}/\textit{drive strength}.^{\dagger}$$

We use 1 for the logical effort of an inverter. More complex logic gates tend to have logical effort larger than 1. Usually, the more complex the logic, the larger its logical effort. For example, assuming that P-type transistors are twice as hard to turn on or off as N-type transistors, the logical effort of a NAND gate, NOR gate, multiplexer, and XOR gate are $\frac{4}{3}$, $\frac{5}{3}$, 2, and 4, respectively.

In some situations one can resize the transistor strengths of a complex logic gate to reduce its logical effort for targeted output transitions. For example, recent work by Swetha Mettala Gilla *et al.* on sizing mutual exclusion elements resizes two frequently used and referenced arbiter designs, giving each a logical effort less than 1 for uncontested grants [4]. The resized designs provide least uncontested grant delay, making the common case—uncontested arbitration—fast.

Similarly, by moving the reset load from the input signal to an amplified output signal, each Proebsting amplifier in Figure 7.6 reduces the logical effort of its first stage inverter. With an input load of $\frac{1}{3} \times 3$ or 1 *each* and drive strengths of 3 for (a) and 2.5 for (b) the first stage inverters in Figure 7.6 have a logical effort of $\frac{1}{3}$ for (a) and $\frac{1}{2.5}$ for (b). Likewise, with input loads of 12 for (a) and 10 for (b) and with drive

†Rather than measuring the *logical effort* of a logic gate as (1) how much more load the gate presents at its input than would an inverter of equal drive strength—as we do here—feel free to measure instead (2) how much weaker its strength is if the gate is allowed to present only the same input load as an inverter, or—alternatively—(3) how many times longer than an inverter it takes the gate to drive a copy of itself. These three views are mathematically the same [3].

strengths of 18 and 15 respectively, the second stage inverters of the two Proebsting amplifiers have a logical effort of $\frac{2}{3}$ *each*. By reducing the logical effort of each stage, the two Proebsting amplifiers achieve outstanding amplification.

We use logical effort to design fast CMOS circuits, with a guiding principle: *The fastest logic equalizes the product of logical effort and amplification in all stages.* This guiding principle translates any cost increase in doing logic, compared to an inverter, into a corresponding decrease in amplification.‡ For more background on the use of logical effort to improve circuit performance, see references [3,5,6].

7.1.5 Summary and conclusion of Section 7.1

Ring oscillators set the pace of self-timed circuits. Although rings of three logic gates are possible, we prefer to use slower but easier to design and more robust rings of five or a larger odd number of logic gates. Rings of five logic gates oscillate at ten gate delays per cycle. Globally clocked networks running as fast as that are unlikely due to the difficulty of amplifying short pulses. Section 7.3 in this chapter describes the Weaver—a self-timed on-chip network manufactured in a 40 nm CMOS technology that operates at the speed of a five-gate ring oscillator. We designed this high-speed self-timed network and the various circuit components in it in accordance with the logical effort guideline described in Section 7.1.4.

The circuit components in the Weaver are partitioned into *Links*, which store and transport local data and state information, and *Joints*, which compute on the data and state information in their Links and control the flow and distribution of the locally computed results and state updates. The information exchange from each individual Link to a Joint and back to the Link forms a ring oscillator. In the Weaver, each Link-Joint pair forms a five-gate ring oscillator. Each such ring oscillator generates a five gate delay low-to-high-to-low pulse signal that is amplified and then used to capture results and state updates computed by the Joint and stored by the Link. The Weaver uses simple pulse generation and amplification techniques.

This section broaches the topic of simple as well as advanced pulse management techniques because different designs require different techniques. The Weaver could use simple techniques because (1) all its ring oscillators operate at the same speed, (2) its data items are only 72 bits wide, and (3) the routing logic in each of its Joints has sufficiently low logical effort to leave adequate amplification to "locally clock" each Link.

Many of the design and test aspects for the Weaver are built in from the bottom up—starting at the level of individual Links and Joints. We therefore added an intermediate section, Section 7.2, on Links and Joints.

‡In Section 7.1.3 on page 119, we gave an alternative but mathematically identical guideline for achieving fastest delay by equalizing the step-up in all stages. The two guiding principles are identical because the logical effort guideline—equalizing the product *(logical effort* × *amplification)* in all stages—is, according to the definitions of logical effort and amplification on pages 123 and 119, respectively, the same as equalizing *((input load/drive strength)* × *(load driven/input load))*, that is, *(load driven/drive strength)*, which—according to the definition of step-up on page 119—amounts to equalizing *step-up* in all stages.

7.2 The Link and Joint model

As do many asynchronous or self-timed circuit designers, we too started out with a favorite set of handshake components that we compiled to and whose function and timing we validated. The moment we started working with two self-timed circuit families, Click and GasP, each with its own handshake protocol, the compilation process got out of hand. Both families had fixed initialization circuitry built into each component to set the initial states of their handshake signals. Differences in initialization, handshake signaling, and static timing required Click and GasP specific code duplication at various levels in the compiler, and made the resulting compiler more complex and less useful [7]. Each additional initialization version of otherwise identical components multiplied our design and validation efforts. While modeling and validating timing constraints for flow control components in Click we noticed that the components used the same handshaking parts for each handshake interface and we were incidentally validating these parts over and over again. Modeling them for each component was useless as well as harmful: we spent significant amounts of time on managing the complexity of the models [8,9]. What went wrong?

Section 7.2.1 analyzes what went wrong and motivates the various steps that we took to make things right [10]. We illustrate these steps on Click and GasP circuits. Figure 7.7 shows the handshake protocols for Click and GasP, and Figure 7.8 shows what the circuits looked like when we started working with them [1,7,11].

7.2.1 Communication versus computation

The fact that validation of different flow control components in the same family results in repeated validation of the same communication circuits suggests that we combine too much in one component. As reference for what is in a component, see Figure 7.8. So, let us separate communication from flow control. We will combine the communication circuits for the same handshake signals, including their initialization circuitry, into a separate component, called *Link*. We will keep the remaining circuits in the original component and call the remainder a *Joint*. By placing the Click and Gasp handshake communication circuits in their own Links, we move their differences from the interface to the internals of each Link. As a result, the interface between Links and Joints can no longer distinguish Click from GasP, and the Joints become the same for Click and for GasP. Thus, by separating communication from flow control, we (1) reduce the complexity as well as the amount of validation required, and (2) make way for a single compilation strategy to Links and Joints that works for both families.

We see Joints as the places where Links meet to exchange information. This information can be dataless to serve as a mere synchronization means, or it can involve data from several Links that the Joint computes on and for which it distributes results to other Links. Thus, in addition to managing flow control, Joints compute. Computations work on data. Who stores the data—the Link or the Joint? In our old design approach, with the communication and computation circuitry

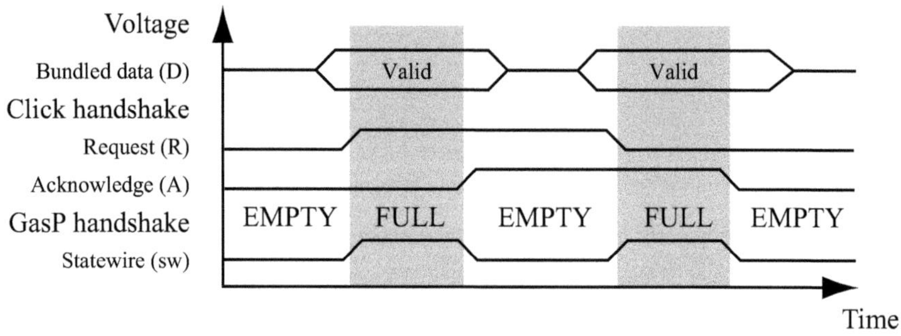

Figure 7.7 Handshakes protocols for Click and GasP. Self-timed circuits in Click and GasP use bundled-data two-phase handshakes. Click has two handshake signals, request (R) and acknowledge (A). GasP has one, called statewire (sw). These signals tell the receiver when the bundle of data wires sent along them carry valid data. Data are valid in Click when R and A differ, and in GasP when sw is high. In the reverse direction, the handshake signals tell the sender when there is space for new data. In the Link and Joint model we view a handshake protocol as a way to encode the presence of data and space, and we focus on ***what*** *rather than* ***how*** *it encodes. So, rather than using R, A, and sw, we use FULL and EMPTY, and we fill a communication Link to make it FULL, and drain it to make it EMPTY. In terms of FULL, EMPTY, fill, and drain, Click and GasP protocols are identical*

residing in the same component, the component stored the data, and the handshake signals were just wires that transferred the data values. In the new design approach with Links and Joints, we made the decision to make the Link both transfer and store the data. As a result, Links and Joints not only separate *communication and storage* from *computation and flow control* but also separate *states* from *actions*.

Figure 7.9 shows what Click and GasP circuits look like in terms of Links and Joints [10]. Except for MrGO, an AND-like gate to be introduced in Section 7.2.2, the two circuits are identical to those in Figure 7.8 for the old design approach—we merely moved the interface! Each Joint in Figure 7.9 acts when its input Link is FULL **and** its output Link is EMPTY. When it acts, it copies the data from input to output Link and it drains the input Link and fills the output Link. The Links respond by changing their FULL or EMPTY states, thus invalidating the conditions for the Joint's action which causes the action to complete its copy, drain, and fill operations—see also Figure 7.11.[§]

[§]Figure 7.11 uses stick figures for Joints and rectangles for Links to illustrate how the Joint acts, how the Links respond, and how the Link responses invalidate the conditions of the Joint's action. The *go* signal used in Figure 7.11 is normally high. It plays a role in initialization and test, as explained in Section 7.2.2.

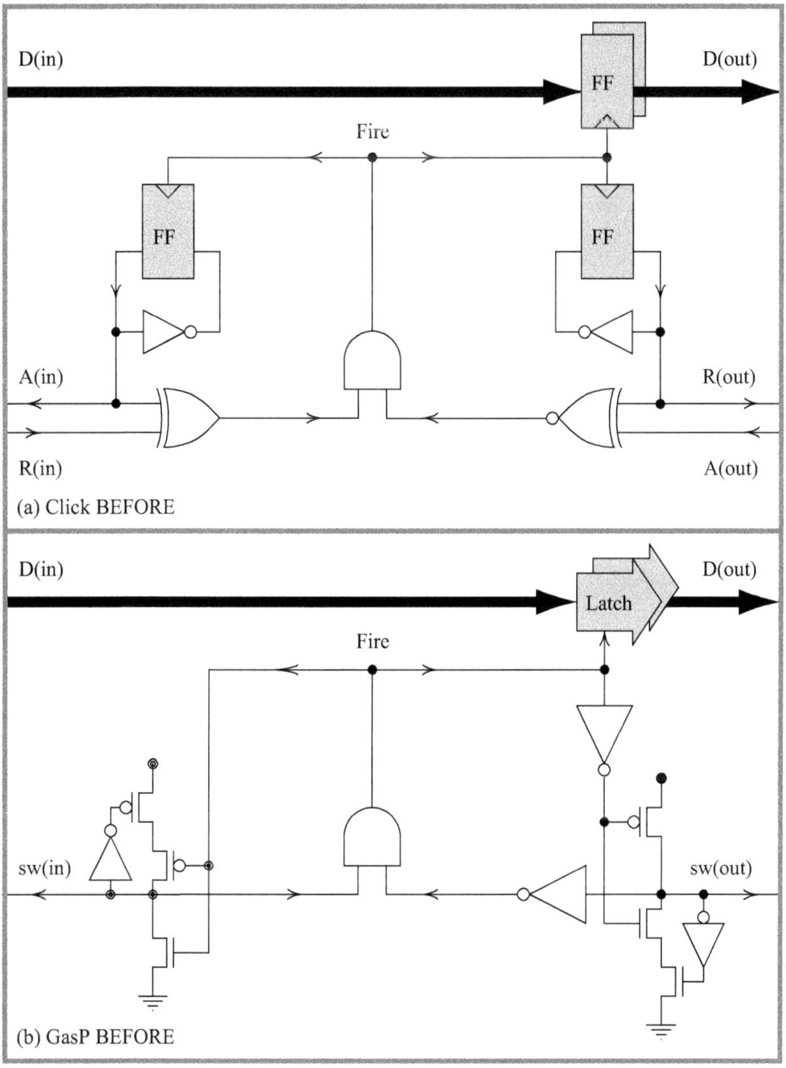

Figure 7.8 *A Click and GasP component before the Link and Joint model. Simple circuits in Click (a) and GasP (b) omitting initialization and amplification. Each acts when (1) its input is FULL, that is, R(in) ≠ A(in) in Click and sw(in) is high in GasP, and (2) its output is EMPTY, that is, R(out) = A(out) in Click and sw(out) is low in GasP. When it acts, it copies and stores D(in) on D(out), makes in EMPTY and out FULL*

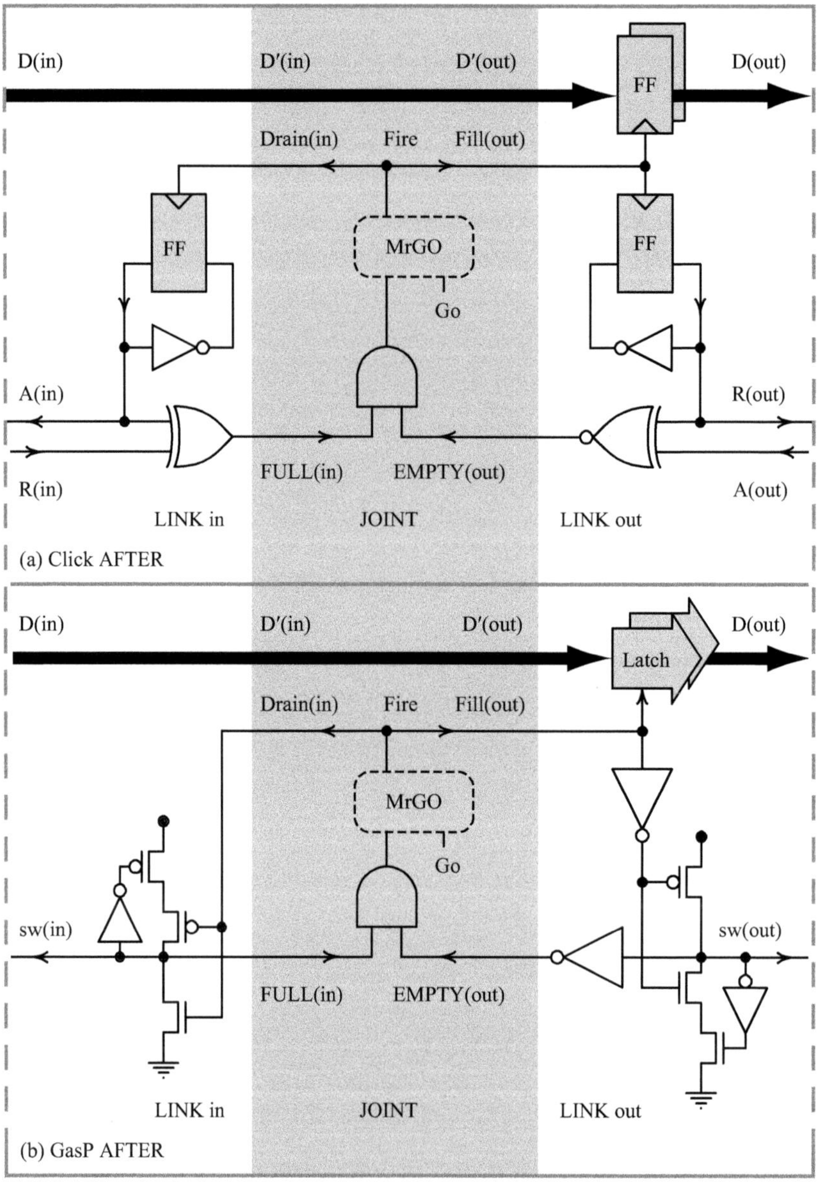

Figure 7.9 *Click and GasP as Links and Joints. We moved the interface from the handshake signals to what they encode, FULL, EMPTY, fill, drain. We stored the data in the Link together with the FULL or EMPTY state. The pictures show only half of each Link. Each complete Link looks like the two pictured half Links put together. The Joint's AND function includes MrGO, an (arbitrated) AND gate—see Figure 7.10*

We describe the behavior of Links and Joints using guarded commands [12,13]. Links and Joints communicate via the FULL, EMPTY, fill, drain, and data signals at their interface. Because the interface signals remain available until the Link changes its FULL or EMPTY state, the communication protocol uses shared variables rather than message passing. As a result, "probes," that check whether or not a Link can communicate [14] and that require special primitives in a message-passing model, are just guards on the Link's FULL or EMPTY state.

7.2.2 Initialization and test

Differentiating states from actions turns out to be key in enabling initialization for design *and* test. For design, fixed initialization circuitry would suffice. But for test, arbitrary circuit initializations may be required, especially if we wish to accommodate unanticipated debug scenarios. So, why bother following the old design approach by building in fixed initialization circuitry that will be used only once, when the circuit starts up? Why not use existing test methods also to initialize the circuit at start-up? There may even be an additional advantage in terms of security if instead of having the circuit initialize itself automatically, initialization takes a separate step—one that is hard to accomplish successfully by accident and would take substantial time to accomplish through trial and error. This section explains how test access to individual actions and states has helped us not only to initialize the Weaver design discussed in Section 7.3 but also to test and debug the Weaver—at speed.

7.2.2.1 Action control: go and MrGO

It is good practice never to let both the design and the test environment initialize the same state at the same time. Letting them initialize different states at the same time may be fine and perhaps even desirable because the self-timed design propagates states faster than the test environment. During initialization, we disable Joint actions that use Link states set by the test environment. Remember that the Links store the states, and the Joints take the actions. To disable an action, we add an extra condition, called *go*, which we control via the test environment. Each Joint in Figure 7.9 now acts when its input Link is FULL **and** its output Link is EMPTY **and** *go* is high. The original condition "input Link FULL **and** output Link EMPTY" no longer suffices: a low *go* signal disables the action—see Figures 7.10 and 7.11.

Thus, by making the *go* signal of every Joint action low, the test environment can disable all circuit actions and safely initialize FULL or EMPTY Link states and data stored in Links anywhere in the design. For the Weaver, we shift the initial values into the chip serially and bit by bit, using a chain of shift registers also known as a *scan chain* [15]. Each shift register is associated with a particular state signal in the design. When all values are shifted into position, the test interface writes them in parallel into the associated Link states and data bits—see Figure 7.12.

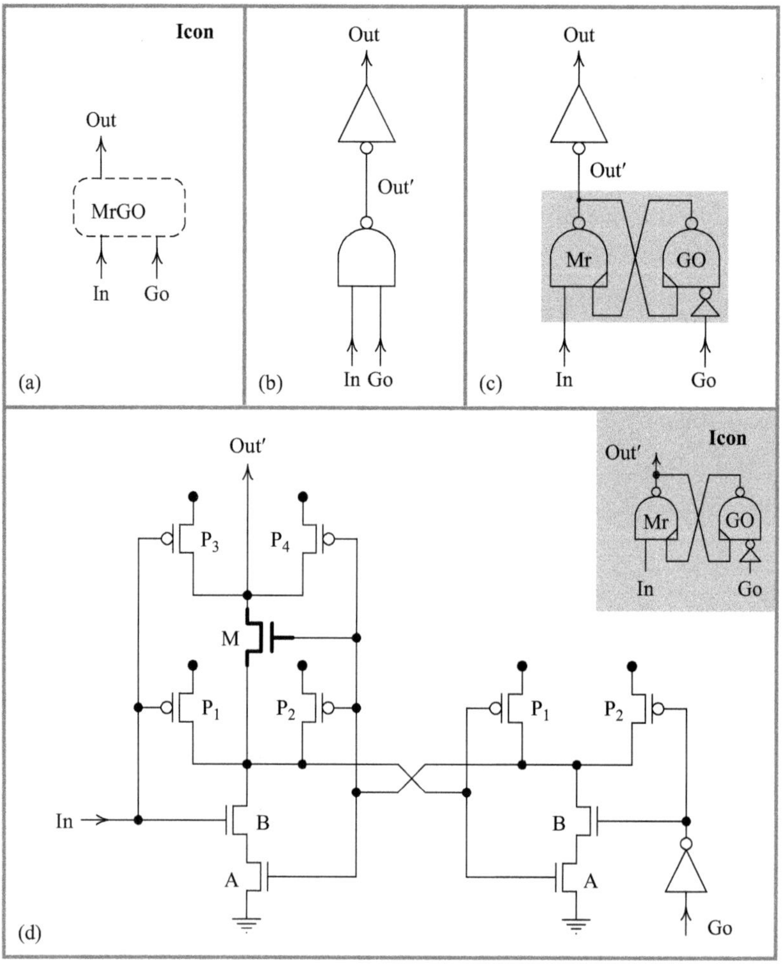

Figure 7.10 *go and MrGO. We use go signals to enable or disable actions. We combine them with the other action conditions through either a simple AND gate (b) or an arbitrated AND gate (c). The arbitrated AND gate, called MrGO and pronounced "Mister GO," is implemented as in (c) and (d). We use the name MrGO with the icon in (a) when the decision which version to use is still open. MrGO arbitrates between a high in signal, to continue the action, and a low go signal, to stop the action. The arbiter's bold transistor in (d) delays active-low grant signal out′ by conducting only after metastability ends. We size the transistors for the common uncontested case to reduce the logical effort from in to out′ [4]. The various pull-up transistors keep out′ from floating. The circuit normally operates with a high go signal*

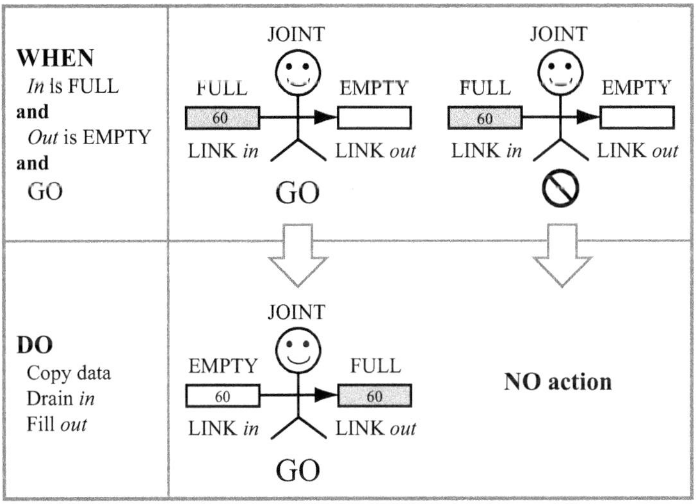

Figure 7.11 Joints act under go control on Link states. This picture illustrates the action of a simple Joint, like the one in Figure 7.9(a) or (b), and the responses of its two Links. The stick figure represents the Joint. The rectangles represent the Links. FULL Links are colored gray, EMPTY Links lack color. The Joint acts only when (1) its input Link is FULL and (2) its output Link is EMPTY and (3) it has permission to act, that is, its go signal is high. When it acts, it copies the data, with value 60, from its input Link to its output Link, drains its input Link, and fills its output Link. The input Link responds by declaring itself EMPTY. The output Link responds by storing the data and declaring itself FULL. Their responses disable and complete the Joint's action, and may enable actions in neighboring Joints. Note that the data in the input Link are unaffected by the drain operation. The data with value 60 will remain in the input Link until a follow-up fill operation or a test write operation changes them

For design initialization, the next step is to make every *go* signal of every Joint action high and thereby start the circuit.

Test methods often use a single *go* signal, typically called *test mode*, to start or stop the circuit. This may work fine when circuit actions are synchronized under global control, but is far from ideal when they are asynchronous, widely distributed, and local. As a thought experiment, try to follow a burst of data items through part of a self-timed design—at speed. Any global control to "walk the burst" will conflict with the "at speed" nature of the experiment. The only candidate qualified to run this experiment at speed is the circuit itself, running self-timed. So, if we could enable all Joints within that part and disable the Joints outside it, then a burst of data would run through that part at speed—and after it has run its course, we could

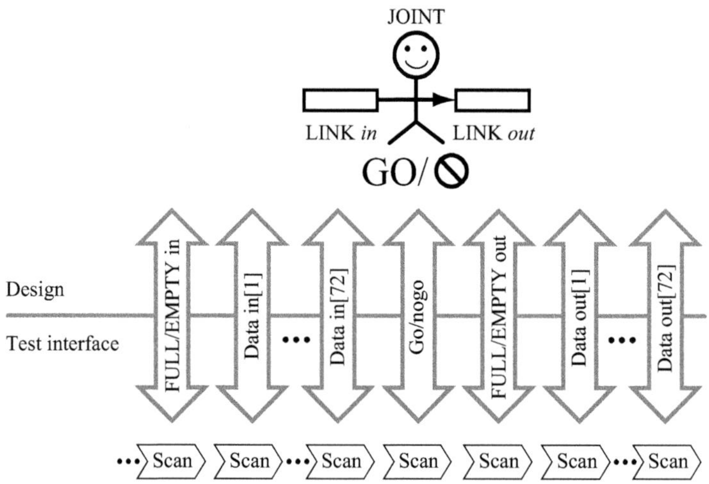

Figure 7.12 Link and Joint scan test interface. A series of shift registers, called a scan chain, shifts bits in and out of the circuit serially. The scan chain (bottom) can shift while the design (top) operates. A scan chain reads or writes the design's data bits, FULL or EMPTY states, and go signals in parallel. To avoid interference when setting up states or actions, we use separate scan chains for states and go signals

read and scan out the states it left in its wake. Guess what. We can do exactly that by giving each Joint action its own *go* signal.

In the Weaver, we use many separate *go* signals, one per Joint. In principle, there are as many *go* signals as actions. To control that many *go* signals, we shift them in using a scan chain—hence the *go/nogo* test interface and scan shift register in Figure 7.12. Note that Figure 7.12 can read as well as write *go* signals, Link states and data bits, and shift their values in as well as out through the scan chain. To avoid interference between controlling actions and reading or writing states, the Weaver has separate mutually exclusive scan chains for *go* signals, for FULL or EMPTY Link states, and for data bits.

Figures 7.13 and 7.14 show examples of the "thought experiment" discussed earlier. The examples run a burst of data items at speed through a part of the design that has a counter attached to it. They test whether the counter can keep up and count the correct number of data items passing by. We did similar test experiments on the Weaver chip, whose counters are located in the NE corner of the floorplan—see Figure 7.16. The test environment can read the Weaver counters, but only reset them to zero. This was a mistake, acceptable in a mere test chip with a transparent circuit-versus-performance relationship, but unacceptable otherwise. Counters are sufficiently important for characterization purposes to justify full scan support for both reading and writing.

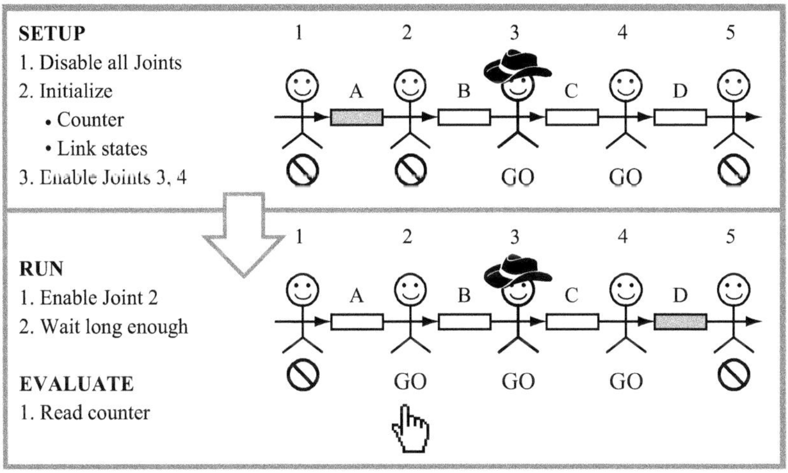

Figure 7.13 *Counting one data item at speed. The counter is attached to Joint 3 with the cowboy hat. The test setup leaves several adjacent Joints enabled to permit data to pass through them at speed. Disabled Joints upstream and downstream of the test part prevent entry of other inputs and escape of results. Enabling "gate keeper" Joint 2 releases the test data to flow through the test part at speed and update the counter*

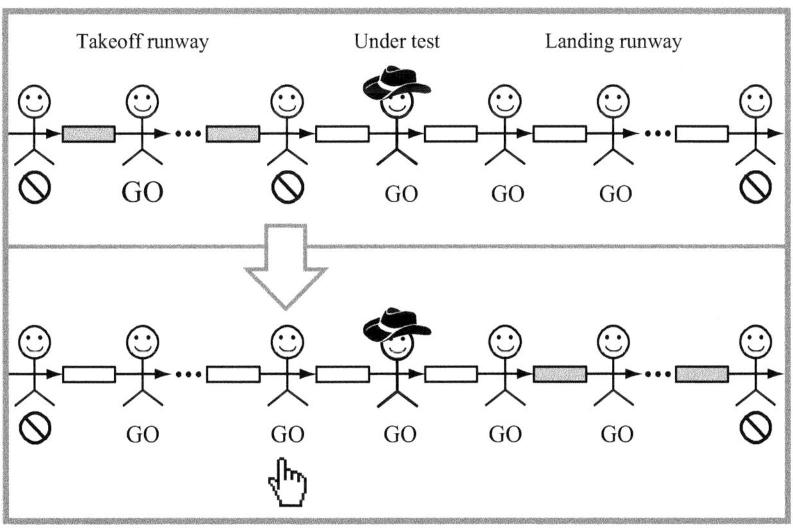

Figure 7.14 *Counting a burst of data items at speed. With longer takeoff and landing runways one can run more data past the counter at speed*

The Links and Joints in Figures 7.13 and 7.14 are similar to those in Figure 7.9 and operate as illustrated in Figure 7.11. Note that the test segments are bounded upstream and downstream by disabled Joints. Most test interface tasks, including circuit initialization and manufacturing testing against, say, stuck-at faults, work on bounded segments. Some though, notably tasks related to performance characterization, require the circuit to run freely. It is for these tasks and our ability to stop them cleanly, without corrupting circuit states, that the *go* signal has its own arbiter. The arbitrated *go* signal, shown in Figure 7.10(c) and (d), is called *MrGO* and pronounced "Mister GO" and arbitrates between continuing or stopping an action.

The arbitrated version of MrGO in Figure 7.10 must "crown" the AND function for the Joint action, that is, all FULL or EMPTY state signals used in either the guard or the command of the action must be AND-ed *before* we AND the *go* signal. If the guard contains data bits, then these may be AND-ed either before or after MrGO—the FULL state signals of their Links already cover them. The Weaver adds the data bits after MrGO, as can be seen from Figures 7.21 and 7.23(a) and (b). Giving MrGO a "crown" position guarantees a nonblocking arbitration that allows the *go* signal eventually to grab the arbiter, either the first time or the next.[¶] If not the first time, because the arbiter favors a high *in* signal over a low *go* signal and thus continues the action, then the action will release the arbiter one cycle later—without blocking. Because the cycle time for new actions is longer than the arbiter's uncontested grant delay, the arbiter will grant *go* next and thus disable further action until released by a high *go* signal. Any MrGO position other than a "crown" position may allow a temporary state to grab MrGO and may inject actions that interfere with the initialization or test run whenever this state changes.

To obtain the throughput and power measurements reported in Section 7.3.5 for the Weaver chip, we let the self-timed circuit run freely for, say, 10 seconds and then read out the counters. The corresponding test setup resembles Figures 7.13 and 7.14: first disable all Joints, then EMPTY all Link states, then enable all but two Joints—the "gate keeper" before the reloader Link, and the reloader Joint after the reloader Link. Reloaders are the only stages in the Weaver where we can scan data bits in and out. They are in the SE corner of the floorplan—see Figure 7.16. We repeatedly scan a data pattern with FULL Link state into the reloader Link, and temporarily enable the reloader Joint to copy the data forward. The data will queue up behind the "gate keeper." When all scan inputs are delivered, we reset the counters, then enable "gate keeper" and reloader, let the circuit run for 10 seconds, disable the "gate keeper"—using MrGO—let operations peter out, and then read the counters.

[¶]We ignore the fact that the arbiter itself can take arbitrarily long to decide which of two contesting inputs to grant. In practice, arbitration time is less relevant for initialization and test. We can avoid arbitration in general for design initialization. We can avoid MrGO arbitration for tests on bounded segments as in Figures 7.13 and 7.14. We can filter out rare arbitration delays by running performance tests multiple times.

7.2.3 Summary and conclusion of Section 7.2

By distinguishing communication, done in Links, from computation, done in Joints, we obtain a simple interface that unifies both Click and GasP as well as many other self-timed circuit families and the compilation and verification tools around them. By also distinguishing states, stored in Links, from actions, performed by Joints, we obtain a simple model of computation that works for computer scientists and electrical engineers alike—a working relation that we intend to explore further.

Traditional scan test access to individual states combined with *go* and MrGO control of individual actions allows clean initialization as well as at-speed test, debug, and characterization. In addition to playing a key role in test, MrGO can also be used to gradually synchronize a self-timed design to a clock domain [16].

Section 7.3 details the Weaver's Links and Joints and their scan connections. The details include amplification needs and cycle times as discussed in Section 7.1.

7.3 The Weaver, an 8 × 8 crossbar experiment

The Weaver is a self-timed 8 × 8 crossbar switch built in 40 nm CMOS by TSMC. The Weaver's crossbar steers individual data items from any of eight input channels to any of eight output channels. Local arbitration throughout the crossbar resolves internal contention without blocking—on a first-come-first-served basis that is fair to the loser—so that only input and output channel capacity limit throughput. Without contention, a data item can pass through the crossbar in less than one nanosecond. Without contention and at nominal 1.0 volt power supply, each channel can pass about 6 Giga data items per second through the crossbar at less than half a watt. Data items are 72 bits wide, giving the crossbar's eight channels a maximum combined throughput of nearly 3.5 Tera bits per second. In the absence of traffic, only leakage consumes energy. The Weaver runs without a clock.

The crossbar on the Weaver chip occupies an area of about a tenth of a square millimeter in a triangle 433 micrometers by 391 micrometers in size. The triangle contains a triangular array of 56 switches, one for each possible channel to channel connection. The Weaver chip places the switches in close proximity and provides recirculating first-in-first-out (FIFO) rings to connect the crossbar outputs back to the crossbar inputs for extended high speed testing. For a network-on-chip application, one would distribute switches like those in the Weaver geographically to form the on- and off-ramps of a freeway-like data network.

The following sections discuss the design and test features of the Weaver from a logical, electrical, and layout floorplanning point of view. Section 7.3.1 discusses the architecture of the Weaver. Section 7.3.2 shows the key circuits used in the Weaver. For consistency, the drawings in both sections show the architecture and circuits from the Weaver's floorplanning point of view. Sections 7.3.3 and 7.3.4 discuss test logistics and Section 7.3.5 reports performance measurements from the Weaver chip. The final Section 7.3.6 concludes this chapter by summarizing where the Weaver's logical, electrical, and layout views differ—and why.

7.3.1 Weaver architecture and floorplan

Figure 7.15 shows a schematic diagram of the Weaver. Eight self-timed channels, each with 48 Links, form FIFO rings that recirculate data from the output of the 8 × 8 *crossbar switch*, also called *crossbar*, back to its input. Two additional channels bypass the crossbar to provide a performance contrast. The two bypass channels flank the recirculation channels. The ten rings are labeled in the floorplan of Figure 7.16. One bypass channel, Ring 0, has 48 Links, but the other, Ring 9, has only 40 Links.

In the floorplan long rectangles represent Links and black dots represent Joints. Arrows connecting the dots and rectangles indicate the direction that data flow. The floorplan arranges the Joints in rows numbered 1 to 20 from bottom to top and in columns with letters *A* to *Z* from left to right. The medium- and dark-gray triangle in the North West (NW) identifies the crossbar switch itself. Data enter the crossbar from the South at row 12, going North, and leave the crossbar at column *K*, going East. The diagonal NW edge of the crossbar (*Double-barrel Ricochet*) folds its datapaths so they turn their data from a Northbound to an Eastbound direction.

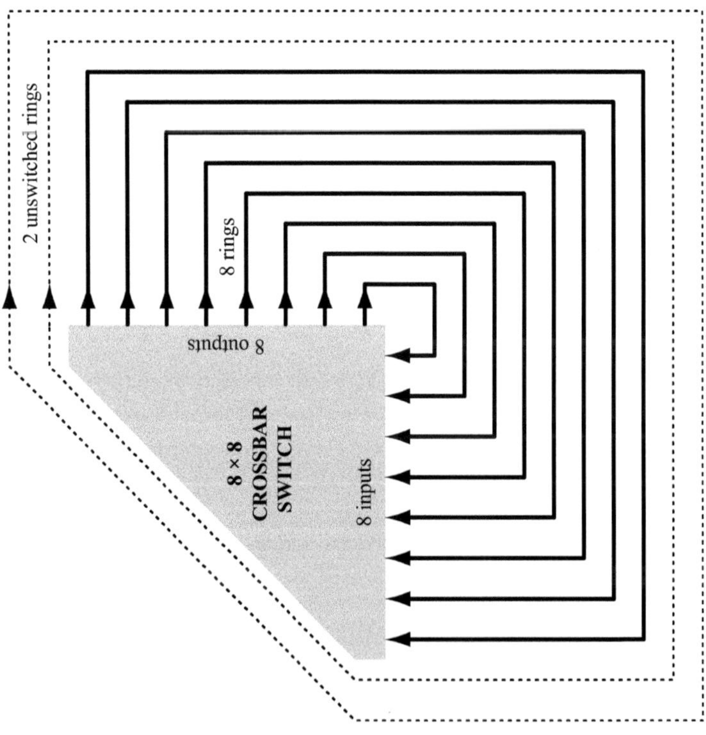

Figure 7.15 Weaver diagram, rotated to match the orientation of Figure 7.16. The crossbar switch is in the gray triangle. Eight rings, 1–8, recirculate data from its output back to its input. Two extra rings, 0 and 9, bypass the crossbar switch to measure relative performance

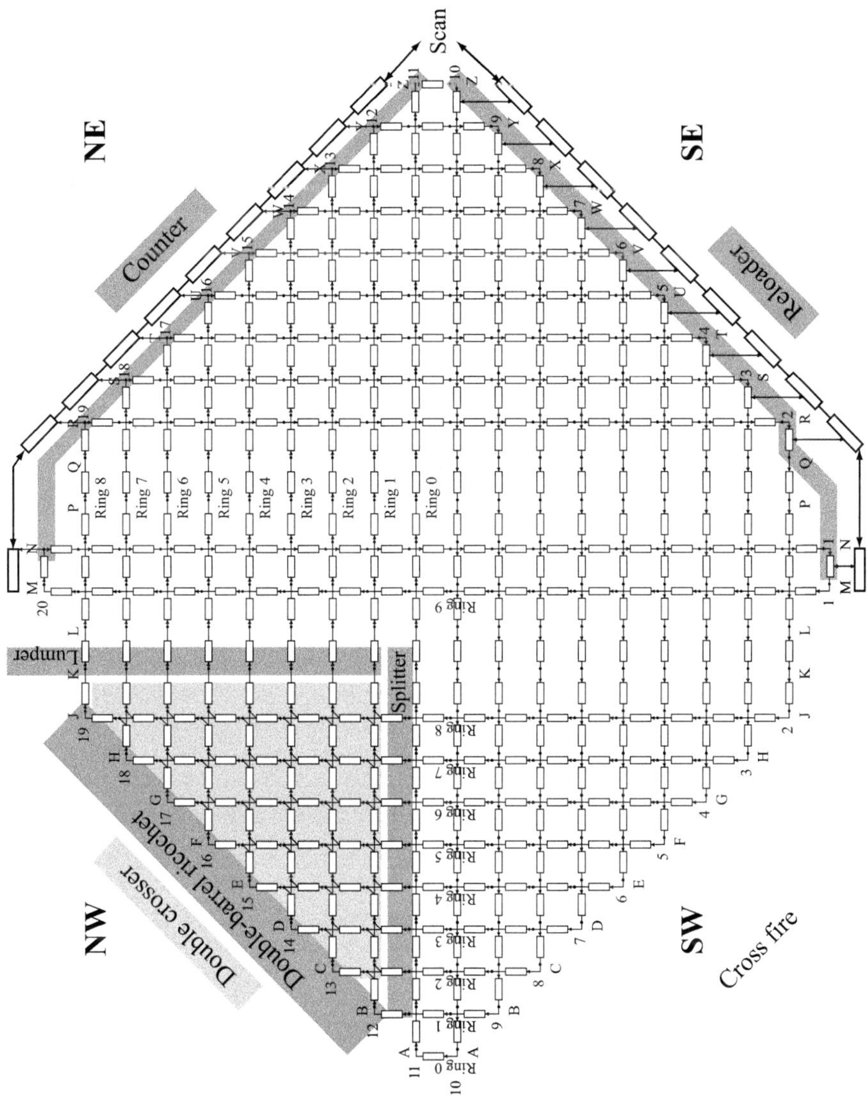

Figure 7.16 *Weaver floorplan, rotated 90 degrees. The Weaver includes ten rings, of which eight recirculate data from the output of the crossbar (NW corner) back to its input. Each of the ten rings has a reloader stage (SE edge) to read and write data from and to one of its Links, and a counter (NE edge) to tally the actions taken by one of its Joints. Scan chains (SE and NE) control the reading and writing of data and the tallying and resetting of ring counters. Other scan chains, omitted here, control the FULL or EMPTY state of each Link and enable or disable the go control signals of each Joint*

Each of the ten rings forms a rectangle through which data circulate clockwise. Ring 0 is wide, stretching between columns *A* and *Z*, and not very tall, going from row 10 to row 11. Ring 1 is a little narrower, stretching between columns *B* and *Y*, and a little taller, going from row 9 to row 12. Successive rings are narrower and taller until Ring 9 is very narrow and very tall, occupying columns *M* and *N* and rows 1 to 20. The rings fold on 45-degree diagonals at the edges of the floorplan—like a ribbon cable. The North East (NE), South East (SE), and South West (SW) ring sections collectively named *Cross Fire* identify simple FIFO pipeline circuits that form most of each ring. Note that even though horizontal and vertical pipelines in the *Cross Fire* sections may cross each other they are entirely independent.

Each ring includes a counter stage to measure throughput. The counters lie along the North East (NE) diagonal edge of Figure 7.16. Each ring also includes a reloader stage to insert, overwrite, or read data values. The reloaders lie along the South East (SE) diagonal edge of Figure 7.16. The Weaver is so named because data items can weave complex paths through its eight middle channels. Many of the part names of the Weaver, such as *Double-barrel Ricochet* and *Cross Fire*, adopt gunslinger lingo from Western movies.

7.3.1.1 Crossbar switch

Instead of a rectangular structure with $N \times N$ switches the Weaver's crossbar switch has a triangular structure with $N \times (N - 1)$ switches and N repeaters. The triangular structure of the crossbar minimizes its wire length and simplifies layout of the rings that recirculate data through the crossbar at high speed.

Figure 7.17(a) illustrates the structure of the Weaver's crossbar. Data items entering from the South on any of eight input channels exit to the East on any of eight output channels. Note that Figure 7.17 shows fewer channel connections. Because the datapath folds diagonally like a ribbon cable, each of the eight input channels crosses every other input channel in the crossbar exactly once. The arrows inside the box at each crossing suggest how the switches at each crossing allow data items to change from one channel to another. The repeaters that fold the datapath at the diagonal edge of the crossbar are *Double-barrel Ricochet* modules described later.

A complete 8×8 crossbar must have 8×7 or 56 individual switches, called *Crossers*. The Weaver arranges these in pairs, called *Double Crossers*, one pair at each of the 28 crossings. Figure 7.17(b) and (c) illustrates one *Double Crosser* while Figure 7.17(d) shows how the *Double Crosser* appears in the floorplan of Figure 7.16. One of its switches serves its North output, and the other one serves its East output. Each *Double Crosser* accepts data items from the South or West and delivers them to the North or East. Conflict in the crossbar happens only if two data items try to leave a *Double Crosser* concurrently by the same exit. Each switch, or *Crosser*, has a mutual exclusion element, or arbiter, with metastability protection [4,17] to resolve exit conflicts on a first-come-first-served basis that is fair to the loser. Although metastability may delay the passage of a data item, the self-timed nature of the Weaver renders such delay harmless. Metastability delays tend to be so rare and short that we expect them to be unnoticeable.

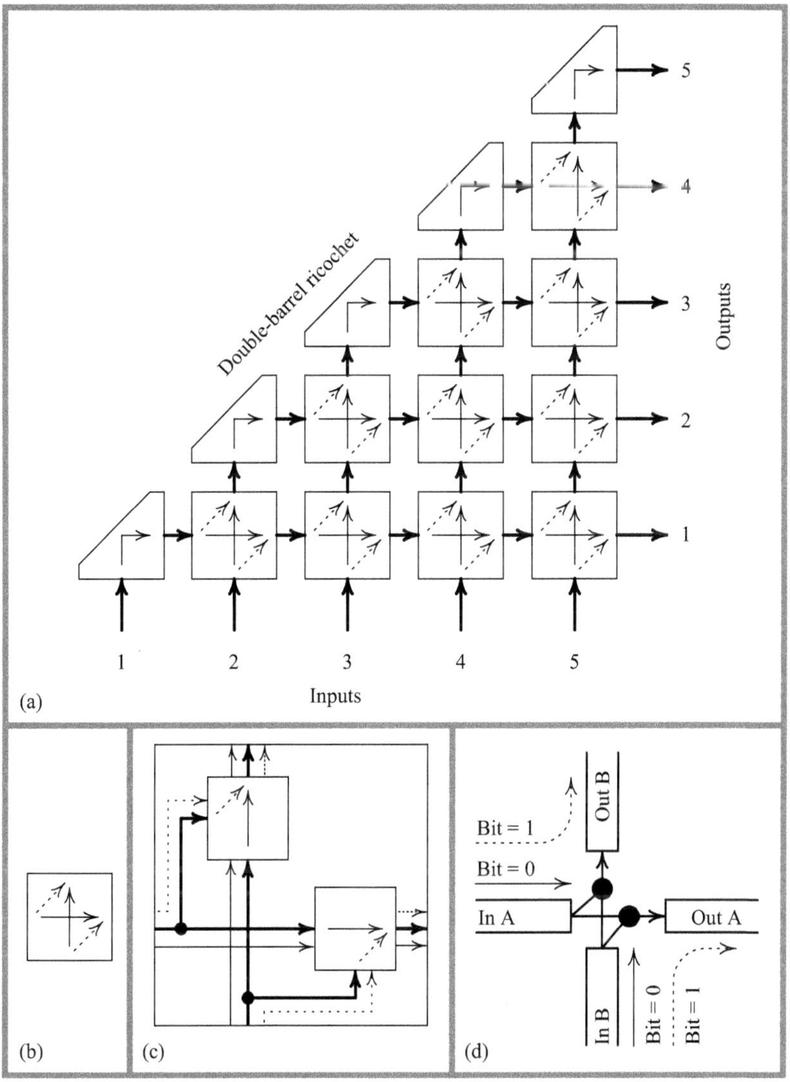

Figure 7.17 *Weaver crossbar switch. The 8 × 8 crossbar switch is a triangular structure with 28 Double Crossers, of which some appear in (a). Each Double Crosser (b) contains two Crossers (c) that arbitrate between competing data items that go in the same direction. The direction is encoded in the data, in a Double Crosser specific steering bit. Data go straight when this bit is 0, and turn otherwise. Link-Joint version (d) of the Double Crosser as used in the floorplan of Figure 7.16 connects four "rectangle" Links and two "black dot" Joints*

7.3.1.2 Steering bits

Steering bits in each data item control how the data item passes through the Weaver. Because each data item carries its own steering bits, each data item weaves its own individual path through the crossbar. To simplify decoding, the Weaver assigns an individual steering bit for each of the 28 *Double Crossers* in the crossbar switch. In other words, 28 of the 72 data bits in each data item may function as steering bits. The remaining 44 data bits are entirely free of assigned meaning in any data item.

Each steering bit applies to a singular *Double Crosser*. Regardless of how it enters a *Double Crosser*, a data item with a 0 in the steering bit position for that *Double Crosser* goes straight through it—West to East or South to North—staying in the same channel. A data item with a 1 in that steering bit position turns either from West to North or from South to East, changing to the other channel.

So, at each *Double Crosser* a data item either remains in its channel or changes to the other channel. Thus, per channel, a data item requires seven steering bits, one for each of the other channels to which it might change. The remaining $72-7$ or 65 bits can carry arbitrary data, though, of course, some of those 65 bit positions may be used as steering bits after the data item switches to another channel.

This choice of rules for steering simplifies testing by forcing every data item to follow a closed path. For instance, data items with 0 in all steering bits keep circulating in their initial ring. Data items with a single 1 in the steering bit position for an intersection of two rings circulate alternately around those intersecting rings. Data items with multiple 1's in steering bit positions weave through several rings in succession, following a closed path that may be less intuitive at first sight. A different application might use other steering rules by decoding a destination address.

7.3.1.3 Test infrastructure: scan, counters, and reloaders

Control of Weaver experiments is entirely through an industry standard low-speed JTAG test interface and scan chains [7,15]. The scan chains serve three purposes.

- First, the scan chains can read or clear the throughput values in each of the ten 54-bit counters, one per ring. These counters appear at the North East (NE) edge of Figure 7.16, and in more detail in Figure 7.18.
- Second, the scan chains can write a data value into or read a data value from a data item in each of the ten reloaders, one for each ring. The reloaders are located at the South East (SE) edge of Figure 7.16.
- Third, because the Weaver implements *Naturalized Communication and Testing* as described in [10] and Section 7.2, the scan chains can stop the flow of data, sense the FULL or EMPTY state of every Link, initialize or reinitialize the FULL or EMPTY state of every Link, and restart the self-timed flow.

In addition to the low-speed JTAG-controlled scan chains, the Weaver has two dedicated medium-speed output pins that deliver reduced-frequency real-time signals from the counters. These reduced-frequency outputs switch at 2^{-20} or approximately one millionth of the throughput rate of the rings. Two series of

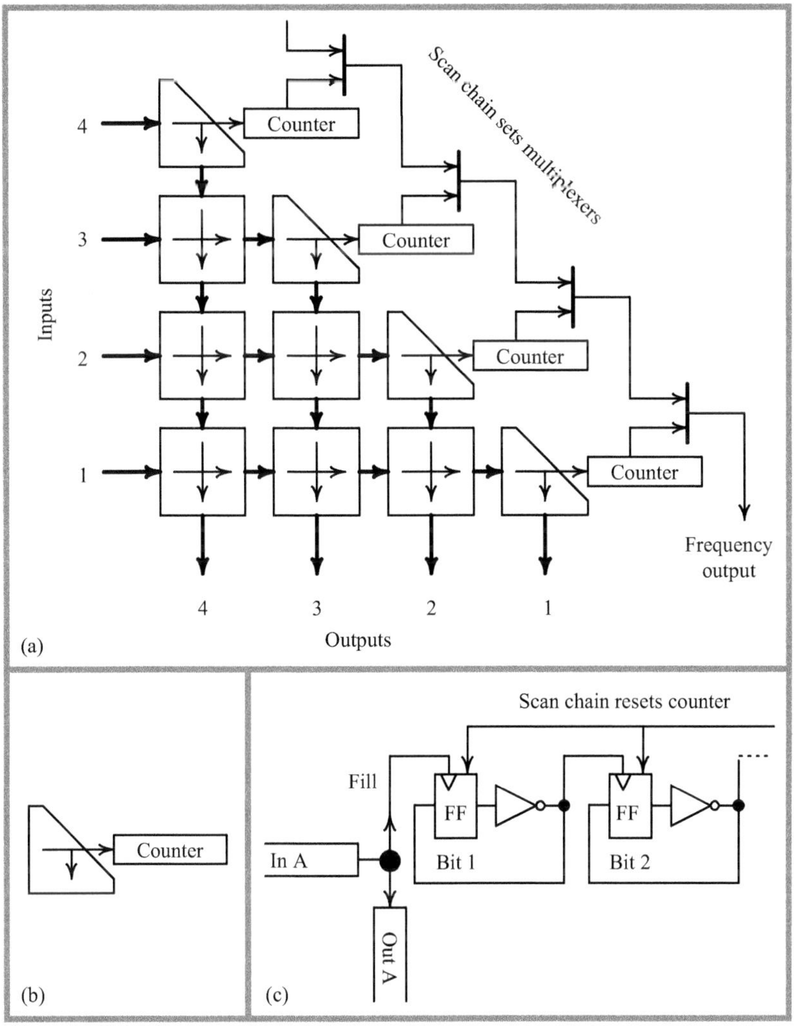

Figure 7.18 *Counters. Each of the ten rings in the Weaver has a counter of which some appear in (a). Bit 19 of each counter provides a frequency output for the ring's monitored fill signal (b) divided by* 2^{20} *or about a million to view in real time on an oscilloscope. The scan chain selects which output is delivered off chip. The counters are implemented as ripple counters (c) that store each bit in a flipflop that uses its inverted output as its input. Each flipflop operates at the rising edge of its clock input. When a bit changes from 1 to 0, its flipflop clocks the flipflop of the next significant bit. The fill signal of the Joint connection to the ring clocks the flipflop of the least significant bit*

multiplexers like those in Figure 7.18(a) allow the scan chain to select two rings for frequency output. The frequency outputs permit real-time observation and comparison of throughput.

7.3.2 Weaver circuits

Motivated by the five- versus three-gate ring comparison presented in Section 7.1.2, the Weaver uses the 6-4 GasP circuit family [10], whose circuits run at the speed of a five-gate ring oscillator, rather than the original 4-2 GasP circuits [1] that run at the speed of a three-gate ring oscillator. The use of GasP is a matter of convenience, given that we have both the expertise and design libraries for GasP, and a matter of target speed, given that designs tend to be faster in GasP than, for instance, in Click.

This section shows the Link and Joint organizations and the 6-4 GasP circuit implementations for various Weaver parts. First, Section 7.3.2.1 describes a simple FIFO circuit that copies, stores, and transports 72 bit wide data. Last, Section 7.3.2.3 describes more advanced parts from the crossbar switch that provide data-driven flow control. Section 7.3.2.2, in the middle, focuses on critical paths and on circuit solutions to manage such paths.

7.3.2.1 First-in-first-out (FIFO) circuits

Figure 7.19 shows the 6-4 GasP implementation for the simple FIFO pipeline circuit in the ring sections collectively called *Cross Fire*—see floorplan of Figure 7.16. The gate-level behaviors of the Joint and the two Links in Figure 7.19 are the same as for the Joint and GasP Links in Figure 7.9 in Section 7.2. The implementation in Figure 7.19 is more detailed in that it specifies the inverting—and amplifying—gates so we know exactly how many inversions there are in each self-resetting loop.

Gates *A*, *B*, *C*, *D*, and *Y* form a self-resetting loop with five inverting gates, as do *B*, *C*, *D*, *E*, and *X*. The two loops share three gates—*B*, *C*, and *D*—to form the Joint's AND function, *FULL(in)* **and** *EMPTY(out)* **and** *go*, and to limit delay variations between the loops. Inverters *D*, *E*, *F*, and *G* amplify the output signal of the AND function to drive the large loads presented by the 72 latches in Link *out* and by the driver-and-keeper gates *X* and *Y* in Links *out* and *in*. Figure 7.19 shows only one end of each Link, the end nearest Joint *fifo*. The far end of Link *in* looks like the near end of Link *out* shown in Figure 7.19—and vice versa, the far end of Link *out* looks like the near end of Link *in* shown in Figure 7.19.

Each time Link *in* is FULL and Link *out* is EMPTY and *go* is high, the AND function is asserted, just like in the Joint of Figure 7.9, Section 7.2, and the two loops generate a low-to-high-to-low *fire* pulse of five gate delays to "locally clock" the latches—that is, render them temporarily transparent—so they store the data copied by Joint *fifo* from Link *in* to Link *out*.

Going around each loop, the *fire* pulse also drives *X* and *Y* for five gate delays. For the Links in the Weaver, a five gate delay drive pulse is long enough to drive the state change at the output of *X* or *Y* from rail to rail across the entire Link length. The state change is sensed at the other end of the Link within a gate delay, just like

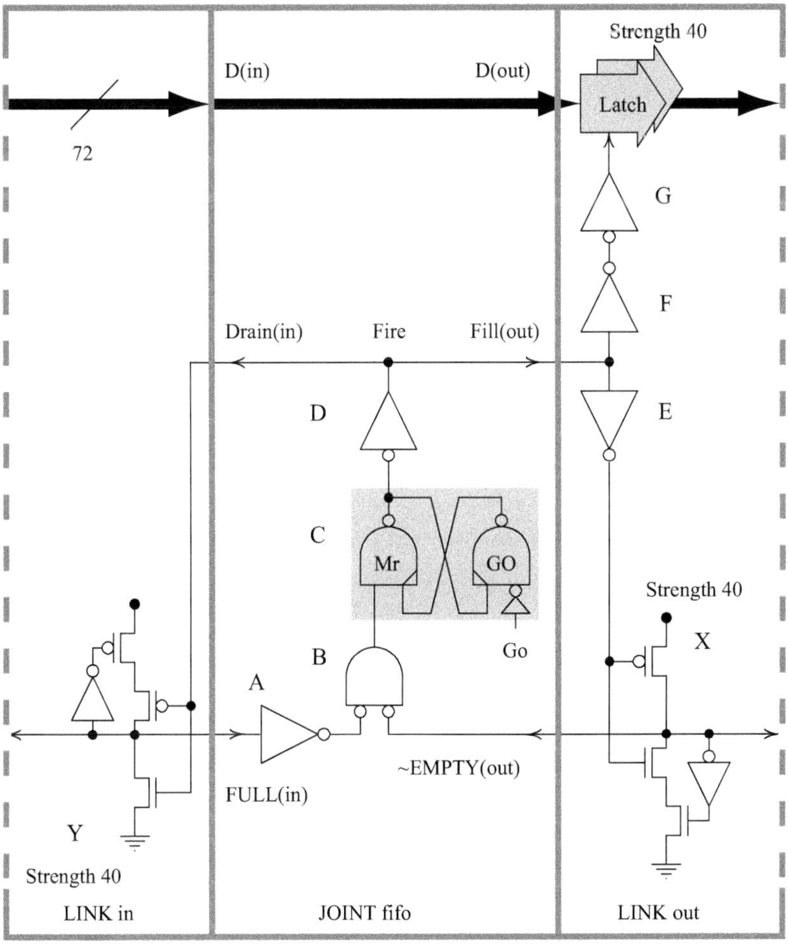

Figure 7.19 FIFO Joint and its two near-end Link connections in 6-4 GasP. This FIFO circuit has the same functional behavior as the Joint with GasP Links in Figure 7.9(b), Section 7.2. With its self-resetting loops, A-B-C-D-Y and B-C-D-E-X, it runs at the speed of a five-gate ring oscillator. We show only one end of each Link, the end nearest Joint fifo. Gray-colored gates are icons—a circuit for MrGO was given earlier in Figure 7.10, the latch follows in Figure 7.20. With the exception of latches all gates invert. To match the inversion count around each loop we use ~EMPTY(out) rather than EMPTY(out) between Joint fifo and Link out. We sized the gates to give each latch and driver-and-keepers X and Y a strength of 40 to help them drive their changes to the other end of their Link within a gate delay

any other gate output change in the FIFO circuit is sensed by subsequent gates within a gate delay. The new FULL or EMPTY Link state is stored by the driver-and-keeper gate at the other end of the Link. To reduce the logical effort of changing a Link's FULL or EMPTY state a driver-and-keeper gate in GasP either drives high and keeps low or drives low and keeps high. This not only creates fast state transitions, it also enables the use of stronger keepers that are more robust to noise.

Note that the asserted AND function is de-asserted after five gate delays, when Link *in* is no longer FULL and Link *out* is no longer EMPTY. The new Link states enable AND functions in neighboring Joints, which in turn fill Link *in* with new data and drain Link *out* and thereby re-assert the AND function in Link *fifo*, and so on.

Fill and drain pulses on the same Link alternate, because one is generated only when the Link is EMPTY and the other only when the Link is FULL. At maximum speed the self-resetting loops run at a cycle time of ten gate delays, barely separating the alternating five gate delay fill and drain pulses on the same Link. Pulse separation is enhanced by the nature of gate *B*—the AND-like gate at the core of each pulse—which turns each pulse on using series transistors and off using parallel transistors. Because parallel transistors are faster than series transistors, each fill or drain pulse turns off before the other turns on. As a result, the driver-and-keeper gate that changes the Link state from EMPTY to FULL or from FULL to EMPTY turns off before the driver-and-keeper gate at the other end of the Link can turn on and change the state back. Separation between fill and drain pulses on the same Link can be enhanced further by threshold shifts in amplifiers *D* and *E* that follow gate *B*.

Because fill and drain pulses on the same Link alternate, latches rather than flipflops can safely hold the data. The pulses play the role of any clock signal that might otherwise have been provided. Because the pulses happen only when needed, "clock gating" is automatic.

The name "6-4 GasP" indicates that (1) the Links are implemented with GasP circuits, that is, with complementary driver-and-keeper pairs and bidirectional state wires, and (2) it takes six gate delays to propagate a FULL state forward from Link *in* through Joint *fifo* to Link *out*—via gates *A*, *B*, *C*, *D*, *E*, and *X*—and (3) it takes four gate delays to propagate an EMPTY state in reverse direction from Link *out* through Joint *fifo* to Link *in*—via gates *B*, *C*, *D*, and *Y*. Together, the forward delay and the reverse delay yield a cycle time of ten gate delays. Note that this cycle time is consistent with the self-resetting of the two five gate delay loops, *A-B-C-D-Y* and *B-C-D-E-X*. We made the forward delay longer than the reverse delay, because the propagation of a FULL state goes together with the propagation of data and thus affects both a Link state and the data stored in the latches of the Link, while the propagation of an EMPTY state affects the Link state only.

7.3.2.2 Critical path: latches to data kiting to double-barrel Links

One might expect that the critical paths in the Weaver are in the crossbar switch where high speed, wide data, and data-driven flow control come together. In this section we explain why the combination of five-gate ring oscillators and 72 latches per Link lead to *data kiting*, why *data kiting* combined with data-driven flow control necessitates *advance decoding* of steering bits, and why *advance decoding* motivates the use of *double-barrel Links* in the crossbar switch.

We designed the circuits in the Weaver using the Theory of Logical Effort [3] discussed in Section 7.1.4. In particular, logical effort helps us determine the amplification required to drive the larger circuit loads. For each gate that drives a large load, we want (1) subsequent gates to sense changes in the gate's output signal within a gate delay, and (2) the gate output signal to change from rail to rail over its entire length within five gate delays. The larger loads in Figure 7.19 are (1) the many latches in Link *out*, which present a large load to gate *G*, and (2) the data and state wires between the two ends of each Link, whose lengths are likely to exceed those of other wires in Figure 7.19 and which present a large load to each latch, *X* or *Y*. Below, we outline how the FIFO circuit in Figure 7.19 supports these larger loads.

- **Provide amplification for *G*:** Each Link has 72 latches to store 72 bits of data, one bit per latch. The series of inverters *D*, *F*, and *G* in Figure 7.19 amplify the output signal of the AND function, *FULL(in)* **and** *EMPTY(out)* **and** *go*, to enable *G* to drive the large load presented by the 72 latches in Link *out*.
- **Limit the load on *G*:** To make the load on gate *G* as small as possible, the latch design—shown in Figure 7.20(a)—uses two complementary and tiny pass gates that are controlled by input *c*, which is the output of *G* in the context of Figure 7.19. One pass gate is transparent when *c* is high and is used to capture new data from *Din* to *Dout*. The other pass gate is transparent when *c* is low and is used to store the captured data and maintain the value on *Dout*. Each pass gate has an N-type and a P-type transistor and a tiny inverter—omitted from Figure 7.20(a)—to invert *c* locally so it can drive both types of transistors.
- **When needed, provide higher-gain amplification for *G*:** The current design with series inverters *D*, *F*, *G* provides enough amplification for *G* to drive the tiny pass gates in each of the 72 latches. Had the FIFO circuit used substantially more latches, say twice as many, it might have "clocked" these with a higher-gain pulse amplifier based on the *post-charge logic* design by Proebsting [2]—for instance by replacing series inverters *F* and *G* with a version of Figure 7.6(b), Section 7.1.
- **Provide amplification for *X* and *Y*:** Just like series inverters *D*, *F*, *G* build up sufficient amplification for *G* to drive 72 latches, so do inverting gates *D*, *E*, *X* and *D*, *Y* provide enough amplification for *X* and *Y* to drive a FULL or EMPTY state change on Link *out* and Link *in*. We gave *X* and *Y* each a drive strength of 40, as indicated in Figure 7.19. We were able to do so partly because *X* and *Y* use small keepers.

  X uses a P-type transistor to drive Link *out* from EMPTY to FULL. Because its keeper is small, driving *X* boils down to driving its P-type transistor, which is two thirds the effort of driving an inverter. The reduced effort makes it

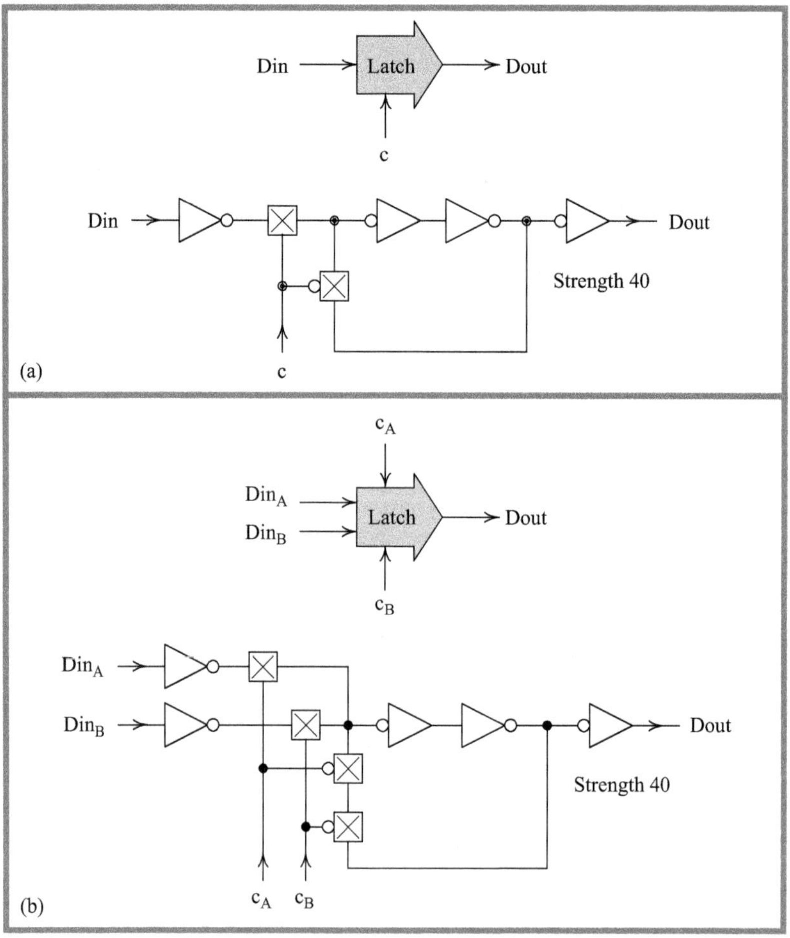

Figure 7.20 *Latch circuits to store and drive one bit of data in a Link. (a) A single-input single-output latch icon (top) and circuit (bottom) uses two complementary and tiny pass gates—the crossed squares—that are controlled by input c. Each pass gate has an N-type and a P-type transistor and a tiny inverter—omitted here—to invert c locally so it can drive both types of transistors. The picture indicates if the pass gate is transparent when c is high or c-inverted is high. With c high, the latch captures data from Din to Dout. With c low, the latch stores the captured data and maintains the value on Dout. The multiplexed latch version in (b) uses two control inputs, c_A and c_B, of which at most one is high at any time. With c_A high, data go from Din_A to Dout. With c_B high, data go from Din_B to Dout. When both c_A and c_B are low, the latch stores and maintains the captured data*

possible to give *X* a drive strength of 40 using three steps of amplification—*D*, *E*, *X*—the first of which, *D*, is shared to amplify *G* and *Y* as well. *Y* uses an N-type transistor to drive Link *in* from FULL to EMPTY. Because the keeper in *Y* is small, driving *Y* boils down to driving its N-type transistor, which takes only one third the effort of driving an inverter. The reduced effort makes it possible to give *Y* a drive strength of 40 with just two steps of amplification—*D* and *Y*.

- **Provide amplification for each latch:** Like *X* and *Y*, each latch has drive strength 40. The design in Figure 7.20(a) achieves this drive strength by amplifying the data signals captured by the pass gates, using a series of three inverting gates directly following the pass gates.

Note that the amplification for each latch, to give it a drive strength of 40, is done inside the latch design. Note too that this amplification comes in addition to the amplification for *G*, to give *G* sufficient strength to drive each of the 72 local latch control signals. Had the data been narrower, say 1 or a few bits, a latch design with two instead of four gate inversions and "clocked" by *fill(out)* instead of the output of gate *G* might have sufficed. In that case, the new data values would have been available at the other end of Link *out* at same time as the FULL state indicator. With 72 bits, however, the new data values will be available three to four gate delays later. In other words, to drive 72 bit wide data at high speed the Weaver must *kite the data*.

Below follows a step by step explanation for why this is the case and how this leads from *data kiting* to *advance decoding* and *double-barrel Links*.

- **Data kiting:** Two gate delays after the start of the *fire* pulse the 72 latches in Figure 7.19 are "clocked" by gate *G*. Likewise, two gate delays after the start of the *fire* pulse driver-and-keeper gate *X* drives the state of Link *out* from EMPTY to FULL.

  In the Weaver, we can assume that the input data for the latch circuits in Figure 7.20 arrive at the pass gates before "clock" control signal, *c*, goes high. Thus, the delay through each latch in Figure 7.20 is determined by the delay from *c* (rising) to *Dout*. With tiny pass gates, the latch delay is closer to three than to four gate delays. As a result, the new data values captured by the latches become available at the other end of Link *out* three to four gate delays after the FULL state. The data are kited—they are tardy by three to four gate delays.

Ring FIFOs in the Weaver can deal with three to four gate delays data kiting. In Figure 7.19, *D(in)* data that arrive three to four gate delays after *FULL(in)* arrives at Joint *fifo* will be at the latches in Link *out* two to three gate delays before *FULL(in)* will have propagated through the six gates *A*, *B*, *C*, *D*, *F*, *G* to "clock" the latches. By the time the "clock" rises, the data will be ready at the pass gates in the latches, having arrived at least one to two gate delays earlier.

If designed with care, *data kiting* can work equally well for Weaver parts with data-driven flow control. Take for instance the *Splitter* circuit in Figure 7.21. Its incoming data contain a steering bit, *D(in[s])*, that must be available in both true and complement forms before the circuit can decide which outgoing Link state to

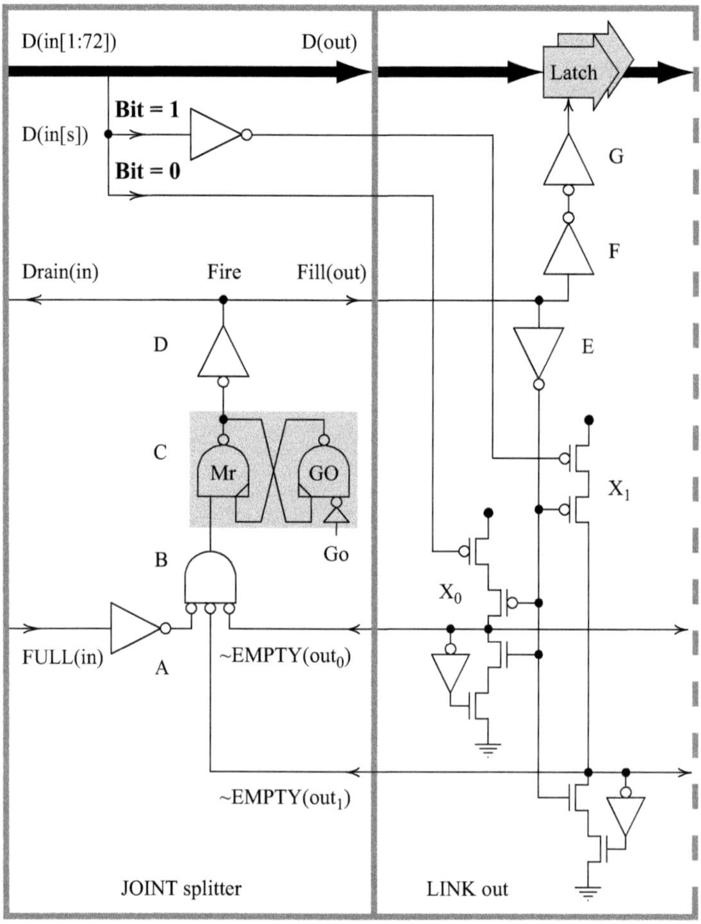

Figure 7.21 Splitter Joint and near-end output Link connection in 6-4 GasP. Gray-colored gates are icons—a circuit for MrGO was given in Figure 7.10, latch circuits can be found in Figure 7.20. Splitters start the advance decoding of steering bits for the crossbar switch by converting a steering bit from bundled data to double barrel form

change—out_0 or out_1. True and complement can be generated within a gate delay. The circuit decides which Link state to change as late as possible by using the true and complement steering signals as an extra selection input to driver-and-keeper gates X_0 and X_1. Thus, *D(in[1:72])* data that arrive three to four gate delays after *FULL(in)* will be at X_0 and X_1 one to two gate delays before *FULL(in)* will have propagated through the five gates *A*, *B*, *C*, *D*, *E* to drive the selected X_0 or X_1.

All circuits in the Weaver have a margin of at least one to two gate delays from the arrival of their data signals to the arrival of their control signals, be it for latching the data or for data-driven flow control. We can get extra delay margins from the wires in the Weaver's layout by allowing different wires to have different widths and different spacings. In particular, data wires in the Weaver use twice the width and twice the spacing used for control wires.

To understand how width and spacing affect the "speed" of a wire, let us consider the real shape of an integrated circuit wire. Wires are relatively thick layers of metal, sandwiched between layers of insulation. Most wires are more tall than they are wide, much as a fence is more tall than it is wide. Wires stand up the full thickness of each layer, like the walls of a room inside a multistory building. Because wires are more tall than wide, most of the capacitance between wires is to adjacent wires on the same layer rather than to wires in layers above or below.

Making a wire twice as wide halves its electrical resistance. Less resistance allows information to get through the wire faster. Doubling the wire width doubles the wire's relatively small capacitance to wires in layers above or below, but preserves the much larger capacitance to adjacent wires in its own layer. Doubling the space between a wire and adjacent wires on the same layer almost halves the capacitive load of that wire. Less capacitance speeds up the transistors that drive the wire.

By doubling both width and spacing, the Weaver's data wires gain an almost four-fold advantage in speed over the Weaver's control wires.

- **From data kiting to advance decoding of steering bits:** The *Splitter* decodes the steering bits one stage in advance, just before the data enter their first *Double Crosser* in the crossbar switch—see Figure 7.16. In the *Double Crosser*, the FULL Link state that accompanies the data is hardwired to the *Crosser* that steers the data in the intended direction—see Figure 7.17.

  The *Crosser* arbitrates between data items that go in the same direction by arbitrating between their FULL Link states—without looking at the data. This is possible because the *Splitter* decoded the direction in advance by making the associated Link state FULL. As a result, each *Crosser* can support arbitration as well as decode its *Double Crosser* specific steering bit one stage in advance.

  The combination of *five*-gate ring oscillators, data kiting, and arbitration would have been impossible without decoding the steering bits in advance.
- **From advance decoding to double-barrel Links:** Note that the *Splitter* circuit in Figure 7.21 as well as the *Crosser* circuit spread over Figure 7.23(a) and (b) take the one-hot Link states that decode the circuit's steering bit and pair these into a single Link with two Link states. Because at most one of these two

Link states can be FULL at any given time, just one set of 72 latches will suffice to store the data sent along each FULL Link state. We call the resulting Link a *double-barrel Link.*

The layout of the Weaver *shields* the narrow FULL or EMPTY state wires in each Link with an adjacent grounded wire on each side, to protect them from capacitance coupled noise. Ordinary Links with only one state wire use a three-wire control bundle: *ground-state-ground*. Double-barrel Links use a five-wire control bundle: *ground-state*$_0$*-ground-state*$_1$*-ground*. In each Link, 36 wider data wires with their wider spacing flank each side of the control bundle. Double-barrel Links are only slightly wider than ordinary Links.

One might view a double-barrel Link as a peephole optimization of the more typical implementation with two separate ordinary Links that can share data because they take the data from the same crossbar source to the same crossbar destination and they operate in mutual exclusion. We prefer to view a double-barrel Link as just a Link with typed interfaces, where the type information conveys the one-hot encoding properties of the two Link states. Adding type information to a Link and Joint interface makes it possible to fine-tune the interface and the tasks on each side of it as well as to graduate the delay sensitivity of the information exchange.

The stages in the crossbar switch communicate through double-barrel Links. Section 7.3.2.3 describes three representative circuits related to the crossbar.

7.3.2.3 Crossbar circuits: Splitter, Double-barrel Ricochet, Crosser

Double-barrel Links appear only inside the 8 × 8 crossbar switch and at its inputs and outputs. Data enter the crossbar from the South and leave toward the East. Just South of the crossbar, a dark gray area in Figure 7.16 holds eight *Splitter* stages that fill the double-barrel Links for the first row of *Double Crossers*. The dark gray area at the North West (NW) boundary holds eight *Double-barrel Ricochet* stages that act as FIFO stages for double-barrel Links. They repeat and fold double-barrel Links, directing data from the South heading North to make an Eastbound turn instead. Each *Double Crosser* stage in the light gray area in Figure 7.16 has two double-barrel input Links, coming from the South and the West, and two double-barrel output Links, going North and East, respectively. A *Double Crosser* also has two *Crossers*—one for Northbound data and the other for Eastbound data. As illustrated in Figure 7.17(c) and (d), the two *Crossers* share data from the double-barrel input Links. Each *Crosser* arbitrates between data items that go into the direction it controls. Just East of the crossbar, another dark gray area in Figure 7.16 holds eight *Lumper* stages that drain the double-barrel Links for the last column of *Double Crossers* and pass their data to the ordinary Links and FIFO rings, for recirculation.

Figures 7.21–7.23 show the 6-4 GasP circuit implementations that the Weaver uses for the *Splitter*, *Double-barrel Ricochet*, and *Crosser*. To emphasize how similar these implementations are to each other and to a simple 6-4 GasP FIFO with ordinary Links, all three Figures borrow the alphabetic gate identifier scheme of Figure 7.19. Below follow brief explanations of the circuit implementations in

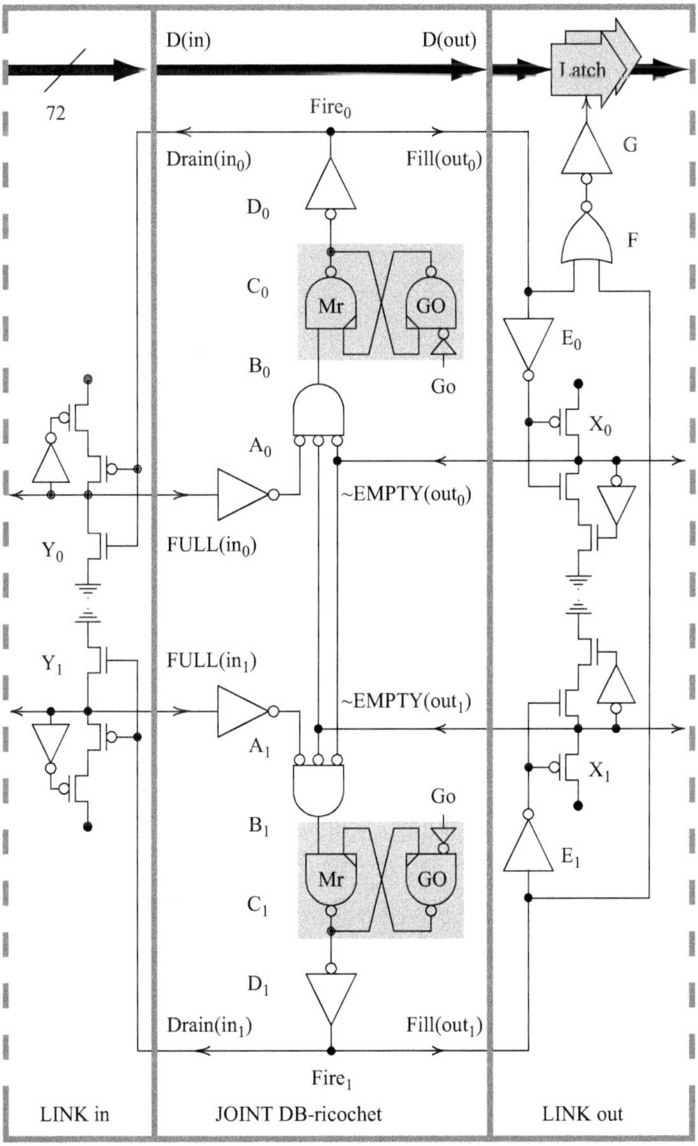

Figure 7.22 Double-barrel Ricochet Joint and near-end Link connections. An advanced FIFO circuit for double-barrel Links

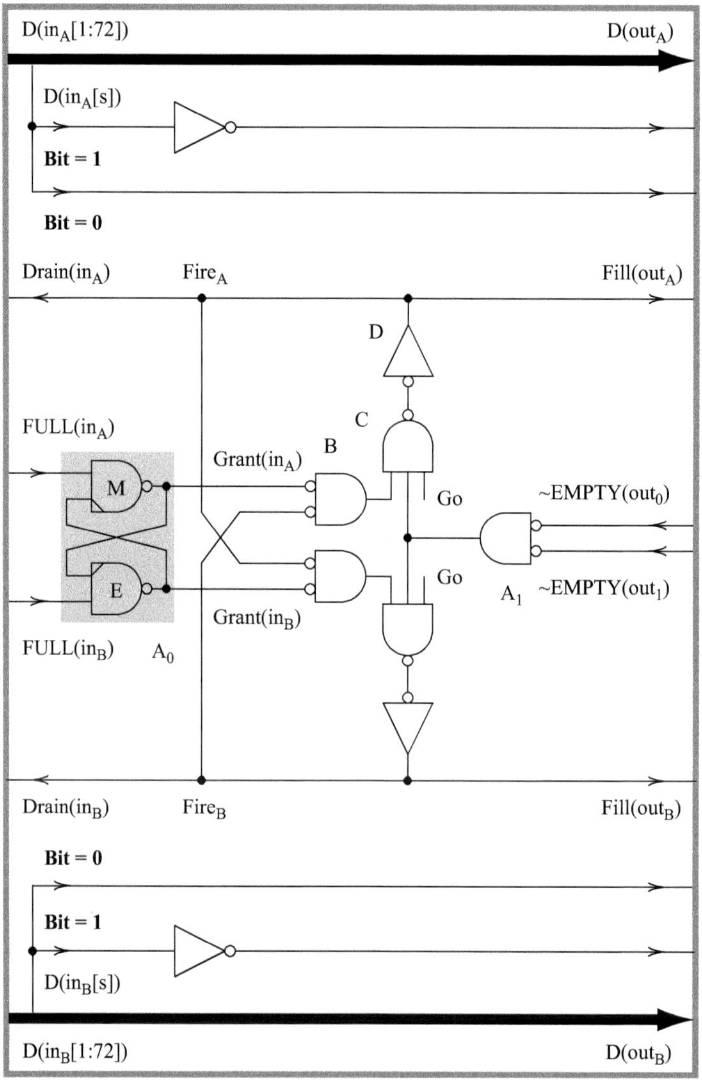

Figure 7.23 *(a) Joint crosser and its input and output interfaces. Gray-colored gates are icons—the mutual exclusion (ME) circuit resembles MrGO in Figure 7.10(c) and (d) but exports both its outputs*

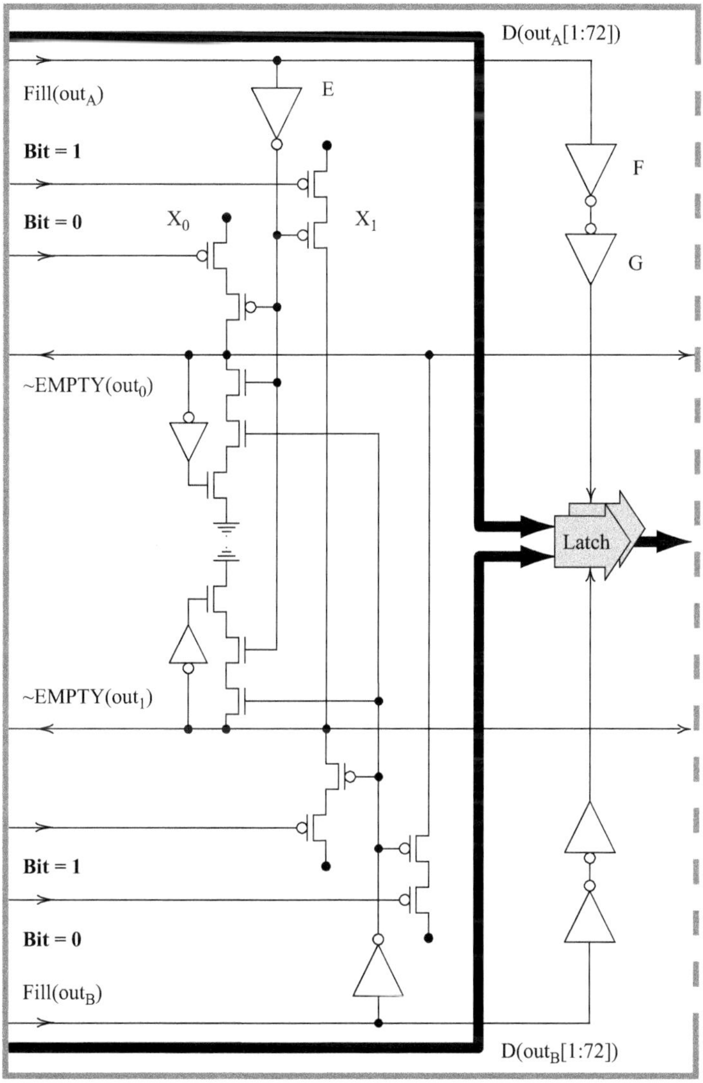

Figure 7.23 (b) Double-barrel output Link connection to Joint crosser. Gray-colored gates are icons—for latch circuits, see Figure 7.20

Figures 7.21–7.23. We focus primarily on new circuit aspects that each subsequent Figure brings in.

- **Splitter:** Figure 7.21 gives a 6-4 GasP implementation of the *Splitter* showing its Joint *splitter* and omitting its input Link *in* and half of its output Link *out*. The Joint receives 72 bundled data input bits labeled *D(in[1:72])*. One of these 72 bits, *Din[s]*, acts as a *Splitter* specific steering bit to select which double-barrel output state to fill—out_0 or out_1. The Joint copies all 72 input bits *D(in[1:72])* to *D(out)*, including steering bit *D(in[s])*. Thus, *Din[s]* remains available in bundled data form to repeat its steering task on a subsequent pass through the *Splitter*. As explained in Section 7.3.2.2, the kited steering bits are used as late as possible, in X_0 and X_1, to compensate for their kiting. Similar to Figure 7.19, inverters *D, E, F, G* amplify the Joint's AND function so it can drive the large loads presented by the 72 latches in Link *out* and by the driver-and-keeper gates X_0, X_1, and *Y* in Links *out* and *in*. Joint *splitter* has the following AND function: *FULL(in)* **and** $EMPTY(out_0)$ **and** $EMPTY(out_1)$ **and** *go*.
- **Double-barrel Ricochet:** Figure 7.22 gives a 6-4 GasP implementation of the *Double-barrel Ricochet*, showing its Joint *DB-ricochet* with half of its input and output Links, *in* and *out*. A *Double-barrel Ricochet* is an advanced FIFO circuit for double-barrel Links. It has two mutually exclusive AND functions. When it fires its first one, $FULL(in_0)$ **and** $EMPTY(out_0)$ **and** $EMPTY(out_1)$ **and** *go*, inverters D_0, E_0, *F*, *G* provide the amplification to drive the latches in Link *out* to copy the data from *D(in)* to *D(out)* and to drive X_0 and Y_0 to fill out_0 and drain in_0. When it fires its second one, $FULL(in_1)$ **and** $EMPTY(out_0)$ **and** $EMPTY(out_1)$ **and** *go*, the inverters D_1, E_1, *F*, *G* provide the amplification to drive the latches and copy the data and to drive X_1 and Y_1 to fill out_1 and drain in_1.

  Note that Joint *DB-ricochet* has two MrGO circuits, one per AND function, but that we tied their *go* input signals together, resulting in one *go* signal that enables or disables both AND functions.
- **Crosser:** Figure 7.23(a) and (b) gives a 6-4 GasP implementation of the *Crosser*, showing its Joint, *crosser*, and double-barrel output Link, *out*. The input interface of the Joint suggests two ordinary incoming Links, in_A and in_B, with 72 bits of data including a steering bit. Links in_A and in_B subset the signals of the double-barrel Links that enter the *Double Crosser* and relate to Link *out*—see Figure 7.17. The *Crossers* handle contention in the crossbar switch. At the heart of Joint *crosser* in Figure 7.23(a) is an *arbiter* or *mutual exclusion (ME)* circuit, gate A_0, patterned after the 1980 design by Charles Seitz [17], and sized to minimize its delay for the common uncontested case [4]. This mutual exclusion circuit grants on a first-come-first-served basis, and waits for metastability to end before it lowers the selected grant signal. Besides handling contention, each *Crosser* also decodes its *Double Crosser* specific steering bit one stage in advance of need, by producing double-barrel outputs—just like the *Splitter* in Figure 7.21. The new aspects in the *Crosser* are:

(a) Joint *crosser* in Figure 7.23(a) has two mutually exclusive AND functions: one for granting $FULL(in_A)$ and the other for granting $FULL(in_B)$. When it fires the first, *~grant*(in_A) **and** *~fire*$_B$ **and** $EMPTY(out_0)$ **and** $EMPTY(out_1)$ **and** *go*, inverters *D*, *E*, *F*, *G* spread over Figure 7.23(a) and (b) provide the amplification to drive the latches in Link *out* to copy *D* (in_A), and drain in_A, and fill either out_0 or out_1 depending on whether steering bit $D(in_A[s])$ is zero or one. The other AND function results in a similar action between in_B and *out*. Note that *~fire*$_B$ is an input to the AND function that generates $fire_A$. Likewise, *~fire*$_A$ is an input to the AND function that generates $fire_B$. Such cross-coupling prevents one AND function from overtaking the other in case of back-to-back grants [13]. Cross-coupling ensures that the signals that drive the data and state changes over the Links have adequate pulse widths—five gate delays wide.

(b) Note that the AND functions in Joint *crosser* both have a *go* input signal, but that both signals come unarbitrated, that is, without a corresponding MrGO circuit. We found the task of adding two MrGO arbiters to an already arbitrated circuit with a tight layout and with tight five gate delay loops and high amplification needs simply too daunting. One *go* input signal serves both AND functions.

7.3.3 Test logistics

The Weaver's rings, including the parts that go through the crossbar switch, can each transfer up to about 6 Giga data items per second (GDI/s). The supporting throughput measurements follow in Figure 7.30, Section 7.3.5. With 72 bit wide data items, this amounts to 3.5 Tera bits per second. Yet, we use a low-speed test interface consisting of only five wires to test the functionality of the Weaver and to debug and characterize its high-speed operations. Moreover, we use this low-speed "test interface" to initialize and start the Weaver. The photo in Figure 7.24 shows a Weaver chip in its ceramic package, mounted on its test board.

In addition to the five low-speed test signals, the Weaver has two dedicated medium-speed outputs that deliver one-millionth reduced ring frequency outputs. These two medium-speed outputs follow the switching frequency of bit 19 in two of the ten 54 bit long ring counters in the Weaver—see Section 7.3.1.3 and Figure 7.18. The two black coaxial cables in Figure 7.24 carry the reduced ring frequency outputs to an oscilloscope for real-time observation.

The ring counters are 54 bits long to accommodate long test experiments. When counting 6 Giga items per second, a 54 bit counter will overflow about every 30 days.

The counters can be reset to zero at the beginning of a test experiment and read out at the end. They are read out over the white flat ribbon cable visible midway the right edge of Figure 7.24. The flat ribbon cable carries low-speed signals between the chip and a computer. The computer contains the test program with instructions for controlling the low-speed test stimuli and observing the low-speed test

Figure 7.24 Photo of a packaged Weaver chip on its test board. The chip contains two experiments, one of which is the Weaver. The other, called Anvil and designed by Chris Cowan, is a case study in radiation hardening—not further discussed here. The two black coaxial cables connected near the middle of the board carry two one-millionth reduced ring frequency outputs to an oscilloscope for real-time observation—see Section 7.3.1.3. L-shape connectors at the top-right corner of the photo bring in power and ground. A white flat ribbon cable visible midway the right edge of the photo carries five low-speed test signals to and from the chip and a computer. The computer contains the test program with instructions for controlling the low-speed test stimuli and observing the low-speed test responses. Board and final chip layout are by the late Jon Lexau of Sun Labs

responses. The chip contains a low-speed JTAG test interface with five test pins, an on-chip test access port, and on-chip scan chains. This low-overhead low-speed test interface is an industry standard for testing manufactured chip designs and printed circuit boards. It was codified by the Joint Test Action Group (JTAG) and the Institute of Electrical and Electronics Engineers (IEEE) in IEEE Standard 1149.1-1990, entitled Standard Test Access Port and Boundary-Scan Architecture [15].

The JTAG test interface in the Weaver runs at 500 kHz. The ten ring counters hold 54 bits each. We use a scan chain to read out all 540 counter bits at once. We then shift the scan bits one by one over the JTAG test interface, which takes on the order of a millisecond. We use a similar approach for reading and writing approximately 500 *go* control signals, one per Joint, approximately 500 Link states, and 720 data bits, 72 for each of the ten ring reloaders.

The JTAG test interface is synchronous and clocked. It has one output signal, *test data out*, and four input signals, *test clock*, *test data in*, *test mode select*, and an optional *test reset* signal. These signals are used to set up and select test operations, to read and write Weaver states, and to enable and disable Weaver actions. Details about setting up test operations can be found in IEEE Standard 1149.1-1990 [15]. Here, we show the Weaver specific parts of the test interface [7]—the scan chains and the transfer circuits to and from the scan chains and the Links and Joints in the Weaver—that read and write (Link) states and enable and disable (Joint) actions.

7.3.3.1 Scan chains and connections to Weaver Links and Joints

The scan chains in the Weaver consist of shift registers connected in series. Each shift register has two small latches that are also connected in series, as illustrated in Figure 7.27(c). The circuit designs for the two small latches follow in Figure 7.25. With two latches, the shift register can store one bit safely. This bit can be shifted in or out serially. To shift bits in or out, the two latches in the shift register are clocked alternately, using two scan clocks, c_1 and c_2. Instead of shifting a bit in through the scan chain, the shift register can read a bit from the Weaver and store it into its second latch, using a special scan signal called *read*. In addition to shifting a bit out through the scan chain, the shift register can write the bit that it stores in its second latch into the Weaver, using a special scan signal called *write*.

Each shift register comes with a bundle of eight scan signals, including shift register specific scan input and output signals, *sin* and *sout*. The other signals in the scan bundle travel the entire length of the scan chain, with regular amplification. These include the two scan clocks, c_1 and c_2, and the scan *read* and *write* signals for interaction with the Weaver. Other scan signals that travel the entire length of the scan chain—c_1*Return*, c_2*Return*, *sReturn*—are the far-out scan clock signals and the scan output signal of the last scan shift register. These travel in reverse direction through the scan chain, back to the first scan shift register and its JTAG test interface.

The far-out scan clock signals that return to the JTAG test interface are important for generating nonoverlapping clocks for shifting data in and out of the scan chain. The clock generator in Figure 7.26 combines the low-frequency JTAG *test clock* with c_1*Return* and c_2*Return* to generate low-frequency scan clocks c_1 and c_2 that are never high at the same time. If the two clocks are never high at the same time, the two small latches clocked by them in Figure 7.27(c) are never transparent at the same time, and neither are any other subsequent latches in the scan chain. The nonoverlapping clocks make the scan chain shift bits properly.

After a bit has arrived in the second latch of the shift register, we can write it into the Weaver signal associated with this shift register. Or we can read the value of the Weaver signal into the second latch and shift it out for inspection. Some writes, specifically those associated with *go* signals, enable or disable circuit actions and can even start or stop them. Other writes merely change circuit states.

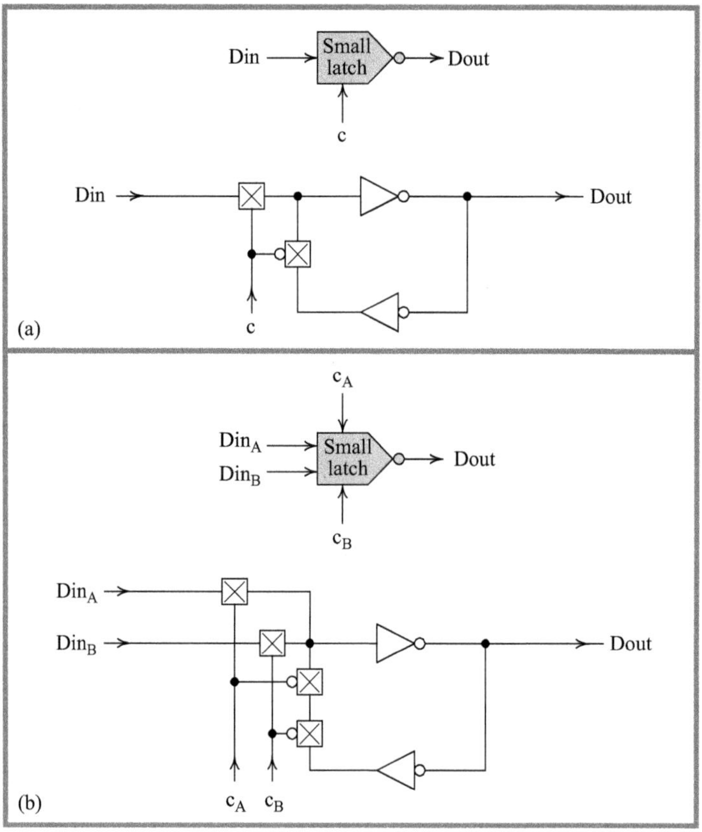

Figure 7.25 Scan latch circuits. Because of their low, 500 kHz, clock frequencies the scan latches can be half the size of the data latches in Figure 7.20

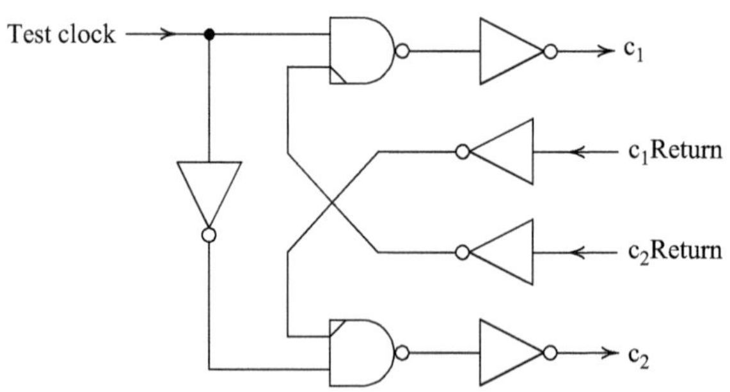

Figure 7.26 Scan clock generation. We use the low-frequency JTAG test clock to generate two low-frequency scan clocks that are never high at the same time. Each scan clock, c_1 or c_2, goes high only after the longest branch of the other scan clock, c_2Return or c_1Return, has gone low

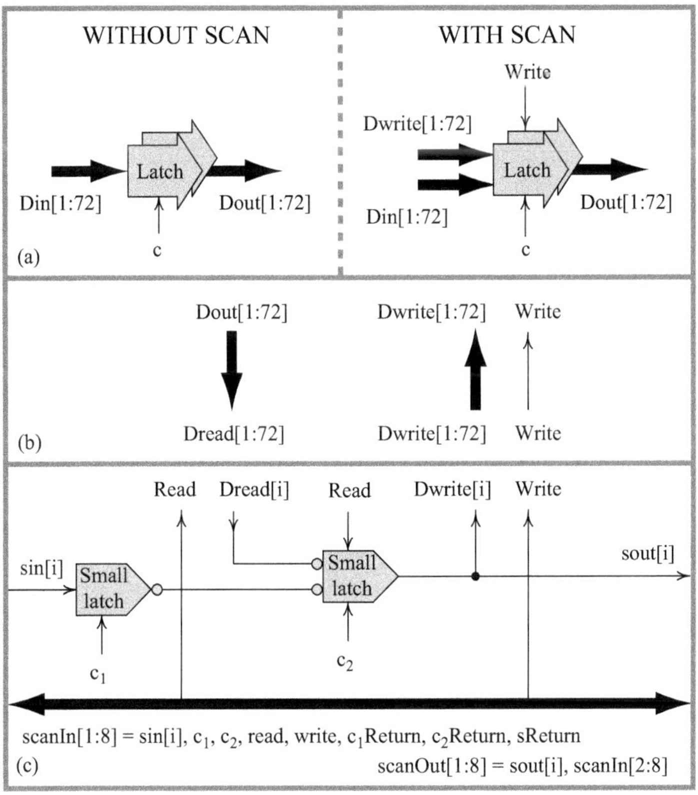

Figure 7.27 Scan connections for reading and writing Weaver data. Weaver data can be read for inspection and written for initialization or test at a reloader stage located in the SE corner of the Weaver floorplan—see Figure 7.16. A reloader stage is just a FIFO circuit, as in Figure 7.19, but one with scan access to the data latches in its output Link. Each data latch is associated with a specific scan shift register (c) which can read the bit stored in the data latch, Dout[i] alias Dread[i], or write its own bit, Dwrite[i], into the data latch (b). To allow the shift register to overwrite latch content, each data latch is replaced by its multiplexed version (a)—see Figure 7.20. Thanks to their small latches, the 72 serially connected scan shift registers and their bundles of scan signals occupy a footprint similar to the FIFO circuit. Reloader FIFO and scan fit in one Weaver layout module

The Weaver uses different circuits to transfer bits to and from the scan chains and its data latches, FULL or EMPTY Link states, and *go* signals. Figures 7.27–7.29 illustrate the differences. The three Figures show similar shift registers (c) but different transfer circuits (b) and different Link and Joint circuit modifications after scan insertion (a). In particular, the Weaver writes and stores the data bits that it receives from the scan chain in its own data latches—see Figure 7.27. Likewise, the Weaver reuses its own driver-and-keeper gates to write and store the FULL (1) or EMPTY (0) bit that it receives from the scan chain. It turns the keepers off while it writes the Link state, and stops driving the Link state when the *write* signal is low—see Figure 7.28. In contrast, the Weaver writes and stores each *go* signal that it receives from the scan chain in a separate small latch—see Figure 7.29. Note that all three Figures enable their shift register to read back what it wrote. To reduce wire capacitance and switching power, we gate the read connections from the high frequency Link state signals to the shift registers when *read* is low.

The Weaver has several scan chains. One scan chain follows the NE edge of the Weaver, reads the ring counters, initializes them to zero, and sets their multiplexers for monitoring frequency outputs in real time—see Figure 7.18. A second and similar scan chain, based on Figure 7.27, follows the SE edge and reads and writes the data latches in each reloader stage. A third one, based on Figure 7.28, reads and writes the Link states of each Link in the Weaver. A fourth, based on Figure 7.29, reads and writes the *go* signals of each Joint. The scan chains for FULL or EMPTY and *go* signals visit each module in the Weaver, and do so in boustrophedonic order—turning like oxen in ploughing a field. Their first shift registers start at the JTAG test interface in the corner where NE and SE edges meet. Their last shift registers end at the opposite corner where NW and SW edges meet. Their scan shift registers and the corresponding transfer circuits to each Link and Joint are combined with the layout modules of the Links and Joints. The Weaver's JTAG test interface operates the scan chains in mutual exclusion. Note that the shift registers in Figures 7.27–7.29 *read* Weaver bits in parallel, *write* Weaver bits in parallel, but *shift* bits serially in and out the scan chain. Note too that the scan chain can shift while the circuit operates, without mutual interference.

7.3.4 How low-speed scan chains test high-speed performance

Asynchronous or self-timed circuits operate as fast as they can—when they can. Externally lock-stepping their operations to, for instance, the JTAG *test clock* would take the "self" out of their timing and run them synchronously and no longer at speed. Once this realization sinks in, it becomes obvious that we need merely identify the borders to where the circuit can run, and allow it to run "flat out" inside these borders. Circuit actions stop at the border. The *go* signals in our circuits give us the necessary control to enable actions within borders and disable actions at borders.

We can test the high-speed Weaver operations at speed because we enabled the JTAG test interface to control actions and states separately. The Weaver's test interface recognizes and controls the individual *go* signals in each Joint's action

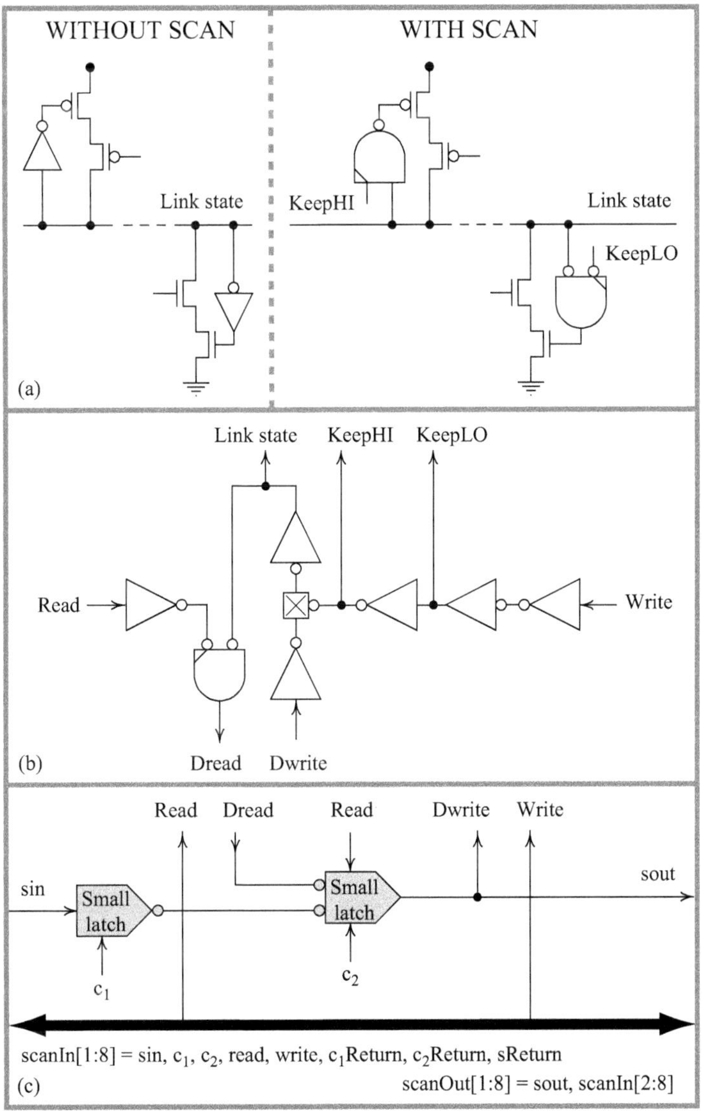

Figure 7.28 Scan connections for reading and writing Link states

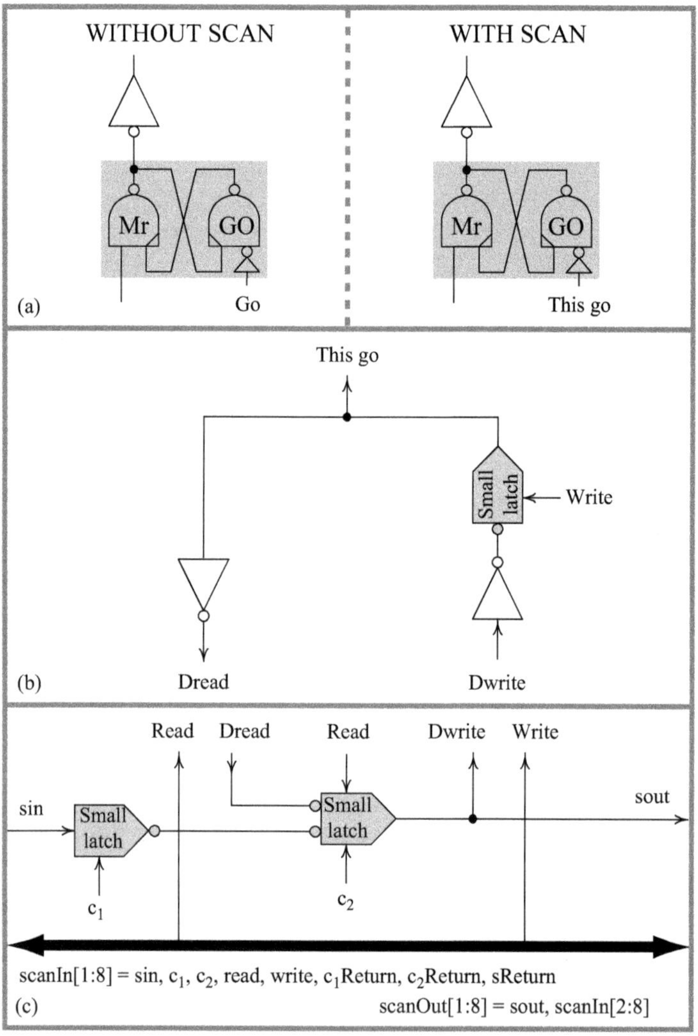

Figure 7.29 Scan connections for reading and writing go signals. Each go signal receives a separate small latch for scan access. The latch isolates scan shift operations from circuit operations, which can be done in parallel without affecting each other's data or control flow

and it recognizes and controls the individual FULL or EMPTY and data signals that are stored in each Link. In Section 7.2, we explained how distinguishing actions from states and controlling them separately accommodate initialization, structural testing, and at-speed testing of parts or the entire design.

In the Weaver, a burst of data items will run *at speed* from one end to the other end of an empty ring segment, with the two ends marked by disabled *go* signals. Any preparation work for running this burst—by (1) first disabling the *go* signals, so we can initialize the Links in the takeoff, under test, and landing parts of the ring, as in Figure 7.14, by (2) then enabling the *go* signals to enable all parts to run freely except for the two ends and the "gate keeper" to the part under test, and by (3) finally enabling the "gate keeper" to release the burst and let it run freely through the ring segment—any of that can be done at low speed, using the JTAG test interface. The at-speed performance follows from letting the circuit run freely from end to end. Similar low-speed preparations let us run data items through an endless ring and stop their circulation by disabling the "gate keeper" in real time, as described at the end of Section 7.2.2. We do this to measure performance—throughput, power, and energy. For throughput, we scan out the ring counters and relate their values to the run time. For power, we use a current probe to measure the average current that the Weaver draws while running, and relate its value to the supply voltage used while running. The energy consumption can be calculated from throughput and power. Section 7.3.5 presents our throughput, power, and energy measurements from the Weaver chip.

7.3.5 Performance measurements

Figures 7.30–7.33 show four collections of canopy graphs with throughput and power measurements from the Weaver chip, measured for various traffic and supply voltage levels. Each graph plots the measured information as a function of ring *occupancy*—the number of FULL Links or valid data items [18,19]. Sections 7.3.5.1–7.3.5.4 analyze the canopy graphs, showing throughputs of about 6 Giga data items per second, and energy dissipation around 3 picojoules to forward one data item one stage.

7.3.5.1 Throughput versus occupancy at nominal power supply

The canopy graphs in Figure 7.30 plot the throughput for four of the ten FIFO rings at nominal power supply voltage. Rings 0 and 9 bypass the crossbar switch and thus omit switching elements. Ring 1 has the highest maximum throughput and Ring 8 has the lowest maximum throughput of all eight rings that go through the crossbar switch. The throughput reflects the count reached in each ring counter stage in the NE corner of the Weaver floorplan—see Figure 7.16. We normalized the count to average one second of run time. Each graph plots the throughput as a function of ring occupancy, that is, of the number of valid data items in the ring.

An empty ring has zero throughput just as an empty freeway carries no traffic. Likewise a completely full ring has zero throughput just as a congested freeway stalls traffic. Therefore, at its left and right ends a canopy graph shows zero throughput. The linear rise in throughput with occupancy at the left of each canopy

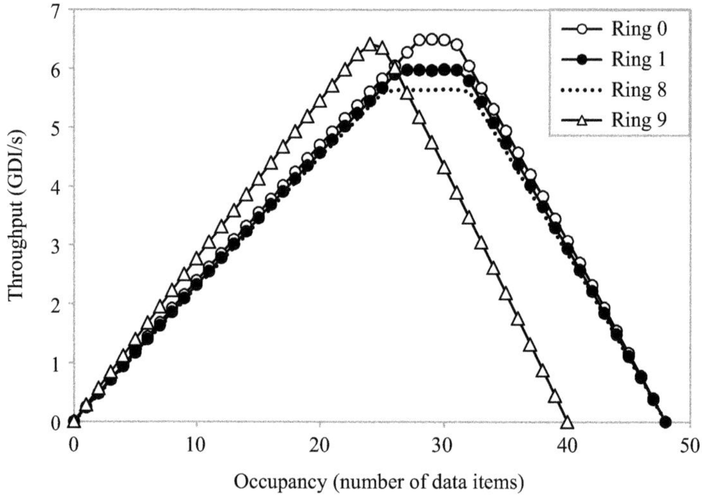

Figure 7.30 Canopy graphs for throughput versus occupancy. Each canopy graph plots the frequency measured at nominal supply voltage as a function of ring occupancy. Ring 9 has 40 stages, all others have 48 stages. Maximum throughput is at 60% occupancy, and around 6 GDI/s. Weaver's layout accounts completely for the differences in maximum throughput shown by the graphs

graph is easy to understand. One data item circulates with a period set by the forward latency around the ring, and—just like a few racecars on a circular track—so do any small number of data items. Because data items cannot overtake each other, throughput increases with the number of circulating data items—as long as congestion is avoided. The right side of the canopy graph shows the impact of congestion. As congestion decreases, more spaces become available for forwarding data items and there is a corresponding linear increase in throughput.

Somewhere between a completely full and a completely empty ring there is an occupancy with maximum throughput. The canopy graph for Ring 9 shows its maximum at 6.4 Giga data items per second (GDI/s) at 60% occupancy, that is, with 24 valid data items in its 40 stages. The 6-4 GasP circuits in the Weaver transport space faster than data: the forward latency of each 6-4 GasP circuit in the Weaver is about 100 picoseconds, and the reverse latency is only about 66 picoseconds. The choice to transport space faster than data is inspired by the relative ease of transporting space. It is easier to declare a Link EMPTY, when transporting space, than it is to declare a Link FULL *and* drive the latches *and* capture an arriving data item, when transporting data. At 60% occupancy the net velocity of spaces and data items match, resulting in maximum throughput.

The Weaver's layout accounts completely for the differences in the shapes of its canopy graphs. These differences can be explained by examining the basic layout of the *Cross Fire* sections outside the crossbar switch and by examining the layout of the *Double Crossers* inside the crossbar switch. We start our explanation by examining the layout of the NE, SE, and SW *Cross Fire* sections in Figure 7.16.

- **Layout modules:** In the *Cross Fire* layout, we pair independent FIFO stages that cross each other at a North-South and East-West ring crossing. Each stage has a FIFO Joint and its two near-end Link connections as shown in Figure 7.19. Crossing stages are paired to form one layout module. Each layout module is approximately square. Layout modules abut. The two FIFO stages fit side by side in the module, each taking a slice that spans the full module height and half the module width. Per FIFO stage or slice, the control circuits—gates *A* to *G*, *X*, *Y* in Figure 7.19—occupy a center row flanked above and below by two groups of 36 latches each that belong to the outgoing Link. The scan circuits for *go* signals and Link state signals occupy the bottom row, below the latches, and part of the center row.
- **North-South module connections are longer:** Per FIFO, the Link state signals extend horizontally in East-West direction, almost all the way across the center row. A Link state signal that connects two FIFO stages in East-West direction must jump horizontally from one layout module's center row to an adjacent module's center row, a slice distance away. The jump requires about half a module width of extra wire. A Link state signal that connects FIFO stages in North-South direction must jump vertically from center row to center row, which requires a full module length of extra wire—twice that of an East-West connection.
- **North-South module connections are slower:** Longer wires are harder to drive than shorter wires. Moreover, the Weaver's throughput depends in part on the control speed of its Links, that is, of its state signals and their driver-and-keeper gates. For design modularity and simplicity, the Weaver uses the same drive strength of 40 for each Link driver-and-keeper gate, independent—within reason—of the wire length of the state signal it drives. Because North-South module connections have longer Link state signals than East-West module connections, using equally strong Link driver-and-keeper gates makes the North-South module connections slower.

The slowness of North-South connections compared to East-West connections is even more pronounced for the *Double Crosser* layout modules in the crossbar switch. Their footprint and organization are similar to those for the *Cross Fire* layout modules: approximately square, with two *Crossers* and their near-end Link connections positioned side by side, with control circuits occupying a center row and groups with 36 latches each above and below. Because the control complexity is higher and the number of gate connections per Link state is higher, *Double Crosser* layout modules have longer Link state signals in and between them than *Cross Fire* layout modules. *Double Crosser* module connections are slower than *Cross Fire* module connections in any direction, and slowest in the direction North-South.

All other circuits, barring the reloader stages, are organized as narrow half-width layout modules, with just one slice instead of two slices side by side. North-

South connections for these are about as slow as for *Cross Fire* layout modules. The reloader stages in the SE corner of Figure 7.16 take an entire layout module each. The FIFO ring stage takes one slice. The other slice contains the scan circuits to read and write the 72 data bits in the FIFO ring stage. North-South and East-West connections to a reloader stage are similar to those for a *Cross Fire* layout module.

We now have enough information to understand the differences between the canopy graphs in Figure 7.30. Consider first the canopy graphs for Ring 0 and Ring 9 that avoid the crossbar switch. The maximum throughputs of Ring 0 and Ring 9 are about the same because both are limited by their slower North-South Links. Next consider the canopy graphs for the switched rings, Ring 1 and Ring 8. Ring 1 passes across the bottom of the *Double Crosser* triangle of the crossbar switch, and is therefore slower than Ring 0 and Ring 9. Ring 1 avoids North-South connections between *Double Crossers* and is therefore faster than Ring 8. The canopy graph for Ring 8 is about the same as for Ring 2 to Ring 7—omitted from Figure 7.30 for this reason—because each is limited by its slower *Double Crosser* North-South Links.

7.3.5.2 Throughput for various power supply voltages

Speed scales with power supply voltage. Figure 7.31 shows canopy graphs that plot the relative throughput of Ring 4 as a function of ring occupancy at different supply voltages. Throughput numbers are scaled relative to the maximum throughput of Ring 4 at nominal supply. The throughput of Ring 4 at nominal supply is about the same as that of Ring 8 in Figure 7.30. The spacing of the flat canopy tops at different voltages indicates a nearly linear relationship between throughput and power supply. The Weaver operates flawlessly between 0.6 and 1.0 volt. In this operating region, throughput is very nearly proportional to the excess of power supply voltage over threshold voltage. Any excess beyond that required just barely to overcome the transistor threshold voltage of about half a volt can be used to charge the wires.

The medium-speed real-time outputs connected to the counters, shown in Figure 7.18, give a vivid demonstration of speed as a function of power supply voltage. Lacking a global clock, it is unnecessary to adjust a clock frequency when changing the supply voltage. Turning the knob to adjust power supply voltage makes the self-timed Weaver automatically speed up or slow down because each part proceeds as fast as the available power supply voltage permits. Turning the power supply voltage knob stretches or shrinks the square wave seen on an oscilloscope attached to Weaver's real-time counter outputs.

7.3.5.3 Power for various power supply voltages

Figure 7.32 shows canopy graphs that plot the relative power of Ring 4 as a function of ring occupancy for five different power supply levels, using a worst-case data pattern.[||] The graphs are normalized in proportion to the maximum power at

[||]The data pattern used in Figure 7.32 is a checkerboard pattern, with alternating bit values for each data item that flip in opposite direction for subsequent data items. For more details, see Section 7.3.5.4.

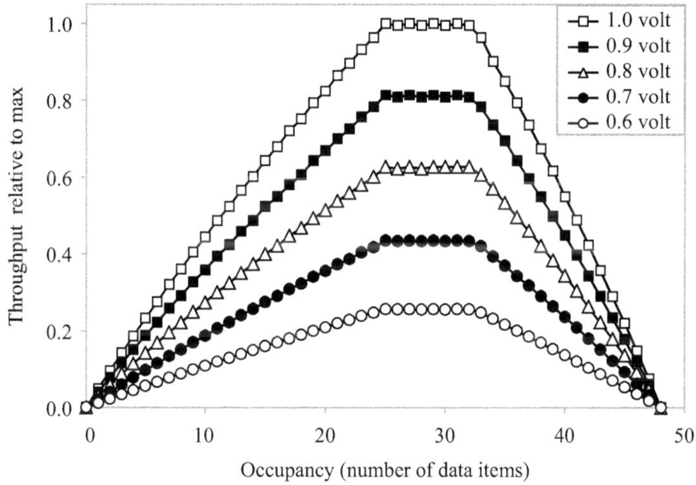

Figure 7.31 *Canopy graphs showing throughput at various supply voltages. The graphs plot the relative throughput of Ring 4 as a function of ring occupancy at five different power supply levels, and indicate that throughput is very nearly proportional to (supply voltage – 0.5 volt)*

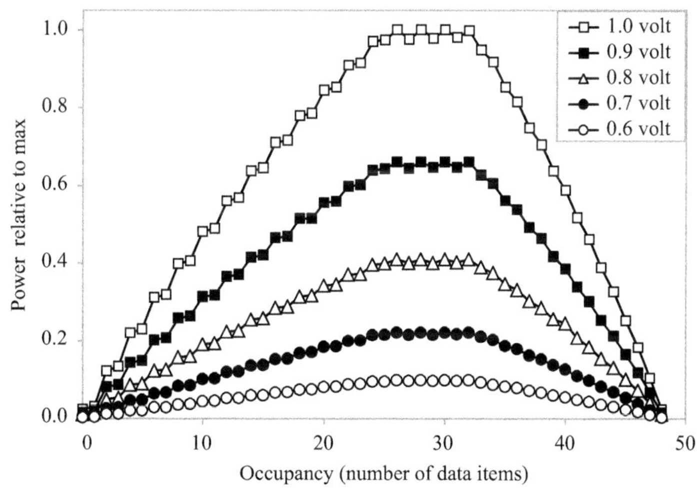

Figure 7.32 *Canopy graphs showing power at various supply voltages. The graphs plot the relative power of Ring 4 as a function of ring occupancy at five different power supply levels, using data patterns like 101010 followed by 010101. The measured power is very nearly proportional to (supply voltage – 0.5 volt) × (supply voltage)^2*

the highest voltage. We measured the power as the product of the current drawn and the power supply voltage.

Upon examination one can see that the measured power is very nearly proportional to $(\textit{supply voltage} - \textit{0.5 volt}) \times (\textit{supply voltage})^2$. Figure 7.31 in Section 7.3.5.2 already noted $(\textit{supply voltage} - \textit{0.5 volt})$ as proportional to the throughput. Thus, the first term, $(\textit{supply voltage} - \textit{0.5 volt})$, relates to how many data items per second pass a given point, for example, the counter stage. The second term, $(\textit{supply voltage})^2$, relates to the energy for forwarding a data item by one stage, which involves charging or discharging the capacitance of the data wires. Power is energy per second, and so the units work out correctly. The graphs in Figure 7.32 serve mostly as a sanity check. The more compelling power measurements follow in Figure 7.33, Section 7.3.5.4.

7.3.5.4 Power for various data patterns

Figure 7.33 shows canopy graphs that plot the active power of Ring 4 as a function of ring occupancy and different patterns of data measured at nominal power supply.

Power is lowest when all data items are identical, because for identical values the data wires need never change. Power is highest for a checkerboard pattern in which data wires adjacent in the layout switch in opposite directions as each data item passes. Power numbers drop slightly if instead of a checkerboard pattern the Weaver alternates all-zeros and all-ones, because of a reduction in side capacitance for adjacent data wires that carry the same value. Thc intermediate graph in Figure 7.33 is for random data and shows random local variation from sample to sample.

In both the checkerboard graph and alternating all-zeros and all-ones graph, the power numbers ripple between even and odd occupancy, up to about 60% occupancy. For N data items circulating, there are either N or $N - 1$ changes in value depending on whether N is even or odd. Adding one more data item to an existing even set of data items maintains the number of data changes, but adding one more data item to an existing odd set of data items introduces another data change with a corresponding increase in active power—barring congestion.

The power numbers in the graphs for the checkerboard and alternating data patterns differ by only about 5%. The data wires in the Weaver are all double width at double spacing to reduce their capacitive load—see Section 7.3.2.2. The graphs confirm that the side capacitance between data wires contributes relatively little load.

The minimum power measured for circulating a constant data pattern is about one quarter of the maximum power reported in Figure 7.33, for the same occupancy. This minimum reflects the power required to "locally clock" the 72 latches in each occupied stage, making them repeatedly transparent and opaque—even though their data inputs and outputs remain unchanged.

The power measured for circulating a random data pattern is more than half the power measured for circulating a checkerboard pattern. Comparing the two after subtracting the fixed power overhead for "local clocking" provided by the constant data pattern gives $(\textit{random} - \textit{allzero})/(\textit{checkerboard} - \textit{allzero})$, which is about 0.54 over a wide range of occupancies. This is consistent with the statistical model that a random data bit changes about half the time.

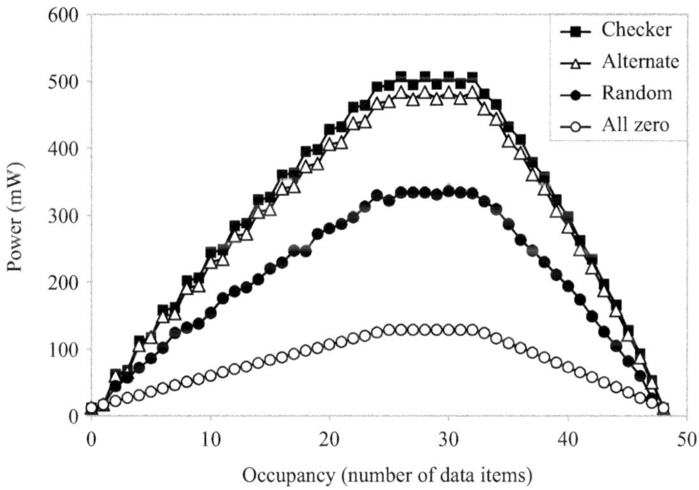

Figure 7.33 Canopy graphs showing power for various data patterns. The graphs plot the power of Ring 4 at nominal power supply voltage as a function of ring occupancy and four different data patterns. Power depends on how many latches change as data items travel through the Weaver. With constant data (All zero) the latched data remain unchanged, resulting in lowest power. Checkerboard patterns like 101010 followed by 010101 (Checker) cause adjacent data wires to change with every passing data item, resulting in highest power. Patterns with all-zeros alternating with all-ones (Alternating) take almost as much power. Random data (Random) give average power. By combining these power measurements with the throughput measurements in Figure 7.30, one can estimate that the energy to forward one data item one stage is at most 3 picojoules—for details, see end of Section 7.3.5.4

The canopy graphs in Figure 7.33 show clearly that the Weaver's power is determined by how many data bits change when data circulate through the rings and the crossbar switch. The maximum power of 500 milliwatts is for circulating 60% $\times$ 48 or around 29 data items in a checkerboard pattern through 48 stages of Ring 4, including eight stages in the crossbar switch. Any stage in the Weaver has the same number of latches driving about the same length of data wires. Each stage therefore has the same power when circulating the same pattern at the same speed. So, if we were to circulate a worst-case data pattern at maximum speed through each switching ring, the 8×8 crossbar would run at $(64/48) \times 500$ or about 667 milliwatts.

All switching rings have a throughput up to between 5.5 and 6 GDI/s. Worst-case, a checkerboard pattern with 29 data items running through a switching ring would run at 5.5 GDI/s and 500 milliwatts. If we call x the energy required to forward one data item one stage, then worst-case $x = 500/(5.5 \times 29)$ or 3 picojoules.

7.3.6 Summary and conclusion of Section 7.3

The Weaver implements a simple logical function: an 8 × 8 nonblocking crossbar switch with recirculating channels connecting its eight outputs back to its inputs. Its simplicity allowed us to push its limits. Wide datapaths of 72 bits stretch the Weaver's layout. A short cycle time based on five-gate ring oscillators and a complex flow control with steering bits and arbitration stretch the Weaver's electrical design. The Weaver's high throughput of 6 Giga data items per second per channel—nearly 3.5 Tera bits per second for the full crossbar—is outstanding.

For initialization and at-speed test and debug, the Weaver has separate *go* control in each and every Joint and FULL or EMPTY state access in each and every Link. Its functional simplicity made it possible to read and write all its data through a single reloader stage per channel. Testing the Weaver was a delight—an experience we intend to cultivate further through the Link and Joint model of computation.

The Weaver's logical design separates communication and states in Links from computation and actions in Joints. The Weaver's electrical design maintains this separation. Although all its circuits use the 6-4 GasP self-timed circuit family, they might equally well have used Click.

The Weaver uses different kinds of Links. For instance, the crossbar combines the steering bits and the fill signals in the driver-and-keeper gates of its output Links—see Figure 7.21. Doing so compensates for kiting delay in the data, as explained in Section 7.3.2.2. A novel feature of the Weaver is its use of double-barrel Links. A double-barrel Link bundles data bits with two state signals that carry steering information in one-hot form. The end of Section 7.3.2.2 considers whether to view a double-barrel Link as a peephole optimization of two mutually exclusive ordinary Links or as a Link with a typed interface that carries data in a different form. Adding type information permits fine-tuning of Link-Joint interfaces.

The Weaver's layout modules package a Joint with the near ends of its Links, cutting the Links where they can be stretched. The layout modules conceal the Link-Joint interface and expose the handshake interface—an unfortunate side effect. All too often, designers let layout guide the way they design. The Link and Joint model guides the Weaver's design. And that has made all the difference.

References

[1] Ivan Sutherland and Scott Fairbanks. GasP: a minimal FIFO control. In *International Symposium on Asynchronous Circuits and Systems*, pages 46–53, 2001.

[2] Robert J. Proebsting. Speed enhancement technique for CMOS circuits. US patent US 5,343,090, assigned to National Semiconductor Corporation, 1994.

[3] Ivan Sutherland, Bob Sproull, and David Harris. *Logical Effort: Designing Fast CMOS Circuits*. Morgan Kaufmann, 1999.

[4] Swetha Mettala Gilla, Marly Roncken, Ivan Sutherland, and Xiaoyu Song. Mutual exclusion sizing for Hoi Polloi. *IEEE Transactions on Circuits and Systems II—Express Briefs*, 66(6):1038–1042, 2018.

[5] Jo Ebergen, Jonathan Gainsley, and Paul Cunningham. Transistor sizing: how to control the speed and energy consumption of a circuit. In *International Symposium on Asynchronous Circuits and Systems*, pages 51–61, 2004.

[6] Ivan Sutherland and Jon Lexau. Designing fast asynchronous circuits. In *International Symposium on Asynchronous Circuits and Systems*, pages 184–193, 2001.

[7] Swetha Mettala Gilla. Silicon compilation and test for dataflow implementations in GasP and Click. PhD thesis, Electrical and Computer Engineering, Portland State University, 2018.

[8] Hoon Park. Formal modeling and verification of delay-insensitive circuits. PhD thesis, Electrical and Computer Engineering, Portland State University, 2015.

[9] Hoon Park, Anping He, Marly Roncken, Xiaoyu Song, and Ivan Sutherland. Modular timing constraints for delay-insensitive systems. *Journal of Computer Science and Technology*, 31(1):77–106, 2016.

[10] Marly Roncken, Swetha Mettala Gilla, Hoon Park, Navaneeth Jamadagni, Chris Cowan, and Ivan Sutherland. Naturalized communication and testing. In *International Symposium on Asynchronous Circuits and Systems*, pages 77–84, 2015.

[11] Ad Peeters, Frank te Beest, Mark de Wit, and Willem Mallon. Click elements: an implementation style for data-driven compilation. In *International Symposium on Asynchronous Circuits and Systems*, pages 3–14, 2010.

[12] Edsger Dijkstra. Guarded commands, nondeterminacy and formal derivation of programs. *Communications of the ACM*, 18(8):453–457, 1975.

[13] Marly Roncken, Ivan Sutherland, Chris Chen, *et al*. How to think about self-timed systems. In *Asilomar Conference on Signals, Systems, and Computers*, pages 1597–1604, 2017.

[14] Alain Martin. The probe: an addition to communication primitives. *Information Processing Letters*, 20:125–130, 1985.

[15] IEEE-SA Standards Board. IEEE Standard Test Access Port and Boundary-Scan Architecture, IEEE Std 1149.1-2001 (Revision of IEEE Std 1149.1-1990), 2001.

[16] Sandra Jackson and Rajit Manohar. Gradual synchronization. In *International Symposium on Asynchronous Circuits and Systems*, pages 29–36, 2016.

[17] Charles Seitz. Chapter 7: System timing. In Carver Mead and Lynn Conway, *Introduction to VLSI Systems*, pages 218–262. Addison-Wesley, 1980.

[18] Gennette Gill and Montek Singh. Automated microarchitectural exploration for achieving throughput targets in pipelined asynchronous systems. In *International Symposium on Asynchronous Circuits and Systems*, pages 117–127, 2010.

[19] Ted E. Williams and Mark A. Horowitz. A zero-overhead self-timed 160-ns 54-b CMOS divider. *IEEE Journal of Solid-State Circuits*, 26(11):1651–1661, 1991.

Chapter 8

Asynchronous network-on-chips (NoCs) for resource efficient many core architectures

Johannes Ax[1], Nils Kucza[2], Mario Porrmann[3], Ulrich Rueckert[2] and Thorsten Jungeblut[2]

The advances in the miniaturization of microelectronic circuits enable the integration of an ever-increasing number of transistors on a single chip. This allows for the realization of massively parallel on-chip multiprocessors (multiprocessor-system-on-chips, MPSoC). Instead of solely increasing the clock frequency to maximize performance, on-chip multiprocessors achieve the required throughput by parallel processing at a comparatively moderate clock frequency, enabling high energy efficiency. Basis for an efficient communication between the compute nodes is a powerful on-chip network (network-on-chip, NoC). The regularity of the architecture leads to a high scalability, which enables the simple realization of MPSoCs scaling to hundreds of processor cores.

With a steadily growing number of processor cores on a single die, it is becoming increasingly difficult to implement the design as a fully synchronous circuit. Synchronous circuits are usually implemented using edge-triggered flip-flops. If the system is based on a clock that is valid everywhere at the same time (see Figure 8.1), design tools must ensure that the clock edge arrives without any significant clock skew at all flip-flops of the system. This results in a large and widely branched clock tree of clock drivers that compensate for a phase shift of the clock on the whole die. In a synchronous circuit, the number and size of the required clock drivers increases with the design complexity. This leads to increased area requirements and higher energy consumption. Furthermore, it is shown in [1] that scaling synchronous MPSoCs results in a strong reduction of the maximum clock frequency—even when using NoCs.

[1]dSPACE GmbH, Paderborn, Germany
[2]Bielefeld University, Bielefeld, Germany
[3]Osnabrueck University, Osnabrueck, Germany

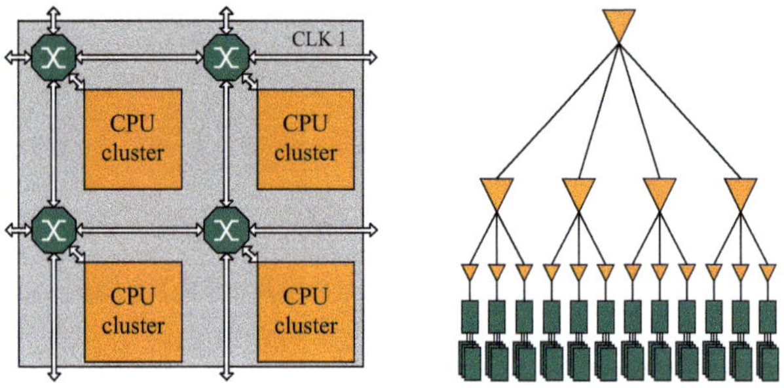

Figure 8.1 MPSoC with one clock domain and the respective clock tree

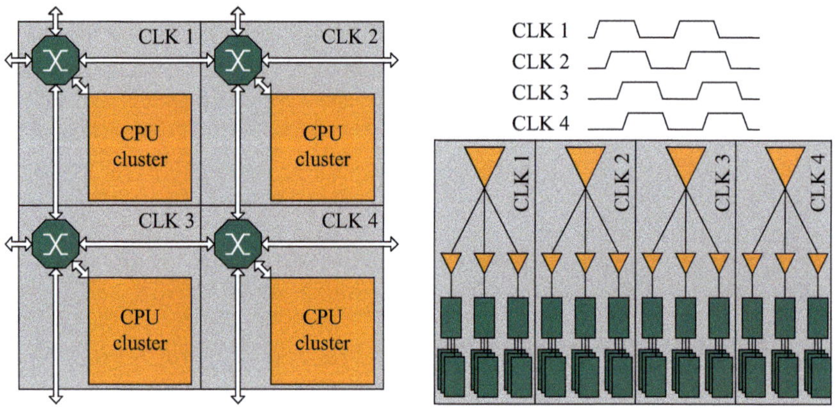

(a) NoC with mesochronous domains (b) Clock trees in a mesochronous design

Figure 8.2 Mesochronous MPSoC: (a) shows the distribution within the NoC and (b) shows the potential corresponding clock trees and the phase shift of the clocks

8.1 Basics of asynchronous NoCs

To take advantage of the scalability of network-on-chips (NoCs), an implementation as a globally asynchronous locally synchronous (GALS) system is often promising [2]. This means that individual modules of the system operate synchronously, whereas the global communication between these modules is asynchronous. The development of a completely asynchronous system is difficult to implement because both the design tools and the IP libraries are strongly based on the synchronous paradigm. The GALS paradigm therefore offers a good compromise between hardware efficiency and development time.

The implementation as a GALS system divides the previously very large clock tree into multiple small independent clock trees (see Figure 8.2). This reduces number and size of required clock drivers, since a phase shift may occur between individual clock trees. Another advantage of a GALS approach is the reduction of electromagnetic interference (EMI). Interference causes noise between the supply voltage and ground, so that critical switching delays can occur for implementations based on small structure sizes and low supply voltages [3,4]. Especially, a synchronous clock tree has a large impact on EMI, since by synchronously switching all registers on the chip during a short period of time a very high surge current occurs [5]. By implementing a GALS system it is possible to set a specific phase shift between different synchronous modules so that the switching time of all registers can be distributed over the entire clock period. This results in a uniform power consumption, which leads to a significant reduction of EMI [6].

Furthermore, in [6] the potential to reduce power dissipation is shown. By implementing a NoC as a GALS system, it is possible to dynamically adapt individual synchronous modules to current performance requirements. This allows nodes with lower performance requirements to be operated at a lower clock frequency. As a result the supply voltage can be adjusted during runtime as well, which is also known as dynamic voltage and frequency scaling (DVFS) [7]. The system is partitioned into several blocks (also called voltage islands) with different supply voltages [8]. According to [9], both the adaptation of the clock frequency and the supply voltage result in a strong reduction of the power dissipation.

Since in a GALS system the different synchronous modules are asynchronous to each other, synchronization is required. This synchronization is necessary to prevent registers in a module to take metastable states.* If a signal arrives at the input register of a receive module, it cannot be guaranteed that the setup or hold times of the register are met. This can lead to a metastable state over an indefinite period of time.

When implementing an MPSoC as a GALS system, two different scenarios can be distinguished. The first scenario is the loosely synchronous system, in which a certain dependency between the different clocks is known at design time. This scenario can be divided into the *mesochronous*, *plesiochronous*, and *heterochronous* approaches. The second scenario is an asynchronous system in which modules are operated by completely different clocks [10].

However, improvements and the use of new design methods in modern tool flows also offer new possibilities for scaling synchronous architectures. Modern design tools offer the so-called clock concurrent optimization (CCOpt) design flow, in which clock tree and combinatorial logic are optimized simultaneously [11]. This makes it possible to use the phase shift of the clock (useful skew) to achieve higher clock frequencies. Nevertheless, GALS methods still allow for a higher resource efficiency because they are almost completely independent of any phase shift. A comparison of the new CCOpt design process with different GALS methods is shown in Section 8.2.5.

*Flipflops can take an undefined state for a certain time before the stable states (voltage levels for logical-0 and logical-1) is taken.

8.1.1 Mesochronous architectures

In a mesochronous architecture, a separate clock domain is defined for each synchronous module (see Figure 8.2(a)), and each clock tree can be divided into several subtrees. All modules are still controlled by the same clock frequency but an unknown phase shift may exist between the clock signals of the individual modules (see Figure 8.2(b)). Usually, all modules are controlled by the same clock source. Since globally no clock drivers are used, a phase shift occurs between the individual modules due to different line lengths across the die [10].

Due to the potential phase shift, synchronization between the modules is necessary to avoid metastability. In addition to bisynchronous FIFOs, special synchronization circuits can be used at the module interfaces, since the clock frequency of the transmitter and receiver modules is identical (see Section 8.2).

8.1.2 Plesiochronous architectures

In plesiochronous GALS systems, all modules operate almost at the same clock frequency. Very small deviations between the frequencies may occur, resulting in a slow phase shift between the modules over time. One way to avoid this behavior is to detect the unstable point of time and return to a stable state. However, this is only possible if the detection is feasible. An advantage of this method is that no synchronization is required while the system is in a stable state. Thus there are no additional latencies during data transfer.

8.1.3 Heterochronous architectures

Heterochronous systems come closest to asynchronous implementations because clock frequencies are completely different. They can be divided into ratiochronous and nonratiochronous systems. The only difference between nonratiochronous systems and asynchronous systems is that in nonratiochronous systems the frequency does not change dynamically. As a special feature different clock frequencies in ratiochronous systems are exactly a rational multiple of each other. Special bisynchronous FIFOs are used for synchronization in heterochronous systems. These can be implemented with less effort, depending on whether the system is ratiochronous or nonratiochronous.

8.1.4 Asynchronous architectures

An asynchronous GALS system is characterized by the fact that the various modules are supplied by completely independent clocks, which can even change if necessary. On the one hand this offers a high flexibility, on the other hand a complex synchronization between the different modules is necessary. Bisynchronous FIFOs at the module interfaces can be used, but this highly increases latency, area requirements, and power dissipation of the system. However, the increase in power dissipation can be compensated by dynamically reducing the frequency or supply voltage of individual nodes. A more efficient method for implementing asynchronous GALS systems is using fully asynchronous routers.

Unlike a synchronous system, a fully asynchronous system responds only to state transitions of certain signals and not to clock edges. A challenge in receiving asynchronous signals is to ensure that the received signal is not evaluated until it has taken a valid stable state (transient). This increases the complexity of the routers and also requires additional circuits when switching from a synchronous to an asynchronous domain. The evaluation of the signals needs to be done accurately and quickly, without triggering unwanted signal transitions. Some implementations are already described in the literature. In [12] and [13], different logic elements for asynchronous circuits are introduced. The different arbiter, C elements, handshake methods, and buffers should be emphasized. Additionally, it is necessary to check error states in asynchronous circuits, which increases the design effort. The publications [14] and [15] deal with this issue and present circuits of a scan test, a self-test, and testing by external test circuits. For synchronous circuits scan tests and self-tests are sufficiently well known. The authors present additional test circuits with which asynchronous circuits can be checked for errors.

8.2 GALS extensions for embedded multiprocessors

For very large microelectronic circuits, a fully synchronous implementation often leads to higher area and energy requirements as well as lower performance [1]. As mentioned in Section 8.1, it is useful to design large circuits as globally asynchronously and locally synchronously. In MPSoCs it is common to implement CPUs synchronously and connect them via an asynchronous communication structure following the GALS paradigm.

In this section, different GALS methods for NoC communication are introduced that allow for a globally asynchronous many-core architecture. First, the state of the art is presented for various GALS-based NoCs. Next, the implementation of a mesochronous and an asynchronous router for the CoreVA-MPSoC is presented. A comparison of these implementations with the synchronous NoC is given in Section 8.2.5.

8.2.1 State-of-the art of GALS-based NoC-architectures

Many different implementation of GALS-based MPSoC NoC architectures are described in the literature. Basically, two different paradigms can be distinguished: a completely asynchronous NoC and a multisynchronous NoC.

In multisynchronous NoCs, routers are operated synchronously based on an internal clock signal. However, direct communication between routers is asynchronous. The most common approach in multisynchronous NoCs is the use of mesochronous links. Although routers still work with the same clock frequency, an unknown phase shift may occur [10]. This simplifies the global clock tree for the local synchronous routers and CPUs. Therefore, smaller clock drivers can be used. However, to prevent violations of setup and hold times, the phase shift between the routers must be compensated. In addition to bisynchronous FIFOs [16], special synchronizers can be used on the links between the routers, since the clock frequency of transmitter and receiver modules is identical. Synchronizers have a smaller footprint as well as lower latency [17]. An example of a mesochronous synchronizer is presented in [18]. This ensures synchronization by means of three data registers, a DLL-based frequency doubler and a decider. Also in [19], synchronization is ensured after detection of the phase shift and by a dynamic change of the signal delay on the links. A similar approach is used in Spidergon-STNoC [20]. Their skew insensitive mesochronous links (SIML) are used for synchronization. With SIML, four data registers are used on the receiver side. In addition, a sampling pulse must be generated in the transmitter. Four registers are used as buffers with two stages, so that SIML implementations have a latency of two clock cycles. The tightly coupled mesochronous synchronizer (TCMS), which is presented in [21], offers a more area-efficient synchronization. This architecture requires only three latches and two 2-bit counters for phase synchronization. In addition, a TCMS has a latency of one to two clock cycles, which is considered to be minimal due to the unknown phase shift. For this reason, the TCMS is also used in the mesochronous NoC of the CoreVA-MPSoC.

In asynchronous systems, handshake protocols are used for the explicit synchronization between two participants. Two different protocols are described in the literature: a four-phase *level signaling protocol* (LSP) and a two-phase *transition signaling protocol* (TSP) [22]. The LSP signals a new data word by a *request.* After the receiver *acknowledges* the request, the transmitter returns to its default level (return to zero transmission). As soon as the receiver has also returned to the default level, a new access can be initiated. The TSP avoids these level changes. New data words are indicated by a change in the state of the request signal (non-return to zero transmission). Examples of these two methods are presented in the following. Thonnart *et al.* [22] propose a framework for an asynchronous NoC. Asynchronous communication is based on the LSP. Via four I/O ports, the individual routers are connected to a 2D mesh with wormhole routing. A fifth port connects local IPs. In order to overcome the limitations of CAD tools, a new design process has been created and adapted to the needs of asynchronous circuits. The system has been manufactured in a 65 nm CMOS technology and offers a throughput of 550 MFlits/s, with a power consumption of 86% compared to synchronous system. In [23], an asynchronous router is presented that operates on a level-encoded-dual-rail protocol. This protocol is similar to the two-phase LSP, but uses a larger set of control signals. In the implementation, the I/O ports of a router are allocated for the entire transmission (circuit switching) and only released after completion. For analysis, a 4 $\times$ 4 2D mesh with 16 Spidergon cores was manufactured in a 130 nm technology. It achieves a latency of 2.74 ns and a throughput of 526 MFlits/s. The Argo-NoC also uses an asynchronous NoC [24,25]. The asynchronous router features special crossbars to connect input and output ports. In this crossbar, so-called *handshake clocks* are used, in which a flit is consumed by each input port and a flit is produced at all output ports. A state bit signals invalid flits, used for synchronization. Consequentially, all incoming flits must be shifted to the same phase, as it is based on a LSP. This is realized by a mousetrap pipeline between the router and the preceding FIFO. Argo-NoC also uses circuit switching, which simplifies arbitration within the routers. However, for the transition between synchronous and asynchronous domains, no precautions are considered for the asynchronous request and acknowledge signal. This can lead to metastable states in the synchronous registers.

Basically, the two-phase TSP offers advantages over the four-phase LSP. When returning to zero state in the LSP, signals are forced to change their level frequently. With the two-phase TSP, a state is only changed when required, resulting in a more compact communication. Therefore, TSP can be more energy-efficient, since significantly fewer switching operations are required.

Different GALS methods and fully synchronous systems have been compared in many publications [26,27]. However, a comparison between mesochronous and completely asynchronous NoCs in the use of hierarchical MPSoCs has rarely been analyzed. To the best of our knowledge, this topic has only been considered in Sheibanyrad *et al.* [28] by comparing a mesochronous NoC (DSPIN) and a fully asynchronous NoC (ASPIN). In the DSPIN, bisynchronous FIFOs between the routers are used to build a mesochronous NoC. The ASPIN, on the other hand, works with fully asynchronous routers that use the four-phase LSP. Furthermore, a dual-rail protocol is used in which two lines represent one bit. This dual-rail protocol is required in the ASPIN architecture to synchronize the bits of a flit. The comparison between DSPIN and ASPIN is performed for maximum throughput, minimum latency, chip area, and power consumption. It should be emphasized that the work also considers the line delay between the routers, which cannot be neglected, especially with hierarchical MPSoCs. Both NoC designs show very similar results for maximum throughput and chip area. The asynchronous ASPIN exceeds the DSPIN with a 2.5× lower minimum latency. In addition, the power consumption in idle mode is three times lower in ASPIN, at the cost of a slightly higher power consumption during active data transmission. Especially for long lines between the routers, the dual-rail protocol loses the advantages of asynchronous NoCs compared to synchronous NoCs, since twice as many lines are required.

In the following, the CoreVA-MPSoC architecture and the implemented asynchronous NoC are presented. The asynchronous NoC uses the two-phase TSP in combination with a so-called mousetrap circuit. To allow for waiving the dual-rail protocol an additional delay of the request signal in relation to the data bits of a flit is added. As a result, the number of lines between routers is almost identical for all three router designs (synchronous, mesochronous, and asynchronous).

8.2.2 The CoreVA-MPSoC

The resource-efficient many-core architecture CoreVA-MPSoC [29] was developed at Bielefeld University. It is mainly optimized for the processing of streaming applications and is based on a hierarchical communication infrastructure (see Figure 8.3). The CPU core is based on the CoreVA-VLIW-architecture and supports VLIW vectorization and SIMD processing. Multiple CPU-cores comprise a CPU-cluster. CPUs within a CPU cluster are connected via a low-latency high-throughput bus-based communication. As shown in [30], however, the bus-based

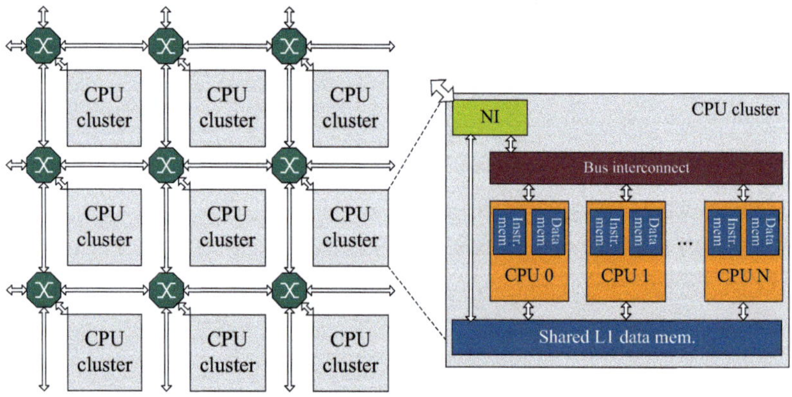

Figure 8.3 The many-core architecture CoreVA-MPSoC

communication is limited when integrating more and more CPUs on a single chip. For this reason the CoreVA-MPSoC introduces an on-chip network (CoreVA-NoC [29]) as the second communication hierarchy.

8.2.3 Mesochronous router implementation

The mesochronous NoC in the CoreVA-MPSoC replaces the input registers of the synchronous router by mesochronous synchronizers (TCMS) [21]. A TCMS is characterized by a very close coupling to the router. Functionality of the TCMS is to compensate a possible phase shift of the clock between two routers so that all data bits of a flit can be stored in a stable state in the FIFO of the router.

The TCMS is divided into the so-called *front end* and the *back end* (see Figure 8.4). The front end is located at the input of the TCMS and receives the clock *CLK_EXT* together with the flit from the external router. The back end forms the output-stage of the TCMS and belongs to the clock domain of the local router. Synchronization takes place between the front and back ends.

The front end contains three latches with a data width of one entire flit, including the request signal. A 2-bit counter determines in which of the three latches data is stored. To avoid a metastable state when writing the data into a latch, the counter is operated with the clock of the external router. In addition, it must be prevented that an enable signal of a latch is active too long so the data of the next data word is stored. This is ensured by an enable signal which is only active during the high level of the external clock. The TCMS-implementation of [21] has been optimized in order to prevent misbehavior in case of glitches in the enable signals of the latches. This misbehavior is caused by the fact that an enable signal switches during the high level, but the value of the counter only changes a short time later due to signal delay. To avoid this problem, the counter used in the TCMS of the CoreVA-MPSoC is operated by the falling edge of the external clock. As a result, the counter value is already stable for half a clock period before the next high level of the clock activates the corresponding enable signal. Since the signal delay of the

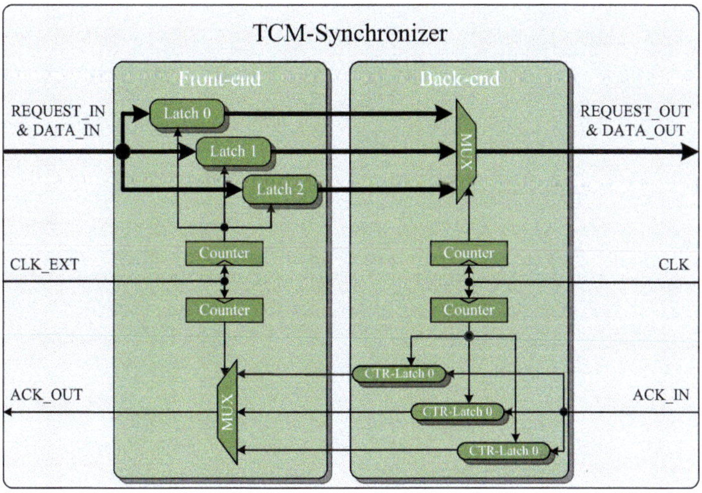

Figure 8.4 Architecture of the tightly coupled mesochronous synchronizer (TCMS) in the CoreVA-MPSoC

counter is always higher than that of the external clock due to a longer path, a correct behavior can also be ensured at the end of the high level.

The back end of the TCMS decides from which latch the data are forwarded. The multiplexer is controlled by another corresponding 2-bit counter operated by the positive edge of the local router clock (CLK). The counter must be set in a way, that it only forwards the data of a latch that have been stored for a longer period of time and are therefore stable. The waveforms shown in Figure 8.5 give an example of the switching behavior within the TCMS.

To avoid metastability of the TCMS at all, the counter in the front end must match exactly the setting of the counter in the back end. A constellation is required which forwards only data of a stable latch for all possible phase shifts between -180° and $+180^\circ$. The correct setting of the counters is ensured by a matched reset of the counters. The global reset of the CoreVA-MPSoC is asynchronous implying that the phase relation of the reset to the clock edge is not defined. Thus the state of the counters is not deterministic and therefore, no suitable default value of the counters can be determined. In order to force a defined initial state, first of all the reset signal must be synchronized with the clock of the local router. To synchronize the reset signal, a *brute-force* synchronizer is used, which is realized by two flip-flops connected in series. With this reset signal synchronized to the local router, all counters in the TCMS can be initialized. However, this can lead to metastability in the counters of the front end, since the reset signal may be too close to the clock edge of the external clock. To avoid this, the clock of the counter in the front end must not be started before the reset has been finished. The delay of the clock is realized by latch-based clock gating to avoid glitches within the clock signal. To avoid a metastable state in the used latch, the reset synchronous to the internal

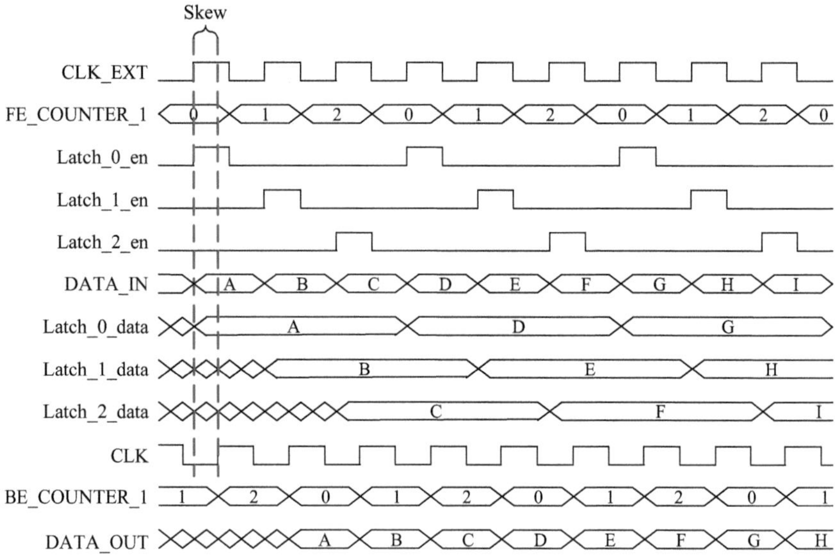

Figure 8.5 Example of the switching behavior of the TCMS

clock is furthermore synchronized to the external clock using a *brute-force* synchronizer. This is important in order to obtain a defined initial state.

After the initialization of the reset, a suitable start value for the counters must be defined. As shown in Figure 8.6, due to the phase shift a window (*Window 1*) exists after the reset in which the first falling edge of the external clock (CLK_EXT_int) can occur. This window extends over a clock period due to a possible phase shift of up to 360°. Also for the counter in the front end there is a window in which it has a fixed value. Window *Window 2* indicates the period in which data can be transferred to the latch based on the counter. In this way it is now possible to determine the starting value for the counter in the back end, which forwards a stable data word from the latch.

Until now, only the data path of the incoming flit was described. In the control path to the external router, however, the ON/OFF signal based on the fill level of the FIFOs must also be transferred to the clock domain of the external router. The same principle is used for the data path. Three additional latches are implemented in the back end of the TCMS to store the ON/OFF signal. Since this is only a single signal, 1 bit wide latches are used here. The enable signals are also controlled by a 2-bit counter which, however, is controlled by the falling edge of the local router clock. For the same reason as in the data path, the enable signals are only active during the high level of the local clock. In the front end there is a multiplexer which is controlled by another counter and transmits the ON/OFF signal from the corresponding latch. This extension within the TCMS ensures that no further synchronization of the acknowledge is necessary in the external router.

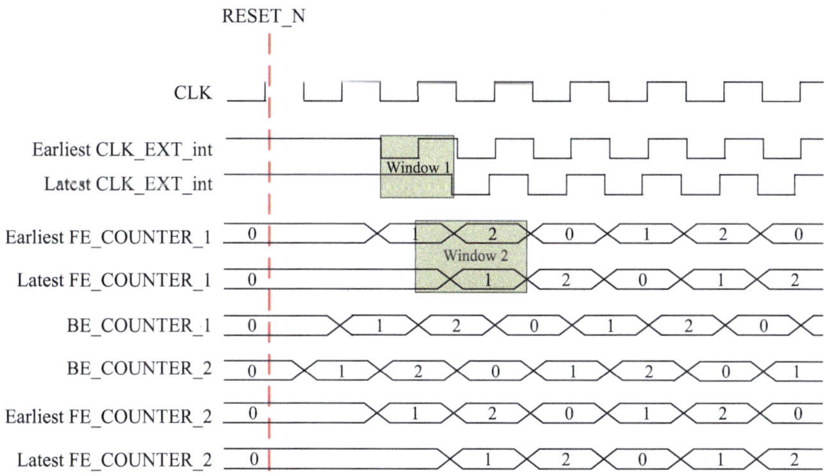

Figure 8.6 Waveform showing the counter initialization of the TCMS

The latency of the TCMS can also be derived from the waveform shown in Figure 8.6. The latency depends on the phase shift between the two clock signals and is either one or two clock cycles. The same applies to the latency of the acknowledgment signal sent back to the external router. However, since this is the opposite case, both delay times always add up to exactly three clock cycles until the acknowledgment signal arrives at the external router. Since this takes one clock cycle longer than in the synchronous case, the *almost-full* signal of the FIFOs must be set if only three free entries are left. This has to be considered for the size of the FIFOs and leads to a minimum size of five flits.

8.2.4 Asynchronous router implementation

For the implemtation of the asynchronous router, the synchronous components of the synchronous router are replaced by an asynchronous implementation. Apart from that, the asynchronous router supports the identical functionality as the synchronous and mesochronous versions. A detailed structure of the asynchronous router is shown in Figure 8.7.

Key component of the router design is a *Mousetrap* FIFO. The Mousetrap (MT), introduced in [31], is a simple but effective asynchronous pipeline stage. At all inputs and outputs of a router, the MT stores flits sent via the NoC and simultaneously supports the two-phase TSP. Figure 8.7 shows the architecture of the mousetrap FIFO. A mousetrap stage consists of latches for the data and request signal, and an XNOR setting the latches to transparent mode. At initialization all latches are transparent. At a transition of the request signal, the mouse trap locks and keeps its value until an acknowledge signal is set. The outgoing request to the next stage is also used as the feedback acknowledgment signal. With this design, the sequence of the signals is important. To avoid metastable states at the latches,

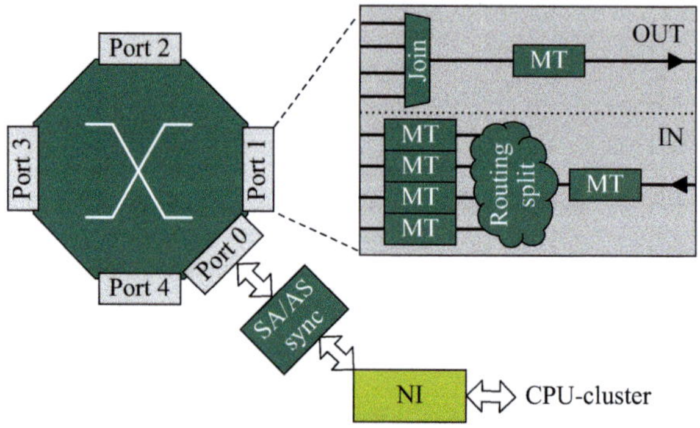

Figure 8.7 Architecture of the asynchronous router

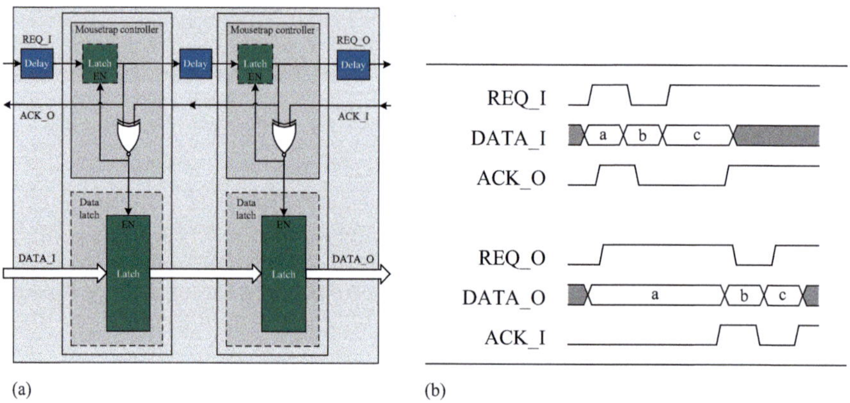

Figure 8.8 Implementation and switching behavior of a mousetrap FIFO of depth two. (a) connection of two mousetrap circuits to realize a FIFO and (b) example of the switching behaviour of the FIFO, the first data word is sampled with a delay.

data signals must be set before the request signal. For this reason, the request signal before a mousetrap stage is always slightly delayed by an inverter chain. For a FIFO component, several of these mousetrap steps are connected in series. The depth of the implemented FIFO can be configured.

The waveform depicted in Figure 8.8(b) illustrates the switching behavior for a mousetrap FIFO of depth two. The incoming request signal (*Req_I*) passes through two latches and an inverter chain with six inverters. In contrast, the data bits only pass through two latches and thus have a lower delay. The correct signal sequence and the TSP approach is ensured by this runtime difference. When implementing a FIFO by combining mousetrap elements, it must be ensured that the feedback acknowledge signal of the previous stage has a defined latency. This is necessary to ensure the hold time of the latches and to avoid any metastability on the signals.

Apart from the propagation delay of gates and interconnects no special precautions have to be considered. Depending on the final chip layout the length of the inverter chain may have to be adjusted.

As shown in Figure 8.7, the routing element must forward the incoming flit to the corresponding mousetrap depending on the destination port. The routing decision itself follows an XY routing analogous to the synchronous router. In order to enable data splitting in the TSP, a split with mutual exclusion is required. The circuit introduced in [32] allows splitting data to two output ports (see Figure 8.9(a)). Data can be routed through this element without modification, since there is only one transmitter and therefore no collision can occur. As only one port may be operated at a time, outgoing control signals must be transmitted unambiguously to a receiver with mutual exclusion. Two latches hold the output values of the request signals if they are not set transparent by the router. An *XOR* gate in front of each of these latches ensures the correct output value since only one of the values is changed at a time. This is essential since in the two-phase TSP a state change always signals a new data word. For the four I/O ports of the CoreVA-MPSoC in a standard 2D mesh topology, multiple split instances are cascaded one after the other.

Analogous to the split, data paths in the TSP have to be merged again. In the router, this is implemented in the crossbar or arbitration before the output ports (see Figure 8.7). The join is based on the implementation from the work in [32] (see Figure 8.9(b)). In this example two input ports are connected to one output port. Four latches keep the outgoing signals at their values and several *XOR* gates set the correct levels. Since both input ports can potentially deliver new data simultaneously, a special arbiter must be used for mutual exclusion. Secure mutual exclusion is a particular challenge in an asynchronous design [33–35]. By the

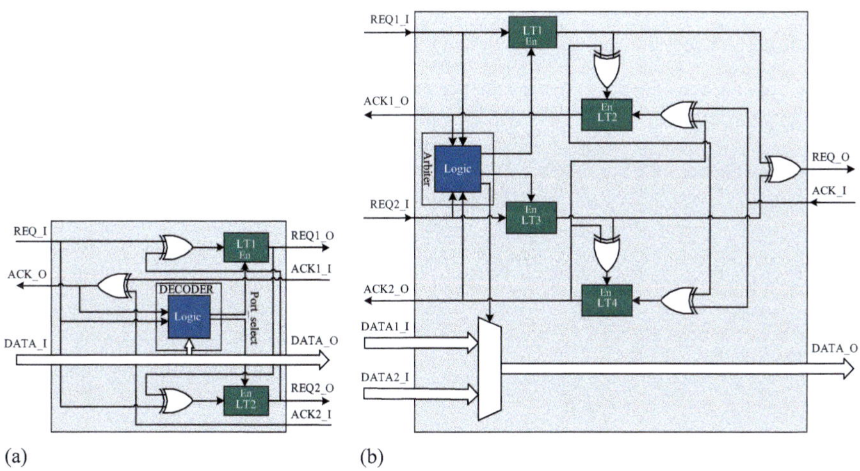

Figure 8.9 Split and join elements for asynchronous data paths. (a) Architecture of a data split with mutual exclusion and (b) architecture of a join

absence of a clock it must be ensured differently that a valid decision between requests of participants can be made at any time and state. Changes of the request signals and particularly metastable states must be avoided at all costs. Possible implementations consist of a part for mutual exclusion and a filter for metastable states. The combinatorial circuit for the mutual exclusion always consists of two cross-connected *NAND* gates, of which only one output may be logical 0.

The schematics in Figure 8.10 illustrate the design used for the CoreVA architecture. It consists of two *NAND* and two *NOR* gates. For mutual exclusion, the outputs of the two NANDs are connected to one input of the other NAND. The other inputs are used for the external input signals. If one of the inputs changes to logic 1, the output of the corresponding NAND gate changes to logic 0. This prevents the other NAND gate from changing to logic 0 and mutual exclusion is given. In the possible case of a simultaneous change of the input signals to logic 1, metastability can occur. In this case, it may take some time until one of the outputs is set to a stable logical 1. Due to the slight variance of the electrical components of the NAND gates, a stable state can always be expected. In order to avoid this metastability, two NOR gates with four inputs each can be used. Due to their design, they are suitable for this operation. With a NOR gate with four inputs, four PMOS transistors are connected in series. These switch more slowly than NMOS transistors and generally have a higher impedance. The structure is shown in Figure 8.10(b). In order to prove the correct behavior of the arbiter, the circuit was simulated in the Virtuoso simulation environment in a 28 nm standard cell technology. In the join element, the decision between the incoming data depends only on the arbiter. The join elements are also cascaded to map the four input ports to one output port.

Another critical issue when using an asynchronous NoC is the interface between router and NI. Since the network interface (NI) is operated synchronously, there is a need to establish a stable signal transition between it and the

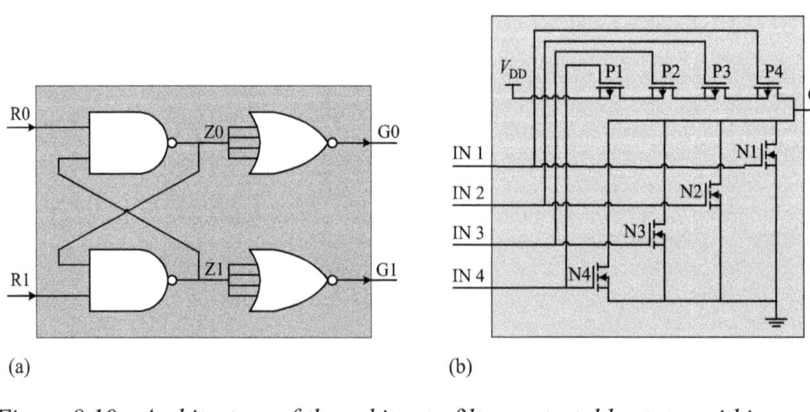

Figure 8.10 Architecture of the arbiter to filter metastable states within a join element. (a) combinatorial implementation of the arbiter and (b) Implementation of a NOR4 in CMOS

asynchronous port 0 (component *SA/AS-Sync* from Figure 8.7). If a single signal between an asynchronous and a synchronous circuit changes, only the transition from asynchronous to synchronous is problematic. To prevent metastable states when disregarding switching times of registers, a synchronizer is required. In the CoreVA-MPSoC, the *brute-force* synchronizer already used in the mesochronous router is used for this purpose. In order to prove the expected switching behavior of this circuit, a SPICE simulation at transistor level was performed.

Between network interface and router not only single bits but whole data words (flits) have to be exchanged. When flits are passed from the router to the network interface, the data bits of a flit and the request signal must be synchronized. The block diagram in Figure 8.11 shows the architecture of this synchronization circuit. Similar to the mousetrap circuit, the request signal is delayed in order to ensure dependency on the data word. The data word always arrives before the request signal and can thus be stored stably in a register. This means that only the request signal must be passed through the brute force synchronizer. Data at the next data register are valid data at least half a clock length earlier. At a target frequency of about 700 MHz this corresponds to a time of more than 700 ps, which in comparison to the setup time of the registers is sufficient to accept the data as valid. If metastability occurs on the data lines, the request signal arrives one clock later and the data word is captured again. This means that only a guaranteed valid word is used. In order to achieve the highest possible throughput, incoming data words are stored in one of two registers. There is one register each for the case of a logical 0 and for the case of a logical 1 of the request signal in the two-phase TSP.

During the transition from the synchronous to the asynchronous domain, the acknowledge signal must be synchronized by the brute force synchronizer. The state of the registered acknowledge signal controls from which registers the data

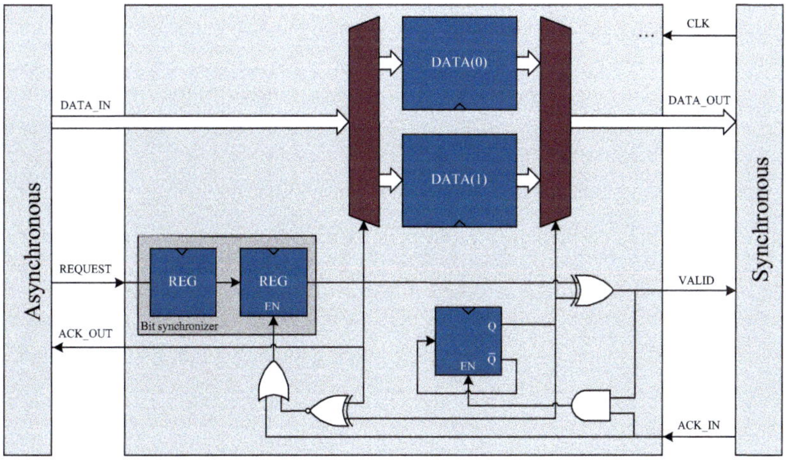

Figure 8.11 Implementation of the synchronizer from asynchronous to synchronous

word is forwarded to the output. Incoming flits on the synchronous side are written to one of two registers. Again, two registers are used to convert to the two-phase TSP of the asynchronous router. A flit has a latency of one clock cycle when passing the synchronization circuits from NI to router as well as from router to NI.

8.2.5 Design-space exploration of the different GALS-approaches

In order to evaluate the different GALS approaches, a design-space exploration was performed, considering measures as area requirement, power consumption as well as latency and throughput. The asynchronous NoC design is compared to the synchronous and mesochronous implementations.

For a fair comparison to the asynchronous design a fully placed and routed design was generated. This is necessary because line delays must be taken into account during the design process of an asynchronous design. Especially with hierarchical many-core systems with large CPU clusters, such as the CoreVA-MPSoC, these cannot be neglected. The following results are therefore based on the place and route of a cluster node. Several of these cluster nodes can be interconnected according to the hierarchical design flow. A cluster node consists of four CPU macros, the cluster interconnect, and 64 kB shared L1 data memory. In addition, it embeds the NoC components, such as the network interface and the router with its I/O port with four links to the 2D mesh. Maximum clock frequency for the synchronous parts of the node is 704 MHz.

In order to examine the effects of the different GALS methods on the entire system, this section concludes with an analysis of the global clock tree resulting from different system configurations. More detailed results are presented in [36].

Figure 8.12 shows the area requirement of a cluster node. The footprint for CPUs, memory, cluster communication infrastructure, and network interface is identical for all three GALS implementations and is shown on the left side of the diagram. The right side of the diagram shows the area requirements of the three different router realizations. "Others" includes all circuit parts that cannot be assigned directly to one instance, such as the clock tree. Due to the lack of a clock tree and the use of latches instead of flip-flops to cache the flits, the asynchronous router requires only 42% of the area of a synchronous router. As expected, the synchronous and mesochronous routers have almost the same area. On the one hand, the area of the mesochronous router increases slightly due to the additional TCMS. On the other hand, this overhead can be compensated by more relaxed timing constraints due to the allowed phase shift between the NoC links.

The layouts for the synchronous and asynchronous realizations are shown in Figure 8.13. The router is highlighted in dark green and shows the different area requirements after place and route. In order to realize the four NoC links that are located at the four sides of the macro in a cluster node, the P&R tool is allowed to route over the CPU macros in the upper two metal layers. In the case of synchronous and mesochronous designs, the entire layout of a cluster node has area requirements of 0.817 mm^2. For the asynchronous implementation, the footprint of

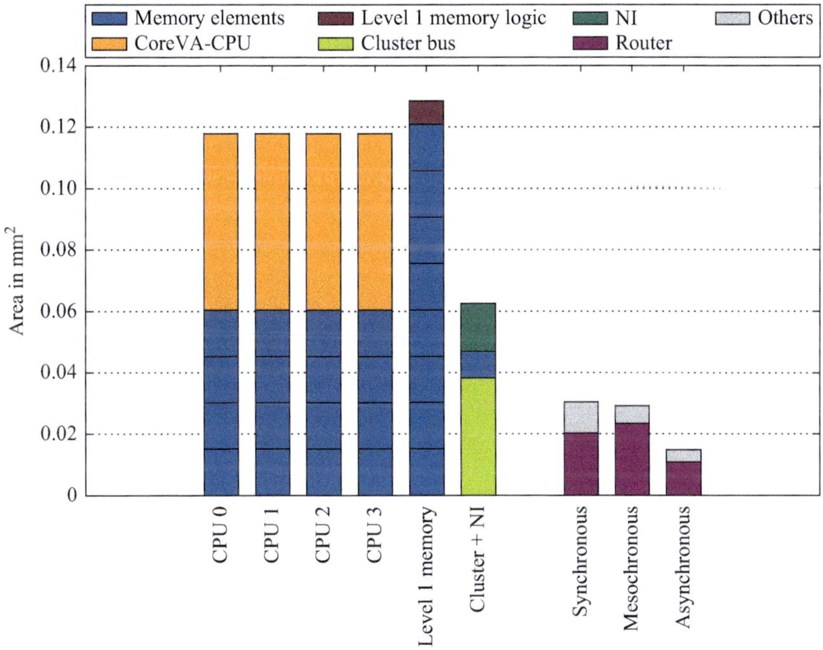

Figure 8.12 Area requirements of CPU, memory, cluster communication infrastructure, network interface, and different router realizations

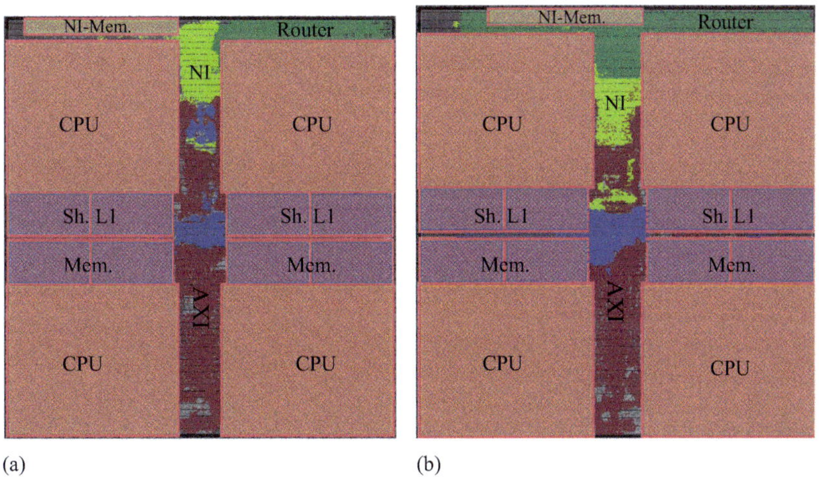

Figure 8.13 Layout of the asynchronous and synchronous CPU cluster. (a) asynchronous NoC and (b) synchronous NoC

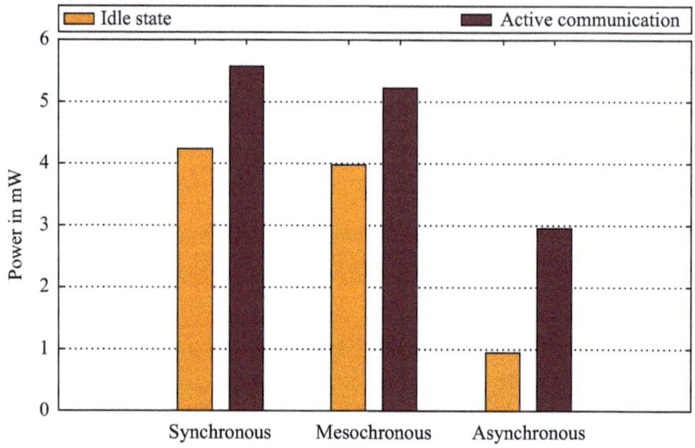

Figure 8.14 Comparison of the power consumption of different router realizations

an entire cluster node is reduced by 3.1% to 0.792 mm^2. For larger clusters (e.g., 8 or 16 CPUs) no significant increase in the area requirements of the NoC components is expected. Just the delay of the request signal by a longer inverter chain (see Section 8.2.4) and additional drivers due to longer lines for the NoC links can lead to a negligible increase of the area requirements.

8.2.5.1 Power consumption

The power consumption of the different router implementations is determined by gate-level simulations of the place and route of two connected cluster nodes. To analyze the dynamic power dissipation, the switching activities during a packet transfer are recorded for different router implementations. These switching activities are then used to perform accurate power simulations using Cadence's Voltus tool.

Figure 8.14 depicts the power consumption of a single router during idle state and active communication. The synchronous router consumes 4.23 mW in idle state and 5.57 mW during active packet transmission. The mesochronous design reduces the power consumption slightly to 3.98 mW or 5.21 mW, respectively. Due to the missing clock signal, the power consumption of the asynchronous router in idle mode reduces to 0.94 mW which is 22.4% of the power consumption of the synchronous router. Most of this power is consumed by the two synchronizers between the router and the network interface, which still contain clock-based circuit elements. Such a synchronizer consumes 0.467 mW, whereas all other components of the router require only 0.056 mW. During active communication, the power consumption of the asynchronous router increases to 2.94 mW, which is 53% of the power consumption of the synchronous router. In this case, the active synchronizer to the network interface consumes 0.7 mW and all other router components 1.75 mW.

8.2.5.2 Latency and throughput

In a synchronous design, both minimum latency and maximum throughput depend solely on the clock frequency of the system. The design has been optimized for a clock period of 1.42 ns, that is, a clock frequency of 704 MHz. In the synchronous router, a flit has a latency of two clock cycles to be passed from one input port to another output port. This results in a minimum latency of 2.84 ns. However, this only applies if there are no collisions with competing flits. The maximum throughput of a unidirectional NoC link is 704 MFlits/s because during a clock cycle only one flit can be sent via one link. The mesochronous design allows the same maximum throughput and in the best case also the same minimum latency as in the synchronous case. However, since the clocks in the mesochronous system are not phase synchronized, the minimum latency of a flit can last almost an entire clock cycle longer in the worst case. The minimum latency in the mesochronous router can therefore vary between 2.84 ns and 4.26 ns.

Compared to the synchronous and mesochronous design, the asynchronous router is completely independent of a clock signal. This means that the minimum latency and the maximum throughput only depend on the switching delays of the logic and memory elements, line delays, and the local handshake. Due to variations in line and gate delays, the results of the I/O ports differ. The minimum latency varies between 1.79 ns and 2.43 ns, and the maximum throughput between 704 MFlits/s and 840 MFlits/s, depending on which I/O ports flits are sent. The minimum latency and maximum throughput that can be achieved for a packet transfer depending on the I/O port of the asynchronous router is listed in Table 8.1.

Maximum throughput (704 MFlits/s) is limited by I/O port 0, due to the synchronous network interface. Overall, it can be observed that both the maximum throughput and the minimum latency of the synchronous and mesochronous implementations are achieved by the asynchronous router in all cases. On average, the router of the asynchronous NoC shows a maximum throughput 15% higher and a latency 25% lower compared to clocked NoCs.

Therefore, the asynchronous router is generally more efficient than the clocked systems and can theoretically achieve an equivalent or higher throughput. However, the maximum throughput is reduced by the synchronous components

Table 8.1 Minimum latency and maximum throughput that can be achieved over the various I/O ports of the asynchronous router

	Minimum latency in ns				Maximum throughput in Mflits/s			
	Port 1	**Port 2**	**Port 3**	**Port 4**	**Port 1**	**Port 2**	**Port 3**	**Port 4**
Port 0	1.81	1.79	2.0	1.97	704	704	704	704
Port 1	—	1.93	2.17	2.20	—	816.4	818.0	797.3
Port 2	1.99	—	2.19	2.21	792.0	—	808.2	801.4
Port 3	2.13	2.11	—	2.34	820.9	817.8	—	825.1
Port 4	2.23	2.21	2.43	—	818.7	821.5	840.0	—

(NI, CPUs). The full benefit of the asynchronous NoC is achieved under full load of the NoC. To analyze the full load condition, one hundred flits were written from each of the three input ports to the remaining fourth I/O port. The synchronous I/O port 0 is omitted during the stress test. With this stress test, inducing a lot of packet collisions, a higher average throughput could be shown for the asynchronous NoC at high load. Both the synchronous and the mesochronous router require 300 clock cycles and thus 426 ns for the transmission of the 3 × 100 flits, since only one flit per clock cycle can be processed by the output port. The asynchronous router needs only 230.95 ns (about 163 clock cycles) and thus only 54.3% of the time to transmit the 3 × 100 flits.

8.2.5.3 Global clock tree

This section discusses the global clock tree of the entire MPSoC. The clock tree results from the interconnection of all cluster nodes and is a measure of the scalability of the system. Traditional synchronous circuit, the number and size of the required clock drivers increases with the number of circuit parts, which lead to an increase in area requirements and energy consumption. Pullini *et al.* [1] have shown that when highly increasing the number of nodes in a synchronous NoC, the maximum frequency is greatly reduced.

Modern design tools, however, also offer new methods for scaling synchronous architectures, called "clock concurrent optimization" (CCOpt) design flow. In CCOpt clock tree and combinatorial logic are optimized simultaneously [11]. This allows for the optimization of the phase shift of the clock, and in some cases the skew can even be used positively (useful skew). Nevertheless, GALS methods still allow better energy efficiency and further advantages, since these are partly completely independent of any phase shifting. This section therefore compares the traditional synchronous design flow (CTS, clock tree synthesis), the new CCOpt design flow, and the different proposed GALS methods.

At global level, many cluster nodes are interconnected to build a 2D mesh NoC. The clock signal must be routed through the global clock tree to all these cluster nodes. In the case of synchronous design, this clock tree must be created in such a way that the NoC links between the cluster nodes meet the required setup and hold times. In the mesochronous and asynchronous designs, however, the clock tree can be more relaxed, since these NoC links and cluster nodes are completely independent of the phase shift of the clock signal. Nevertheless, clock drivers are needed to provide a stable clock signal for all cluster nodes.

In order to show the scalability of the various design methods, a full place and route at top level is used. The previously mentioned cluster node with four CPUs is used here. Table 8.2 shows the power consumption of the global clock tree for different MPSoC sizes and design methods. The results refer exclusively to the power consumption of the global clock tree excluding the internal clock trees within the cluster nodes. For the synchronous design the traditional design flow (trad.-CTS) and the new CCOpt flow are compared. The results show a relatively similar power consumption for the design using the CCOpt flow and the mesochronous implementation. However, the global clock tree of the asynchronous

Table 8.2 Power consumption of the global clock tree for different MPSoC sizes and GALS approaches

# CPUs (NoC-mesh)	Synchronous (mW)		Mesochronous (mW)	Asynchronous (mW)
	Trad.-CTS	CCOpt flow		
16 (2 × 2)	0.39	0.28	0.27	0.27
64 (4 × 4)	2.55	1.79	1.47	1.35
256 (8 × 8)	—	7.83	7.55	5.78

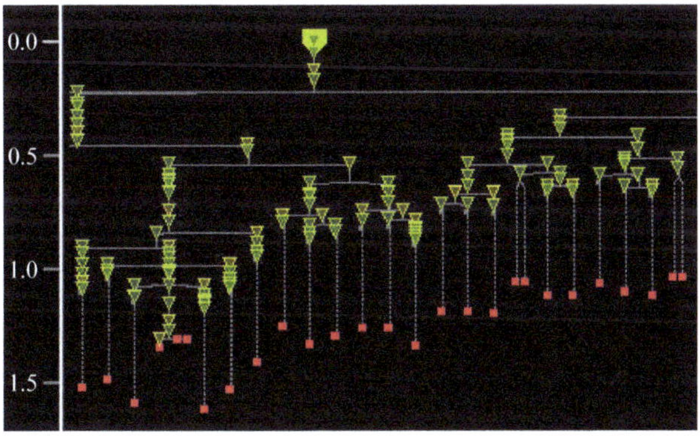

Figure 8.15 Phase shifts (ns) of the extracted clock tree for a 8 × 8 2D-mesh of a CoreVA-MPSoC with 256 CPUs

design has significantly lower power consumption (25% lower for an MPSoC with 256 CPUs). This is due to the fact that the internal clock trees of the cluster nodes are smaller, since the synchronous circuit parts of the NoC are replaced by asynchronous ones. Due to the smaller internal clock trees, the clock tree requires less clock drivers on a global level.

If the traditional CTS flow is used for the synchronous design, this leads to a significantly higher power consumption of the clock tree. Furthermore, maximizing the clock frequency is becoming increasingly difficult, so that MPSoCs the design with more than 256 CPUs was not possible. The CCOpt flow allows for a further scaling of the MPSoC, since the phase shift of the global clock tree can be positively used ("useful clock skew"). An extract of the global clock tree, using the CCOpt flow, is shown in Figure 8.15 for an MPSoC with 256 CPUs and an 8 × 8 mesh NoC (64 cluster nodes). In addition to the clock drivers (yellow) and the connected cluster nodes (red), the maximum phase shift of the clock (in ns) can be seen on the Y-axis. The clock tree does not have to be fully balanced even in synchronous designs. In this example, a phase shift of 0.5 ns can occur between the most distant cluster nodes. This means that using the CCOpt flow, a certain phase shift across the chip is allowed, similar to mesochronous and asynchronous designs.

Apart from the synchronous design using the traditional design process (trad.-CTS), the maximum clock frequency of 704 MHz of the cluster nodes could be achieved for all design variants analyzed. Only the synchronous design using the CCOpt flow reduced the maximum clock frequency slightly by about 1%.

Finally, it can be said that the use of the CCOpt flow allows for the scaling of a synchronous MPSoC, whereby a slight reduction of the maximum clock frequency has to be expected. For this reason, the mesochronous and especially the asynchronous NoC are suitable for a more efficient scaling of the MPSoC. However, they require more design effort.

8.3 Conclusion

In this chapter, different GALS approaches for the implementation of embedded NoC architectures were presented. The GALS approach allows for the reduction of the resource requirements at an increased scalability of the NoC without sacrificing performance. The three approaches of synchronous, mesochronous, and asynchronous NoCs were compared. For the mesochronous NoC special synchronizers between the links were implemented. For the asynchronous NoC, the routers were completely realized as an asynchronous circuits. The results have shown that modern design methods (CCOpt design flow) allow a good scaling of MPSoCs even for synchronous NoCs. Nevertheless, the asynchronous NoC showed lower area and energy requirement compared to the mesochronous and synchronous implementation, while still providing a comparable performance. When comparing a place and route of an MPSoC, the asynchronous NoC leads to 3.1% less area requirements. The power consumption of an asynchronous router is only 22.4% (0.94 mW in idle state) or 53% (3.94 mW during communication) of the power consumption of a clock-based router. In the last section of the chapter, the global clock tree for an MPSoC with 256 CPUs was examined. The synchronous and mesochronous NoC show almost the same power consumption of about 7.7 mW. Using the asynchronous NoC reduces the power consumption by about 25% (5.78 mW). In addition, the mesochronous and asynchronous variants achieve a 2.6% higher clock frequency.

References

[1] Pullini A, Angiolini F, Murali S, *et al.* Bringing NoCs to 65 nm. *IEEE Micro.* 2007;27(5):75–85. Available from: http://dl.acm.org/citation.cfm?id=1320302.1320839.

[2] Hemani A, Meincke T, Kumar S, *et al.* Lowering power consumption in clock by using globally asynchronous locally synchronous design style. In: *Proceedings 1999 Design Automation Conference (Cat. No. 99CH36361).* IEEE; 1999. pp. 873–878. Available from: http://ieeexplore.ieee.org/document/782202/.

[3] Chen LH, Marek-Sadowska M, Brewer F. Buffer delay change in the presence of power and ground noise. *IEEE Transactions on Very Large Scale Integration (VLSI) Systems*. 2003;11(3):461 473. Available from: http://ieeexplore.ieee.org/document/1218219/.

[4] Samanta R, Venkataraman G, Jiang H. Clock buffer polarity assignment for power noise reduction. *IEEE Transactions on Very Large Scale Integration (VLSI) Systems*. 2009;17(6):770–780. Available from: http://ieeexplore.ieee.org/document/4837872/.

[5] Yow-Tyng N, Shih-Hsu H, Sheng-Yu H. Minimizing peak current via opposite-phase clock tree. In: *Proceedings. 42nd Design Automation Conference, 2005*. IEEE; 2005. pp. 182–185. Available from: http://ieeexplore.ieee.org/document/1510316/.

[6] Stanisavljevic M, Krstic M, Bertozzi D. Advantages of GALS-based design. 2010. Available from: http://www.galaxy-project.org/files/D24{_}EPFL{_}R002{_}AdvantagesofGALS-baseddesign.pdf.

[7] Leung LF, Tsui CY. Energy-aware synthesis of networks-on-chip implemented with voltage islands. In: *Proceedings – Design Automation Conference*. 2007; pp. 128–131.

[8] Ogras UY, Marculescu R, Choudhary P, *et al.* Voltage-frequency island partitioning for GALS-based networks-on-chip. In: *Proceedings – Design Automation Conference*. 2007; pp. 110–115.

[9] Clermidy F, Miermont S, Vivet P. Dynamic voltage and frequency scaling architecture for units integration within a GALS NoC. In: *Second ACM/IEEE International Symposium on Networks-on-Chip*. 2008; pp. 129–138.

[10] Teehan P, Greenstreet M, Lemieux G. A survey and taxonomy of GALS design styles. *IEEE Design & Test of Computers*. 2007;24(5):418–428. Available from: http://ieeexplore.ieee.org/document/4338461/.

[11] Cunningham P, Swinn M, Wilcox S. Clock concurrent optimization: rethinking timing optimization to target clocks and logic at the same time. 2010; pp. 1–20. Available from: https://www10.edacafe.com/link/Clock-Concurrent-Optimization-Timing-Clocks-Logic-Same-Time/34504/link download/No/ClockConcurrentOptWPV2.pdf.

[12] Martin AJ, Nystrom M. Asynchronous techniques for system-on-chip design. *Proceedings of the IEEE*. 2006;94(6):1089–1120. Available from: http://ieeexplore.ieee.org/document/1652900/.

[13] Oliveira CHM, Moreira MT, Guazzelli RA, *et al.* ASCEnD-FreePDK45: an open source standard cell library for asynchronous design. In: *2016 IEEE International Conference on Electronics, Circuits and Systems (ICECS)*. IEEE; 2016. pp. 652–655. Available from: http://ieeexplore.ieee.org/document/7841286/.

[14] Miorandi G, Celin A, Favalli M, *et al.* A built-in self-testing framework for asynchronous bundled-data NoC switches resilient to delay variations. In: *2016 Tenth IEEE/ACM International Symposium on Networks-on-Chip (NOCS)*. IEEE; 2016. pp. 1–8. Available from: http://ieeexplore.ieee.org/document/7579332/.

[15] Zeidler S, Krstic M. A survey about testing asynchronous circuits. In: *2015 European Conference on Circuit Theory and Design (ECCTD)*. IEEE; 2015. pp. 1–4. Available from: http://ieeexplore.ieee.org/document/7300128/.

[16] Tran AT, Truong DN, Baas BM. A GALS many-core heterogeneous DSP platform with source-synchronous on-chip interconnection network. In: *2009 3rd ACM/IEEE International Symposium on Networks-on-Chip*. IEEE; 2009. pp. 214–223. Available from: http://ieeexplore.ieee.org/document/5071470/.

[17] Jungeblut T, Ax J, Porrmann M, *et al.* A TCMS-based architecture for GALS NoCs. In: *2012 IEEE International Symposium on Circuits and Systems*. IEEE; 2012. pp. 2721–2724. Available from: http://ieeexplore.ieee.org/lpdocs/epic03/wrapper.htm?arnumber=6271870.

[18] Mesgarzadeh B, Svensson C, Alvandpour A. A new mesochronous clocking scheme for synchronization in SoC. In: *2004 IEEE International Symposium on Circuits and Systems (IEEE Cat. No. 04CH37512)*. IEEE; 2004. pp. II–605–8. Available from: http://ieeexplore.ieee.org/document/1329344/.

[19] Mu F, Svensson C. Self-tested self-synchronization circuit for mesochronous clocking. *IEEE Transactions on Circuits and Systems II: Analog and Digital Signal Processing*. 2001;48(2):129–140. Available from: http://ieeexplore.ieee.org/document/917781/.

[20] Saponara S, Vitullo F, Locatelli R, *et al.* LIME: A low-latency and low complexity on-chip mesochronous link with integrated flow control. In: *2008 11th EUROMICRO Conference on Digital System Design Architectures, Methods and Tools*. IEEE; 2008. pp. 32–35. Available from: http://ieeexplore.ieee.org/document/4669216/.

[21] Ludovici D, Strano A, Bertozzi D, *et al.* Comparing tightly and loosely coupled mesochronous synchronizers in a noc switch architecture. In: *Proceedings - 2009 3rd ACM/IEEE International Symposium on Networks-on-Chip, NoCS 2009*. 2009; pp. 244–249.

[22] Thonnart Y, Vivet P, Clermidy F. A fully-asynchronous low-power framework for GALS NoC integration. In: *2010 Design, Automation & Test in Europe Conference & Exhibition (DATE 2010)*. IEEE; 2010. pp. 33–38. Available from: http://ieeexplore.ieee.org/lpdocs/epic03/wrapper.htm?arnumber=5457239.

[23] Onizawa N, Matsumoto A, Funazaki T, *et al.* High-throughput compact delay-insensitive asynchronous NoC router. *IEEE Transactions on Computers*. 2014;63(3):637–649. Available from: http://ieeexplore.ieee.org/document/6495457/.

[24] Kasapaki E. An asynchronous time-division-multiplexed network-on-chip for real-time systems. PhD thesis, Technical University of Denmark (DTU); 2015.

[25] Kasapaki E, Sparso J. The argo NOC: combining TDM and GALS. In: *2015 European Conference on Circuit Theory and Design (ECCTD)*. IEEE; 2015. pp. 1–4. Available from: http://ieeexplore.ieee.org/articleDetails.jsp?arnumber=7300101.

[26] Yaghini PM, Eghbal A, Asghari SA, *et al.* Power comparison of an asynchronous and synchronous network on chip router. In: *2009 14th International CSI Computer Conference*. IEEE; 2009. pp. 242–246. Available from: http://ieeexplore.ieee.org/document/5349422/.

[27] Gebhardt D, You J, Stevens KS. Comparing energy and latency of asynchronous and synchronous NoCs for embedded SoCs. In: *2010 Fourth ACM/IEEE International Symposium on Networks-on-Chip*. IEEE; 2010. pp. 115–122. Available from: http://ieeexplore.ieee.org/document/5507554/.

[28] Sheibanyrad A, Panades IM, Greiner A. Systematic comparison between the asynchronous and the multi-synchronous implementations of a network on chip architecture. In: *Design, Automation and Test in Europe*. Nice, France: IEEE Computer Society; 2007. pp. 1090–1095. Available from: http://dl.acm.org/citation.cfm?id=1266601.

[29] Ax J, Sievers G, Daberkow J, *et al.* CoreVA-MPSoC: a many-core architecture with tightly coupled shared and local data memories. *IEEE Transactions on Parallel and Distributed Systems*. 2017; pp. 1030–1043.

[30] Sievers G, Daberkow J, Ax J, *et al.* Comparison of shared and private L1 data memories for an embedded MPSoC in 28 nm FD-SOI. *International Symposium on Embedded Multicore/Many-core Systems-on-Chip (MCSoC)*. 2015. Available from: https://pub.uni-bielefeld.de/publication/2760622.

[31] Singh M, Nowick SM. MOUSETRAP: high-speed transition-signaling asynchronous pipelines. *IEEE Transactions on Very Large Scale Integration (VLSI) Systems*. 2007;15(6):684–698. Available from: http://ieeexplore.ieee.org/lpdocs/epic03/wrapper.htm?arnumber=4231890.

[32] Gibiluka M. Design and implementation of an asynchronous noc router using a transition-signaling bundled-data protocol. PhD thesis, Pontifícia Universidade Católica Do Rio Grande Do Sul; 2013.

[33] Oliveira CHM, Moreira MT, Guazzelli RA, *et al.* ASCEnD-FreePDK45: an open source standard cell library for asynchronous design. In: *2016 IEEE International Conference on Electronics, Circuits and Systems (ICECS)*. IEEE; 2016. pp. 652–655. Available from: http://ieeexplore.ieee.org/document/7841286/.

[34] Plummer WW. Asynchronous Arbiters. *IEEE Transactions on Computers*. 1972;C-21(1):37–42. Available from: http://ieeexplore.ieee.org/document/1672022/.

[35] Zhang Y, Heck LS, Moreira MT, *et al.* Design and analysis of testable mutual exclusion elements. In: *2015 21st IEEE International Symposium on Asynchronous Circuits and Systems*. IEEE; 2015. pp. 124–131. Available from: http://ieeexplore.ieee.org/document/7152700/.

[36] Ax J, Kucza N, Vohrmann M, *et al.* Comparing synchronous, mesochronous and asynchronous NoCs for GALS based MPSoC. In: *IEEE 11th International Symposium on Embedded Multicore/Many-Core Systems-on-Chip (MCSoC-17)*. 2017; pp. 45–51.

Chapter 9

Asynchronous field-programmable gate arrays (FPGAs)

Rajit Manohar[1]

Field-programmable gate arrays (FPGAs) are chips that can be electronically programmed to function as an arbitrary digital circuit or system. They were originally used to replace discrete gates in interface electronics, and over the past three decades have evolved to being used in the place of application-specific integrated circuits (ASICs) in low volume and cost-constrained situations. Modern commercially available FPGAs are sophisticated integrated circuits capable of implementing digital chips with millions of gates. In addition, some of them also have special-purpose I/O macros to support memory interfaces, as well as serial links to support high-throughput communication.

FPGAs are widely used to prototype digital logic. This chapter discusses some of the challenges with using standard FPGAs to prototype asynchronous logic and summarizes research efforts that have created alternate FPGA architectures for asynchronous logic.

9.1 Why asynchronous FPGAs?

To understand why there is a need for a different type of FPGA to support asynchronous logic, we first summarize the architecture of typical synchronous FPGAs. Most synchronous FPGAs consist of a mix of four building blocks:

1. The programmable logic that consists of a configurable architecture of combinational logic gates and flip-flops. This is the part of the FPGA that is used to map general-purpose logic. We refer to this as the *logic block*.
2. Multiplier-intensive blocks, sometimes referred to as digital signal processing (DSP) blocks. Many application domains that leverage the benefits of FPGAs perform computations that can take advantage of multiplier/multiply-accumulate hardware. These applications are important enough for FPGA vendors to devote dedicated hardware resources to optimize the execution of such computation.

[1]Computer Systems Lab, Department of Electrical Engineering, Yale University, New Haven, CT, USA

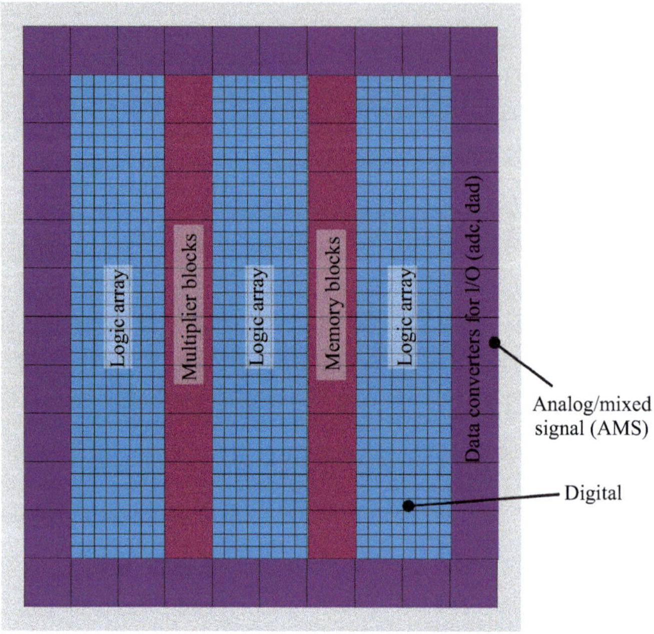

Figure 9.1 Block diagram showing the major components of a typical FPGA architecture

3. Memory blocks, which provide single and/or dual-ported memory blocks. Since memories are ubiquitous in designs and their efficiency often determines overall system performance, FPGAs provide custom blocks that implement memories that can be used for design mapping.
4. Programmable input/output and other hard macros are used so that the FPGA can communicate off-chip with industry standard signaling levels and protocols. This functionality can range from circuits that can conform to various digital signaling standards (e.g., LVDS, SSTL, HSTL, etc.) to circuits that are capable of communicating in excess of 10 Gb/s/pin.

Each block contains configuration bits that determine the detailed functionality used by a design, and they are immersed in a flexible routing architecture that can be used to implement the connections between gates/blocks. The overall block diagram of a typical FPGA architecture is shown in Figure 9.1 [1,2].

9.1.1 Mapping synchronous logic to standard FPGAs

Synchronous digital designs to be mapped to FPGAs are specified in a hardware description language (HDL) such as Verilog or VHDL. The first step in mapping to the FPGA consists of logic synthesis, which converts the HDL into a netlist consisting of gates and their interconnections. The HDL is synthesized into combinational logic and flip-flops, and together they implement the original synchronous circuit description. To map these elements of the design to an FPGA, the FPGA uses two basic building-blocks:

- Look-up tables (LUTs) are used to implement arbitrary combinational logic as a truth table for the logic function.
- Flip-flops are used to implement flip-flops in the synthesized design.

As a concrete example, FPGAs containing four-input LUTs are capable of implementing any Boolean function of up to four inputs using one LUT. Complex combinational logic is remapped into networks of four-input LUTs, and each LUT in the network is mapped to the FPGA directly. The connection between the LUTs is implemented using the programmable routing network. Connections between LUTs and flip-flops can also use the routing network. If the logic synthesis tools recognize structures like multipliers and memories, those are mapped to the efficient built-in hardware resources rather than LUTs and flip-flops. This process of mapping the HDL into hardware resources that are native to the FPGA is referred to as technology mapping.

Once the design is mapped to the logic resources that exist in the FPGA, the design has to be placed and routed onto the FPGA architecture. This process maps each component of the technology mapped netlist to a physical location in the FPGA hardware and allocates the appropriate routing resources necessary to correctly map the connections in the mapped netlist.

Once a design has been mapped to physical locations on the FPGA, timing analysis is performed to determine if the design will operate correctly. In synchronous logic, the two basic constraints that a design must satisfy are the setup and hold time constraints of flip-flops. The setup time is determined by the worst-case delay of the *post place-and-route* logic, and this sets the maximum clock frequency. The hold time is determined by the minimum delay of the combinational logic. Historically hold time has been less of an issue for FPGAs, simply because mapping the design and adding the delays from the routing network typically increases the minimum delay through the mapped design to be large enough to satisfy hold time constraints.

9.1.2 Mapping asynchronous logic to standard FPGAs

Mapping asynchronous logic to commercially available synchronous FPGAs poses a number of challenges. First, many asynchronous logic families use state-holding gates that consist of circuits other than flip-flops. The most commonly used example of this is the C-element, a two-input gate whose output is the *consensus* of its two inputs—if the two inputs agree, the output is the same as the input; if the two inputs disagree, the output maintains its previous value. Such state-holding gates must be mapped to the existing FPGA resources. In cases when flip-flops cannot be used, the state-holding gate must be mapped using a collection of combinational logic gates that have a cyclic topology—a structure that is incompatible with standard static timing analysis tools. An example of a proposed approach to map C-elements to standard FPGAs is shown in Figure 9.2 [3].

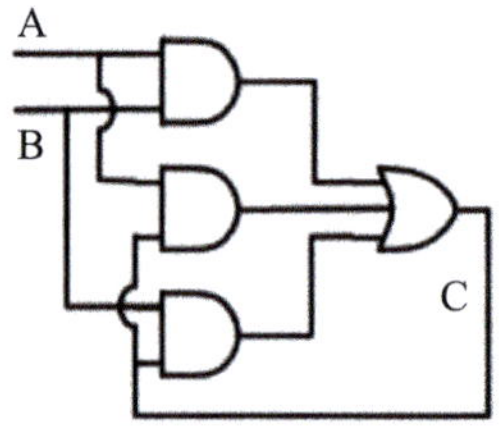

Figure 9.2 Implementation of a C-element with combinational logic and feedback

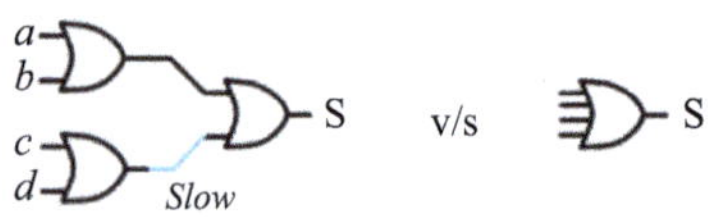

Figure 9.3 Combinational logic decomposition. If the initial state of the circuit on the left is a = 1, b = 0, c = 0, d = 0 (with s = 1), and the input undergoes changes where c = 1 and a = 0, with the path highlighted being very slow, then the circuit on the left will exhibit a hazard on the output, whereas the circuit on the right will not

Second, asynchronous logic typically requires that most control signals be hazard-free. This has implications in mapping combinational logic to an FPGA. Conventional synchronous logic allows combinational logic functions to be implemented using any gate topology that is logically equivalent to the desired function. For instance, since a two-input NAND gate is universal, a valid mapping of general combinational logic would be to only use two-input NAND gates. However, if the combinational logic network is required to be hazard-free, this general remapping strategy is no longer valid. An example is shown in Figure 9.3, which has two different implementations of four-input OR gates. Suppose the path from the lower two-input OR gate is much slower than the upper two-input OR gate. Consider the scenario where initially $a = 1$ and all the other inputs are 0. Both gates have their output high. Now if c is set to 1, after which a is set to 0, then the gate on the right will have its output stay high. However, if the slow path highlighted is very slow, the implementation on the left will exhibit a hazard on the output. Hence, a much more restricted approach and complex analysis that involves state-space exploration is necessary to determine if a particular gate decomposition is valid [4].

A third issue arises because of the different set of timing constraints necessary for the correct operation of asynchronous logic. Different asynchronous logic families have different timing constraints for correct operation. For example, the quasi delay-insensitive (QDI) circuit family has timing constraints on the fan-out of some of its signals. These signals, known as *isochronic forks*, have to satisfy a relative delay timing constraint [5]. Speed-independent circuits assume that the relative wire delay between the fan-out of any signal is small. Circuits with bundled-data logic require that control signals used to indicate that the results are ready must be slower than their corresponding datapath logic—a relative path delay timing assumption. In the FPGA context, these timing constraints have to be satisfied by the design *after* placement-and-routing is complete on the FPGA.

All of these issues are exacerbated by the fact that the circuit details of the underlying FPGA implementation tend to be undocumented, and because the constraints needed to successfully map asynchronous logic to an FPGA are different (and hence, unsupported) from those necessary for mapping synchronous logic. There have been successful efforts to use features of existing design automation tools to map asynchronous logic to standard FPGAs [6,7], although they are tailored to specific FPGA implementations.

9.2 Gate-level asynchronous FPGAs

To directly address the needs of asynchronous logic, several proposals have been made for new FPGA architectures. The primary focus of these proposals is to address the issue of mapping asynchronous logic gates to programmable logic. There were three approaches investigated to achieve this goal: (i) the first, and most direct approach was to enhance the logic block for a synchronous FPGA with additional features that made it suitable for mapping asynchronous logic; (ii) the second approach was to design an FPGA by examining the requirements of asynchronous logic, without considering the needs of synchronous logic; and (iii) the third approach was to examine common asynchronous modules, and develop an FPGA architecture where the logic block could be configured to implement any commonly used module.

9.2.1 Supporting synchronous and asynchronous logic

The first proposed FPGA architecture that added functionality to support asynchronous logic was the Montage architecture [8]. The core logic functionality in the Montage FPGA was similar to the baseline synchronous FPGA architecture used, with logic mapped to a three-input lookup table implemented with pass transistor logic. The goal was to create an FPGA that supported both synchronous and asynchronous logic. To do so, an internal feedback path was provided in the logic block so as to be able to implement state-holding asynchronous logic gates using a feedback structure similar to the one shown in Figure 9.3.

There are scenarios when asynchronous circuits require circuits that have nondeterministic execution. A classic example would be in circuits to implement a network router, where multiple input packets might be routed to a single output port. In such a scenario, asynchronous circuits use an *arbitration* circuit that consists of cross-coupled NAND gates followed by a metastability filter [9]. For this circuit to be well-behaved, the feedback in the cross-coupled NAND gates must be fast compared to the output of the circuit. The Montage architecture has a special logic block that has dedicated support for arbiters as well as synchronizers so as to provide this functionality. Since arbiters are not that common in asynchronous circuits, it was suggested that a 15:1 ratio of nonarbiter blocks to arbiter blocks would be appropriate for a large FPGA [8].

Montage also presented an approach to handling isochronic fork timing constraints. In cases when the only requirement is one branch of a fork is faster than the other, the proposed approach was to route the fork *through* the logic block that implements the gate connected to the faster branch, and from there to the gate for the slower branch. This guarantees that the isochronic fork constraint is met. For other forks, placing the destination gates on a shared routing track was proposed as a way to control the relative delay between the end-points of the fork.

9.2.2 Supporting pure asynchronous logic

The self-timed array of configurable cells (STACC) architecture is an FPGA designed solely for asynchronous logic [10]. The architecture draws its inspiration from a popular asynchronous circuit design style called *micropipelines* [11]. The basic structure of a micropipeline is shown in Figure 9.4. The circuit can be divided into two basic components: the *control* section, consisting of the C-elements and delay lines that synchronize the operation of adjacent pipeline stages and generate a timing signal for the *datapath*, where computation is performed. The STACC architecture therefore contains two different programmable planes—a control plane, that generates timing signals, and a data plane, responsible for the computation.

Part of the timing cell in STACC is shown in Figure 9.5. The cell contains a programmable C-element so that multiple signals from the datapath can be combined to generate a single control signal. The sense of some of the signals can be changed using multiplexors; this functionality enables a signal to serve as a request or an acknowledge signal in a standard handshake protocol. Delay elements are also integrated into the cell, and it should be clear by inspection that the timing cell can be viewed as a generalization of the micropipeline control circuitry.

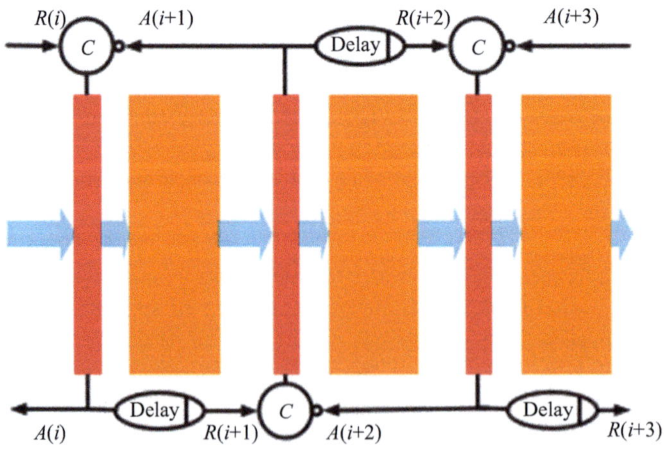

Figure 9.4 Circuit topology of an asynchronous micropipeline

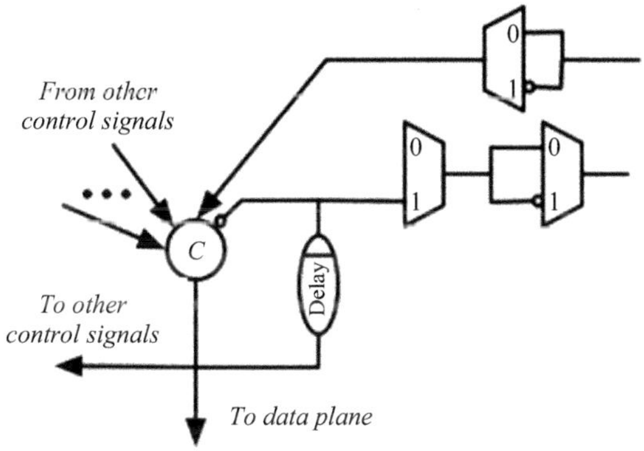

Figure 9.5 Part of the programmable timing cell used by the STACC architecture

The data array in the STACC architecture was envisioned to be similar to a standard synchronous FPGA architecture. The primary difference was that the configurable clock network present in an FPGA is replaced with timing signals from the control plane. This is because bundled data pipelines use datapath components that can be made identical to those used in synchronous logic, and delay lines are used to filter any hazards in the combinational logic.

The first physical realization of an asynchronous FPGA architecture was the plastic cell architecture (PCA) design [12]. In this architecture, the FPGA was partitioned into two components: the *built-in part*, a bidirectional two-dimensional mesh router using dimension-order wormhole routing; and the *plastic part*, an eight by eight array of cells used to implement logic. Each built-in part is connected to a plastic part by two unidirectional bundled data four-bit data channels so that data can be transferred to and from the plastic part.

The cells in a plastic part consist of four four-input LUTs, connected in a fixed topology as shown in Figure 9.6. Each cell has four inputs (from the north, south, east, and west) and four outputs (to the north, south, east, and west), and the logic for those signals is determined by four input LUTs. The eight by eight array of cells also connects to its nearest neighbors, resulting in a "sea of LUTs" architecture for the overall programmable fabric. Each plastic part can also communicate with the built-in part, and packets can be transmitted to other plastic parts through the routing network.

The PCA architecture was fabricated in 0.35 μm CMOS technology, and an array of six by six tiles (9,216 total four-input LUTs) used a die area of 100 mm^2. The performance of the built-in part was found to be above 20 MHz during testing.

9.2.3 Supporting asynchronous templates

The approaches for asynchronous FPGAs outlined so far focused on the development of programmable gates to support the timing/communication protocols, while using standard LUT-based architectures for computation. In PCA, the plastic part

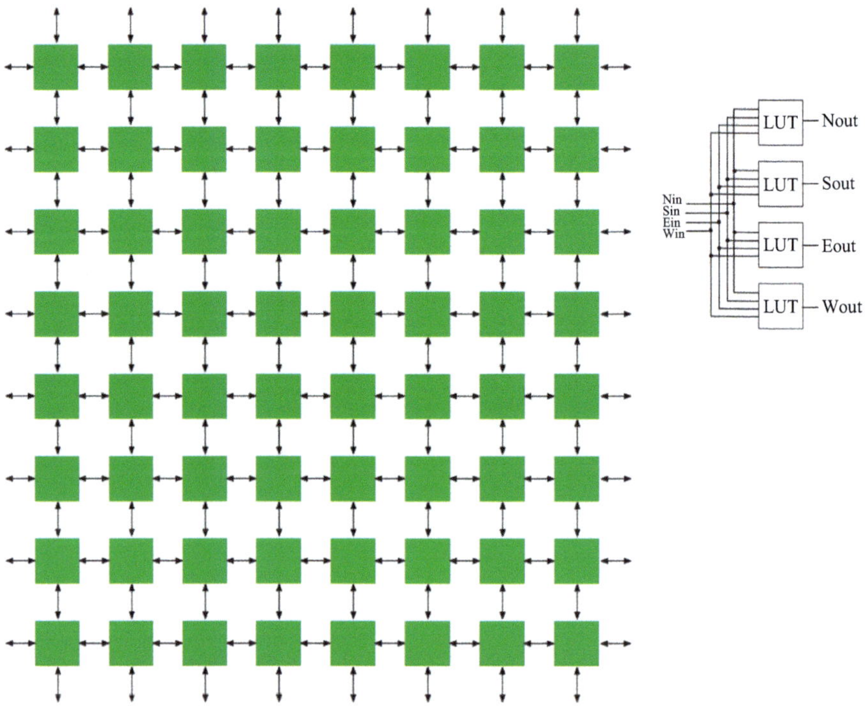

Figure 9.6 The plastic part in the PCA architecture, showing the array of cells connected in a nearest-neighbor configuration

used to map general logic was designed as a sea of four-input LUTs. Instead of this, an alternate strategy is to develop a configurable building block that corresponds to common circuit building blocks for asynchronous logic.

The MiniMIPS asynchronous microprocessor developed at Caltech used the notion of templated modules, where each module was implemented as an asynchronous pipeline stage [13]. These templated modules had the following structure:

- Receive inputs on some or all of a set of input channels.
- Produce an output that is a function of the data received on some or all of a set of output channels.

In addition, local state could be added to the template either as a local variable or via connecting an output channel back to the input, storing the state as a data token in a feedback loop. Since these templates were sufficiently expressive to design a complete microprocessor, a follow-on project developed an FPGA architecture using programmable templates [14]. In what follows we refer to this as the T-FPGA for template-based FPGA.

The basic building block of the T-FPGA is a logic cell, consisting of three one-bit input channels and a single one-bit output channel. Each channel uses a dual-rail encoding, resulting in three wires per channel (two for data and one for acknowledge). This logic cell is analogous to a three-input LUT, except that the logic cell performs handshakes on its input and output channels rather than implementing combinational logic. The logic cell can be configured to skip inputs, or even to skip inputs depending on the value of other inputs. This functionality enables the logic cell to implement building blocks for routing data in asynchronous logic. The logic cells are grouped into clusters of four cells that interface to the global interconnect.

Since the logic cell implements a fixed collection of templates, gates such as C-elements, etc. used for completion detection do not have to be mapped directly. Instead, the logic cell has a custom implementation of a programmable completion detection network that includes custom circuits for C-elements where necessary. Designs are mapped to a collection of templates using a technique known as data-driven decomposition, and these templates are directly supported by the logic cells in the architecture [14].

In terms of routing, the architecture assumes a standard pass-transistor and buffer style programmable wires, with the observation that bundles of wires (the three wires used for one-bit data) can share a configuration bit to reduce the overhead of routing.

SPICE simulations of the logic cell in a 0.18 μm process technology showed a range of throughput depending on the communication pattern configured, with a peak throughput of 235 MHz.

9.3 Dataflow asynchronous FPGAs

A different approach to asynchronous FPGA design is to raise the level of abstraction used to map an asynchronous design to a programmable chip. Instead of mapping digital gates from the user design to the FPGA, the dataflow approach changes the primitive building blocks in the FPGA from gates to static dataflow elements.

Static dataflow is a well-understood computational framework [15]. In this model, the computation is specified as a graph where edges represent information flow, and vertices correspond to computation elements. Data *tokens* flow through the graph and are transformed as they pass through vertices based on the computation implemented at the vertex. Figure 9.7 shows a dataflow graph corresponding to a multiply-accumulate unit with two inputs "*a*" and "*b*." When data tokens arrive on both inputs, the two data values are multiplied together. The feedback edge "*x*" contains the initial value of the accumulator. The output of the multiplier is added to the feedback input, producing the accumulator output on the primary output and a copy back along the feedback loop. The flow of data tokens through the graph performs the computation and new computation results when new input tokens arrive. This static dataflow model is a natural fit to the way pipelined asynchronous circuits operate.

A complete set of dataflow building blocks that is sufficient to implement any deterministic computation is shown in Figure 9.8. The simplest dataflow element is the function computation block, which receives one data token from all of its

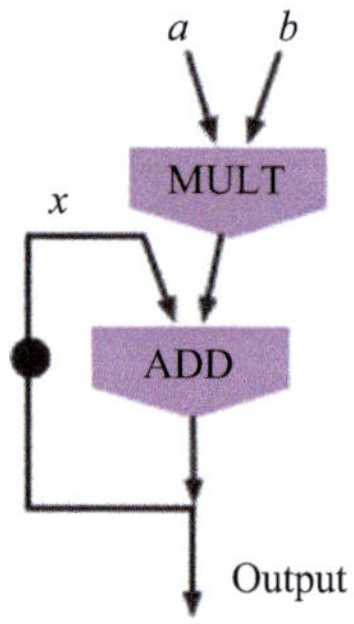

Figure 9.7 A dataflow graph corresponding to a multiply-accumulate unit

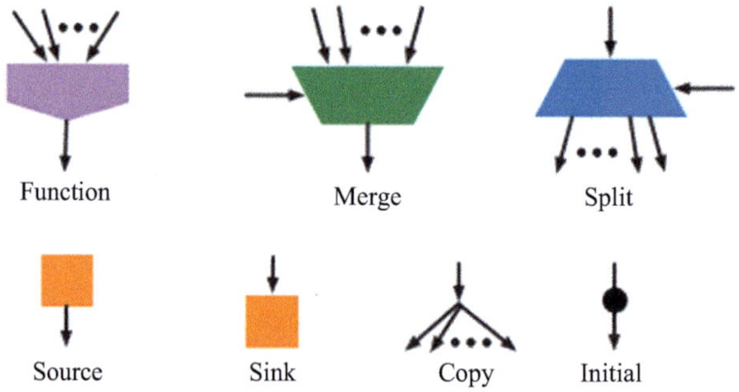

Figure 9.8 A complete set of dataflow primitives that can implement any deterministic computation

inputs, and computes a function of those inputs to produce a single output token. The multiplier from Figure 9.7 is an example of a function block. The copy block simply replicates an input token to all of its output links. The initial block is needed for initial values in feedback loops, as in Figure 9.7. The merge block has a control input token (the horizontal arrow), and the value received on this control input determines which input token is copied to the output. The split block is the dual of the merge, where the control input determines how the other input is routed to one of the output channels. Finally, the source and sink are used to insert constant tokens and discard tokens. A simple example that uses all the building blocks is shown in Figure 9.9, which is an extension of the multiply-accumulate block shown earlier, but with an extra control input "*c*" that can be used to reset the value of the accumulator.

An asynchronous computation specified in a high-level language can be compiled into a collection of dataflow elements using a variety of techniques [16]. Most of the techniques used mimic those used by standard software compilers that use control-flow graphs and data-flow graphs in a variety of forms as their intermediate representation prior to code generation. The biggest departure from

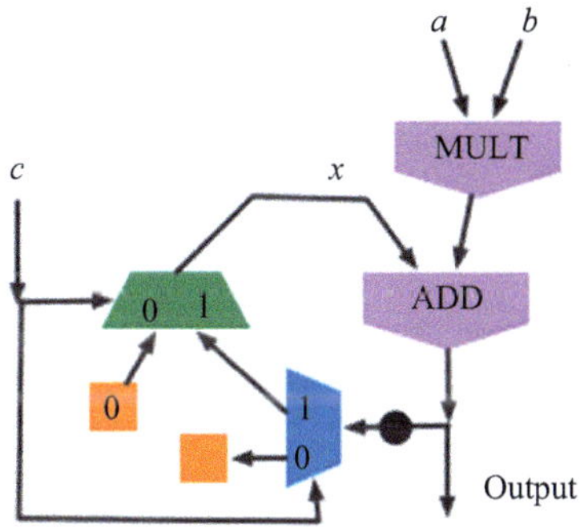

Figure 9.9 Multiply-accumulator with a control input "c" that can reset the accumulator

traditional compiler flows is that dataflow graphs for asynchronous dataflow FPGAs have to be expanded into bit-level operations—a 32-bit addition has to be decomposed into operations on one-bit values. This is similar to transformations that occur during traditional logic synthesis.

An FPGA architecture implemented to support dataflow graphs is a programmable array of dataflow elements. The logic block for an asynchronous dataflow FPGA (AFPGA) must contain configurable elements that can implement each of the building blocks shown in Figure 9.8, so as to be able to implement arbitrary dataflow graphs. Figure 9.10 shows an example of one such logic block from [17]. The logic block contains token sources, token sinks, copy blocks, a function unit, as well as a conditional unit that supports both split and merge functionality.

Instead of routing individual wires, dataflow FPGAs view the routing resources in terms of tracks of *channels*—communication links that carry both data and handshaking information. The channels correspond to the edges in the dataflow graph. This has a significant impact on the performance of the overall FPGA architecture. To understand why, we describe some of the complexities involved in FPGA routing.

Programmable routing is one of the major sources of area overhead in any FPGA architecture. More than 80% of the area of an FPGA is devoted to routing resources [18]. Routing resources can be divided into three major components: (i) the global connectivity, corresponding to the number of routing tracks in the horizontal and vertical direction per row/column of the FPGA and the connections supported at points where the routing tracks intersect; (ii) the local connectivity between components in an individual logic block; and (iii) the supported connectivity from the primary inputs and outputs of the logic block to/from the global routing tracks. FPGAs typically implement connectivity options using multiplexor (MUX) circuits, where the configuration memory is used to control the MUX. The MUX circuits are often implemented using pass-transistor logic and buffers to minimize area. The large number of routing tracks necessary for supporting realistic designs is what makes programmable routing the dominant component of an FPGA architecture.

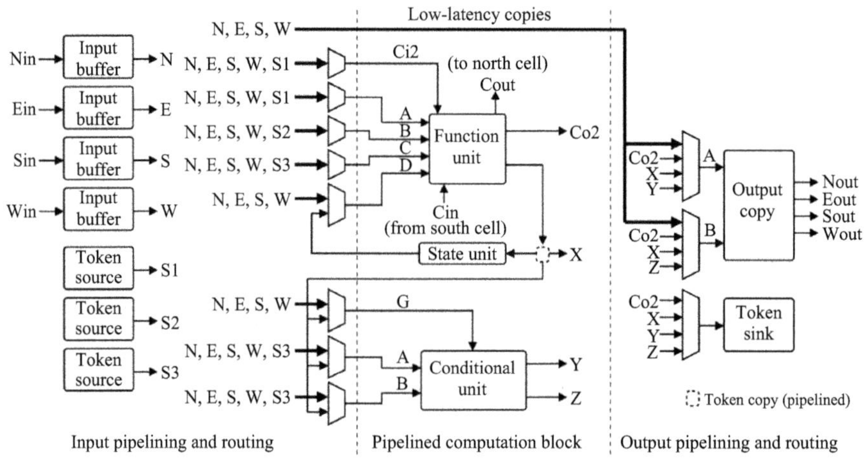

Figure 9.10 Logic block details for an asynchronous dataflow FPGA architecture from [18]

If routing resources are viewed as programmable dataflow *channels* rather than individual wires, then the FPGA architecture has the flexibility to include pipelining in the programmable interconnect. This is because adding buffering to a dataflow channel in a deterministic asynchronous computation does not impact the functionality of the overall design [19]. Unlike traditional synchronous pipelining, where the introduction of a pipeline stage might require the entire circuit to be retimed, an asynchronous pipeline stage can be inserted into a design without impacting the global design. This means that this pipelining is transparent to the place-and-route algorithms as far as correctness of the implementation is concerned [17]. Hence, standard place-and-route tools can be used to map designs to an asynchronous dataflow FPGA—something that had not been the focus of any previous architecture.

SPICE simulations of both the interconnect and logic block in a 0.25 μm process technology result in a peak performance of about 400 MHz for this style of asynchronous FPGA. Measured peak performance in 0.18 μm was in excess of 650 MHz, significantly higher than standard synchronous FPGAs in the same feature size [20]. Mapped (small) designs were shown to achieve a frequency that was close to the peak performance of the FPGA [17]. This FPGA architecture was the technology that launched Achronix Semiconductor Corporation.

9.4 Discussion

A variety of approaches to mapping asynchronous logic to FPGAs have been proposed in the literature. These range from extending a synchronous FPGA design to include building blocks that make it possible to map unique aspects of asynchronous logic, to an architecture that only maps static dataflow graphs implemented with asynchronous circuits.

Asynchronous FPGAs, especially those that support interconnect pipelining can demonstrate significantly higher throughput for designs that are amenable to pipelining. This presents an opportunity for asynchronous FPGAs in application domains such as signal processing and networking, where dataflow-style pipelining is compatible with the typical computation. However, there are a number of challenges with the design and implementation of asynchronous FPGAs that need further study and exploration.

Even though dataflow FPGAs demonstrate high throughput, their performance also comes at a significant area cost [17] compared to synchronous FPGAs that support similar functionality. More detailed research is needed on architectural optimization of asynchronous FPGAs, taking both the interconnect design and logic block design into account simultaneously. Doing this well is a challenge even for synchronous FPGAs; there has been a long history of research in industry and in academia since the 1980s that has resulted in the current, highly optimized, commercial synchronous FPGA architectures. While we can leverage this body of research, many of the studies are not applicable to asynchronous FPGA design due to the significantly different performance and power characteristics of asynchronous logic.

There is a gap between hand-designed and optimized circuits versus designs mapped with standard tools [21]. Hence, it is important that asynchronous FPGA studies use a realistic design flow when analyzing the impact of various architectural features of FPGAs. There are very few quantitative studies that benchmark realistic asynchronous FPGAs with design automation tools, and more such studies are needed so that we can gain a better understanding of the design space of asynchronous FPGAs. While academic FPGA studies in the synchronous domain have similar shortcomings, industry has been driving the evolution of synchronous FPGA architectures with realistic assumptions, thereby advancing the field.

Overall, asynchronous FPGAs can benefit from the average-case behavior of asynchronous circuits. Since optimized routing architectures have highly utilized routing resources, it is more likely that there are a small number of paths that have a significant impact on the overall clock frequency of a synchronous design. Asynchronous FPGAs would only pay this cost when those communication tracks are used, not on every cycle. However, whether this observation can be exploited remains to be seen. The way pipelining can be integrated into an asynchronous FPGA without adding to the complexity of the mapping software is a significant advantage when it comes to mapping pipelined designs. However, this comes at an area cost. What is clear is that asynchronous FPGAs represent a different point in the design space in terms of area, power, and performance compared to their synchronous counterparts.

References

[1] Xilinx, Inc. http://xilinx.com [Accessed April 2019].

[2] Intel (formerly Altera). https://www.intel.com/content/www/us/en/products/programmable/fpga.html [Accessed April 2019].

[3] Ho, Quoc Thai, Jean-Baptiste Rigaud, Laurent Fesquet, Marc Renaudin, and Robin Rolland. "Implementing asynchronous circuits on LUT based FPGAs." In *International Conference on Field Programmable Logic and Applications*, pp. 36–46. Springer, Berlin, Heidelberg, 2002.

[4] Burns, Steven M. "General conditions for the decomposition of state holding elements." In *Advanced Research in Asynchronous Circuits and Systems, 1996. Proceedings, Second International Symposium on*, pp. 48–57. IEEE, 1996.

[5] Rajit Manohar and Yoram Moses. "Analyzing isochronic forks with potential causality." In *IEEE International Symposium on Asynchronous Circuits and Systems (ASYNC)*, May 2015.

[6] E. Brunvand. "Implementing self-timed systems with FPGAs." In *International Workshop on Field-Programmable Logic and Applications*. Oxford, 1991.

[7] Christos P. Sotiriou. "Implementing asynchronous circuits using a conventional EDA tool-flow." In *Proceedings of the 39th Annual Design Automation Conference (DAC'02)*, pp. 415–418, ACM, New York, NY, USA, 2002.

[8] Scott Hauck, Gaetano Borriello, Steven Burns, and Carl Ebeling. "MONTAGE: An FPGA for synchronous and asynchronous circuits." In H. Grünbacher, and R.W. Hartenstein, editors, *Field-Programmable Gate Arrays: Architecture and Tools for Rapid Prototyping*. FPL, 1992, 1993.

[9] C. L. Seitz. "Ideas about Arbiters." LAMBDA, First quarter, 1980.

[10] R. Payne. "Asynchronous FPGA architectures." *IEE Computers and Digital Techniques*, 143(5), 1996.

[11] I. E. Sutherland. "Micropipelines." *Communications of the ACM*, 32(6) 1989, pp. 720–738.

[12] Ryusuke Konish, Hideyuki Ito, Hiroshi Nakada, *et al.* "PCA-1: a fully asynchronous, self-reconfigurable LSI." *Proceedings of the Seventh International Symposium on Asynchronous Circuits and Systems*, pp. 54–61, March 2001.

[13] Alain J. Martin, Andrew Lines, Rajit Manohar, *et al.* "The design of an asynchronous MIPS R3000 microprocessor." In *Proceedings of the 17th Conference on Advanced Research in VLSI (ARVLSI)*, pp. 164–181, September 1997.

[14] Wong, Catherine G., Alain J. Martin, and Peter Thomas. "An architecture for asynchronous FPGAs." In *Proceedings of the IEEE International Conference on Field-Programmable Technology*, 2003.

[15] Jack B. Dennis. "The evolution of 'static' dataflow architecture." In J.-L. Gaudiot and L. Bic, editors, *Advanced Topics in Data-Flow Computing*. Prentice-Hall, 1991.

[16] Song Peng, David Fang, John Teifel, and Rajit Manohar. "Automated synthesis for asynchronous FPGAs." In *13th ACM International Symposium on Field Programmable Gate Arrays (FPGA)*, February 2005.

[17] John Teifel and Rajit Manohar. "An asynchronous dataflow FPGA architecture." *IEEE Transactions on Computers (Special Issue on Field-Programmable Logic)*, November 2004.

[18] Ian Kuon, Russell Tessier, and Jonathan Rose. "FPGA architecture: survey and challenges." *Foundations and Trends^R in Electronic Design Automation*, 2(2), pp. 135–253, 2007.
[19] Rajit Manohar and Alain J. Martin. "Slack elasticity in concurrent computing." *Proceedings of the 4th International Conference on the Mathematics of Program Construction (MPC)*, June 1998.
[20] David Fang, John Teifel, and Rajit Manohar. "A high-performance asynchronous FPGA: test results." In *2005 IEEE Symposium on Field-Programmable Custom Computing Machines (FCCM)*, April 2005.
[21] Brian Von Herzen. "Signal processing at 250 MHz using high-performance FPGA's." In *Proceedings of the 1997 ACM Fifth International Symposium on Field-Programmable Gate Arrays (FPGA'97)*, pp. 62–68. ACM, New York, NY, USA, 1997.

Chapter 10

Asynchronous circuits for extreme temperatures

Nathan W. Kuhns[1]

In our world today, the complexity and variability of circuit applications is ever increasing, which in turn increases the demand for flexibility and capability of the integrated circuits (ICs) themselves. With those applications, comes several new challenges to face, especially when integrating circuitry into various types of extreme temperature environments. For example, the auto industry is pushing to integrate devices with modern-day vehicles to transform them into smart cars. The Internet of things (IoT) requires many new and unique hardware implementations to be developed in order to support the kind of connectivity that is needed. Also, in the aerospace field, circuit technologies can be exposed to wide temperature swings that have the potential to cause critical system failures.

The primary goal for computer hardware in harsh environmental conditions is always to achieve the highest level of performance and functionality while minimizing the cost, or tradeoffs, of doing so. Typically, engineers implement complex climate-controlled cases to house circuitry and ensure critical failures are avoided. These structures are large and bulky and can consume a great deal of power when compared to the devices they house. Another commonly implemented technique is to install long resilient connections between the circuitry and the control logic they communicate with, effectively removing the vulnerable hardware from the harmful environment. This approach, though more energy and cost efficient, causes a severe penalty in performance due to the inherent drawbacks of utilizing long transmission lines.

Whether it be in the cold of space, in a cramped engine compartment, or on the surface of a neighboring planet, it is critical that ICs function properly and perform optimally regardless of the environment they exist in.

10.1 Digital circuitry in extreme environments

There are many factors to take into consideration when designing an electronic system for extreme environments. These factors include but are not limited to the

[1]The Design Knowledge Company, Fairborn OH, USA

behavior of power sources and passive devices, the physical integrity of materials used in packaging and for signal transmission such as solder joints (due to mechanical stress caused by the wide temperature swings), the operating conditions of which the system will be exposed to (i.e., for how long will the circuit need to operate at any given temperature), and power/performance specifications to be met. Though all these factors play a major role in the integrity of IC components, this chapter will focus on the effects of extreme temperatures on ICs themselves and the benefits of utilizing asynchronous logic, as opposed to synchronous logic, under these conditions.

In order to fully describe the inherent benefits of asynchronous paradigms, it should be stated that all semiconductor devices function through the ability to control the movement of charged carriers, referred to as electrons and holes, through the desired regions of the device. The flow of carriers in devices can be controlled by changing the electrical properties of the semiconductor regions through a physical process known as doping. The most common example of this is a p–n junction, which is simply composed of a p-type region (electron acceptor) and an n-type region (electron donor) abutted, for which carriers experience far less resistance in one direction of travel as opposed to the other. When these devices are subjected to a high-temperature environment, the doped regions are inundated with energy particles and create ions. When this happens, the separately doped regions become more alike and are no longer capable of effectively controlling the flow of carriers. This leads to undesired current flow, effectively turning the device into a simple resistor. During this process the threshold voltage of each device fluctuates wildly, meaning their behavior now steps outside the bounds of the timing models that were used to design the circuit they compose. For synchronous systems, this will likely lead to critical failures because they are built with setup and hold time requirements at each register stage. The digital design flow used to create industry-standard ICs utilize specific gate-level timing libraries (with some flexibility due to common process variation, voltage fluctuation, and temperature changes typically between $-40\ ^{\circ}\mathrm{C}$ and $125\ ^{\circ}\mathrm{C}$) and must meet strict timing requirements before they are fabricated.

There are secondary concerns with respect to synchronous logic implemented in advanced processes, as well. As device sizes and operating voltages continue to decrease, there are inherent disadvantages that arise that greatly affect the timing in synchronous circuits. For example, although nominal operating voltages are decreasing as process technology advances, other aspects that affect the on-chip power domains (such as noise, ground-bounce, and cross-talk at high speeds) are not scaling at the same rate. So, the ratio of the effect for these well-known IC aspects becomes greater. Also, the heat generated by the circuits themselves has a greater impact on a larger number of surrounding devices due to the smaller scale and becomes more difficult to dissipate due to there existing less surface area for heat sinks.

Conversely, for extreme cold temperature environments, it is more difficult for individual devices to switch due to a decrease in carrier generation. In these circumstances the energy required for carriers to move (ionization energy) cannot be met, resulting in a condition referred to as "freeze out." Though there is this inherent risk to the ICs as environmental temperatures drop, the performance of silicon (Si) synchronous circuits generally improves due to factors such as a decrease in leakage power, noise, and parasitic capacitance/resistance values. Also, an increase in gain, speed, and heat transfer has been observed.

There are many existing techniques designed to alleviate the effects of extreme temperatures on electronic devices such as heat sinks/sources, thermal shielding, and enhanced MOSFETs (capable of operation near absolute zero due to gate electric field). Also, there was a vast improvement in high-temperature capability with the introduction of silicon-on-insulator (SOI) technologies due to the decrease in potential for voltage to be leaked to the substrate. Despite all this effort, the fact remains that extreme temperatures affect threshold voltage enough to cause timing violations in synchronous systems. This is the reasoning that motivated the use of asynchronous paradigms (which implement handshaking protocols in lieu of a global clock signal) in extreme temperature environments. The inherent nature of "self-timed circuits" gives them immunity to the complications created by fluctuating threshold voltages over wide temperature ranges. As stated in previous chapters, NCL is a quasidelay insensitive, correct-by-construction architecture. This means the circuit will continue to function properly as long as the devices themselves continue to switch properly, regardless of the time it requires for them to do so. Essentially, the performance of the asynchronous circuit will fluctuate in relation to the effects of the environment it is being subjected to, rather than separate stages of the design producing incorrect data due to timing violations like in a synchronous circuit.

For the reader's reference, Table 10.1 has been supplied to visualize and compare points of the wide range of temperatures in everyday applications to the temperatures and that data have been collected for the asynchronous circuits described later in this chapter. The remainder of this chapter reviews the design methodologies and techniques used in recent years to perform research of NCL for both high and low extreme temperature environments. These circuits were fabricated in silicon germanium (SiGe) and silicon carbide (SiC) process technologies and underwent rigorous physical testing, the results of which are presented.

10.2 Asynchronous circuits in high-temperature environments

The work described in this section represents the effort to develop a proven and viable design methodology capable of producing real-world asynchronous ICs in a developing SiC technology process [1]. The final goal was to produce physical testing results of NCL circuits functioning in a wide temperature range for the purpose of demonstrating the paradigm's performance and resiliency in extreme environments.

Table 10.1 Range of reference temperatures for circuit applications and technologies

Fahrenheit	Celsius	Kelvin	Reference point
1,040	560	833	Maximum functional temperature achieved by SiC circuit before breaking down[1]
1,004	540	813	
968	520	793	
932	500	773	Average breakdown temperature for most SiC circuits that were stress tested[1]
896	480	753	
860	460	733	Average temperature on the surface of Venus
⋮	⋮	⋮	
572	300	573	Maximum temperature rating for SiC devices
⋮	⋮	⋮	
338	170	443	Temperature at which Si devices begin to break down
⋮	⋮	⋮	
104	40	313	
68	20	293	Room temperature
32	0	273	Freezing point of water
⋮	⋮	⋮	
−388	−233	40	Absolute zero—theoretical minimum temperature of space environment (lowest recorded SiGe device[2] physical results at 2 K)
−424	−253	20	
−460	−273	0	

[1]References work described in Section 7.2.
[2]References work described in Section 7.3.
Note: The values in red correspond to the reference items listed in the right-most column.

10.2.1 High temperature NCL circuit project overview

Substantial effort was needed to create the tool-flow capable of producing verified SiC NCL circuit designs ready to be submitted for fabrication. The project was performed in two phases, or fabrication runs. It was necessary to develop more accurate device models than was available at the time to allow for high confidence large circuit designs in future work. To accomplish this, the first die fabricated consisted primarily of test circuits, while the second die was composed of larger blocks that more closely aligned with designs intended for real-world applications. The majority of the circuits fabricated on both runs were designed as individual blocks consisting of a logic core with a dedicated I/O ring and include (with justification):

Project phase 1

- **NCL 8 + 4 × 4 multiply accumulate unit (MAC):** A moderately complex combinational NCL circuit (with a large footprint) intended to demonstrate design flow and process capability.
- **4-bit NCL counter:** A moderately sized sequential NCL circuit.
- **Boolean finite state machine (FSM):** A moderately sized synchronous circuit for comparison.
- **4-bit NCL ripple carry adder (RCA):** A small combinational logic design chosen for measuring process capability.

- **4-bit Boolean RCA:** A circuit used to compare performance with its NCL counterpart.
- **11-stage ring oscillator (RO):** A simple switching circuit used to measure process variation and performance across temperature.
- **11-stage ring oscillator with probe pads:** Identical design to previously mentioned oscillator but with the addition of probe pads for physical testing prior to packaging.
- **8-bit Boolean shift register (SR) [NAND gates D-flip flops (DFFs)]:** Three different SR architectures were chosen to implement and compare performance results because it is such an integral part of the NCL paradigm. This variant exhibits basic DFFs that were constructed with NAND gates.
- **8-bit Boolean SR [transmission gate DFFs]:** SR implemented with transmission gate variant DFFs.
- **8-bit Boolean SR [optimized static DFFs]:** SR implemented with static DFF variant.
- **Boolean library:** Basic Boolean gates used in the project that are individually accessible, enabling individual gate performance testing over temperature.
- **Complete NCL library [two circuits]:** The NCL standard cell library also accessible from external I/O on the individual gate level.
- **Transistor sizing test cells [three circuits]:** With such a young process, it was hypothesized that an optimal balance of performance, power, and yield (in relation to device sizing) wasn't established yet. These circuits were implemented to gather data on these metrics over temperature.

Project phase 2

- **Flyback controller:** The flyback controller, fed by the output of an analog to digital converter (ADC), serves as the digital control logic for an intelligent gate driver meant whose specifications were outlined for use in a real-world high-temperature scenario.
- **DAC controller:** A synchronous FSM needed to drive the digital to analog (DAC) phase of the closed feedback loop system this fabrication run produced. This variant of the design included an I/O ring in order to test it individually.
- **DAC controller [placement macro]:** This variant of the DAC controller was the version to be used in the larger system and did not include an I/O ring.

At the inception of the project, no standard asynchronous logic compatible flow was available in the SiC process (developed by Raytheon) and very few SiC circuits were in existence. As a result, the following data and procedures were necessary to develop: the NCL standard cell library (balanced schematics and layouts), a Cadence Encounter setup to enable routing with only a single metal layer, and a parasitic extraction (PEX) flow compatible with the process design kit (PDK). The voltage shifting printed circuit board (PCB) and the apparatus used to perform the high temperature physical testing at an incremental pace were also designed as a part of this work.

The existence of only one metal layer for signal routing provided a unique challenge. First, it meant that the poly layer was necessary to route under the metal layer often, which presented complications when following the typical physical implementation procedures for digital circuitry. Inherently, poly is a higher resistance material which translates to a decrease in performance, so its use as a routing layer was avoided as much as possible. Also, there was a limiting factor on how long any single poly wire could exist due to a design rule check (DRC) violation. The use of one metal and one poly layer for routing requires the use of an older floor planning technique, referred to as "channel routing." When implementing channel routing, every other standard cell row in the circuit core is left empty to allow for the space required by the internal signals to reach their destination. If not for this, the metal and poly that make up the standard cell structures would essentially create a routing blockage. It is also necessary to leave space between the standard cells, so signals traveling vertically through rows can reach their destination. All these stipulations resulted in top level layouts that were inefficient in terms of area utilization. So, in the second fabrication run, the standard cell layouts were optimized for the channel routing technique. This was done by moving the power and ground rails to the center of the cells, effectively separating the pull-up region from the pull-down region. Also, the cell I/O pins were modified from single-point side placements to dual-point top and bottom placements. This allowed signals to be routed from either the top or bottom of the cell and eliminated the need for the space required to route to the sides of each cell. Effectively, this approach reduced the horizontal space required between the standard cells by moving the pins away from the sides, and reduced the vertical space required in the routing channels by giving the automatics placement and routing tool the flexibility to make connections on both the top and bottom of the cells. The area of the resulting top-level core layouts was reduced by 30%–40% when compared with their counterparts [2].

10.2.2 High temperature NCL circuit results

In order to demonstrate NCL's innate flexibility and its viability in high-temperature environments when compared to its synchronous counterparts, a series of physical tests were performed across temperature with the circuits previously mentioned. While performing physical testing of asynchronous circuitry, it is essential to read the output of the device under test (DUT) in order to generate the input to the DUT at the appropriate time. This is due to the inherent nature of the handshaking protocol, and the simplest method for achieving this process is to utilize a FPGA. The physical tests for this work were conducted by generating and receiving the I/O signals from a Xilinx Virtex-7 FPGA that were passed through a voltage shifter PCB. The PCB shifted the voltage between the FPGA's 1.8 volt value to the necessary 12–15 volts needed to operate the SiC circuits. For each individual circuit a VHDL testbench was written that would render the DUT's max performance and simultaneously verify correct functionality across a wide range of input vectors. This was true for both asynchronous and synchronous designs. The

Figure 10.1 Cross temperature physical testing setup

test setup leveraged for these high-temperature physical tests is pictured in Figure 10.1.

The testing procedure for each circuit began with power up and initialization at room temperature. Once correct logical functionality was exhibited by the DUT, the hot plate surface temperature was increased by 20 °C and the user waited until the temperature sensor (placed at the point of contact between the circuit package and the custom heat sink) output the desired value. At this point the test stimuli was presented again and the results were recorded. This process was repeated until the DUT no longer exhibited correct logical behavior. Figure 10.2 displays the average ring oscillator frequency across temperature. The results show a gradual increase in performance as the temperature rises, and then a gradual decrease beyond 200 °C. This trend was seen across most of the circuits on which high temperature physical tests were performed. For added reference, Figure 10.3 displays the average transmission gate operating frequency across temperature and Figure 10.4 displays the NCL counter's average operating frequency versus propagation delay.

10.3 Low temperature NCL circuit project overview

The work described in this section was performed to demonstrate and examine NCL circuit's performance, implemented with the IBM SiGe5AM 0.5 μm process, in extreme low temperatures [3]. The design leveraged for this project was the well-known 8051 microcontroller, made popular by Intel. The 8051 was chosen because

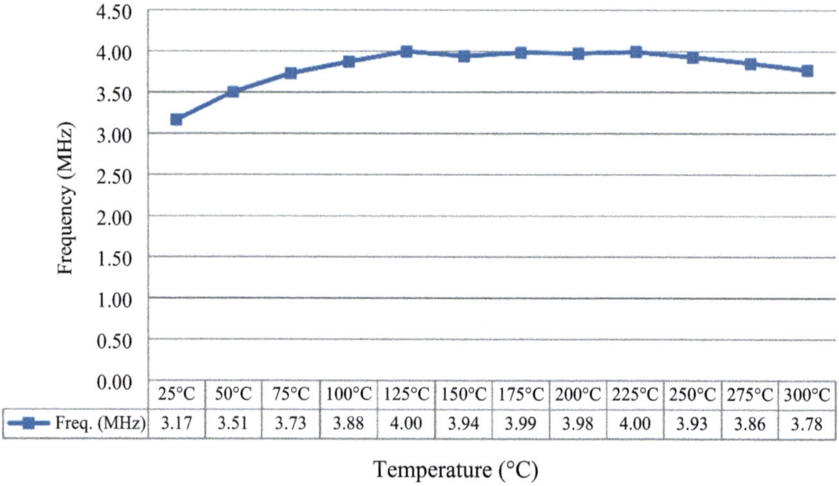

Figure 10.2 Average ring oscillator operating frequency

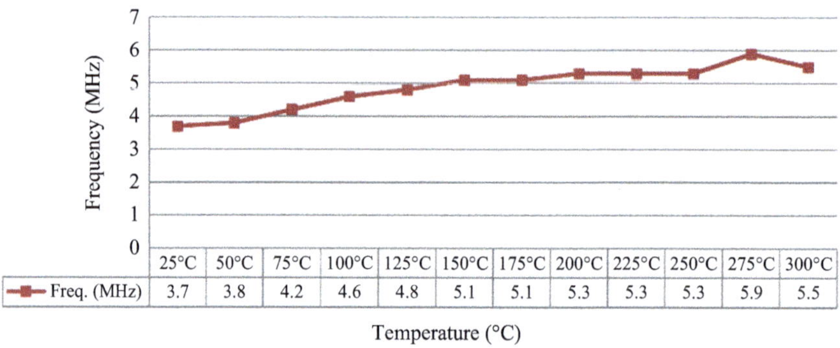

Figure 10.3 Average transmission gate shift register operating frequency

it is one of the most widely implemented microcontrollers in history, and its complexity met all aspirations for the project. A secondary objective for the project was to create an attractive alternative to a synchronous circuit in common industrial applications. In order to accomplish this, the asynchronous 8051 core was combined with a synchronous logic "wrapper" that would make the final design capable of interfacing with a fully synchronous control system. The result would be a user-friendly design capable of directly replacing a synchronous 8051 circuit but would also possess the operational flexibility of an asynchronous circuit across temperature. Also, for further related reading, this asynchronous 8051 design was built upon by implementing a similar circuit (in the previously mentioned SiC process) in which the large dual-rail data transmission bus was replaced with a mux-based

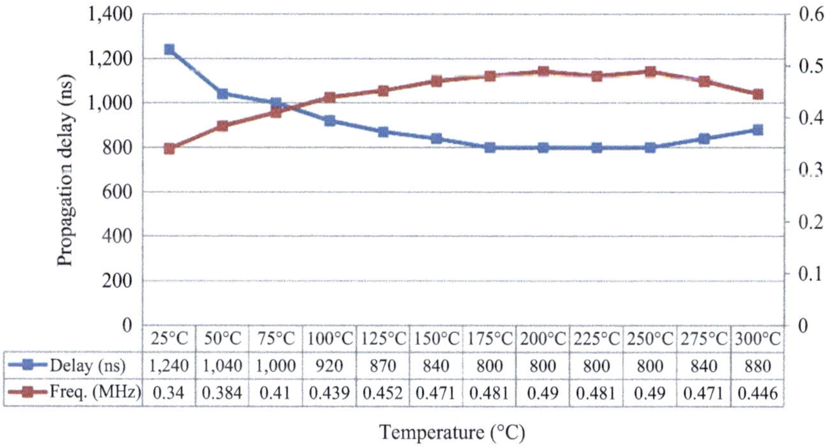

	25°C	50°C	75°C	100°C	125°C	150°C	175°C	200°C	225°C	250°C	275°C	300°C
Delay (ns)	1,240	1,040	1,000	920	870	840	800	800	800	800	840	880
Freq. (MHz)	0.34	0.384	0.41	0.439	0.452	0.471	0.481	0.49	0.481	0.49	0.471	0.446

Figure 10.4 NCL counter operating frequency versus propagation delay

system for the purpose of greatly reducing the switching and leakage power consumed [4].

10.3.1 Low temperature NCL circuit project overview

The 8051 is primarily composed of four 8-bit I/O ports, a configurable combination of on-chip/off-chip RAM (which houses the register banks used in the micro-controller's operations), control circuitry, special function registers, and an ALU. Each of these components were individually evaluated for the process of transforming to asynchronous functionality, and then a holistic approach was taken for the system in order to improve overall performance. In order to satisfy the asynchronous functionality on the system level, the following modifications were made to the basic logical structure for the individual components:

- **I/O ports:** In order to satisfy the constraint for an 8051 synchronous compliant design, the I/O ports were modified to allow for single to dual-rail conversion and the generation of the appropriate handshaking signals.
- **Program counter, register file, accumulator:** These individual blocks were reconstructed using the basic NCL 3-ring register architecture in order to store their respective data values through a data-NULL cycle. This architectural decision was made to avoid the more straightforward approach of implementing a block-to-block bus system that would have a large impact on area and performance of the design.
- **ALU:** The ALU block was redesigned from the ground up as a wholly NCL combinational logic block utilizing large multirail logic encodings in order to optimize performance and area.

This work was an iterative project with many consecutive goals in mind, and as a result there were variations of the design that were fabricated. The first iteration

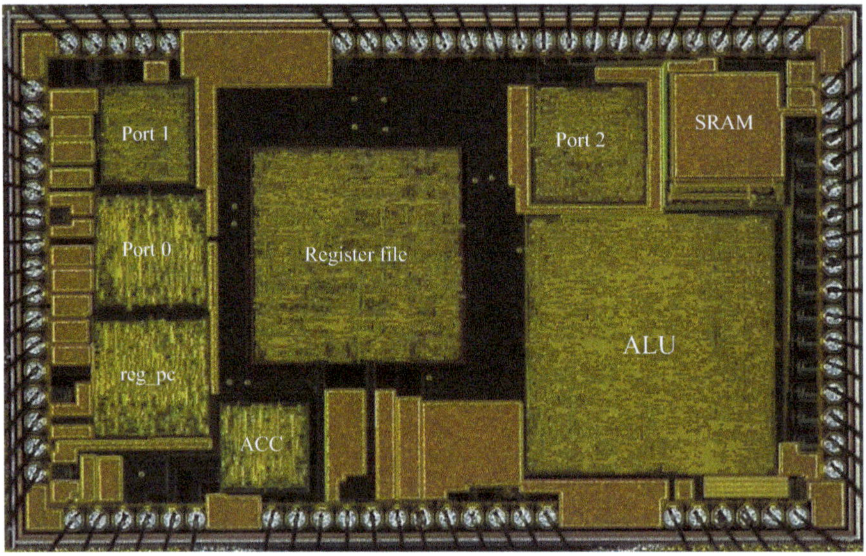

Figure 10.5 8051 microcontroller component die micrograph

consisted of the components labeled on the die micrograph in Figure 10.5. These individual components, most notably the ALU, were chosen for this fabrication run due to their unique functionality in the overall design. Successful physical tests for these components would, first, increase confidence in the complete 8051 design integrity as well as the physical testing setup, and second, produce performance results of the combinational logic components in the critical path of the design.

Once high confidence in the individual components of the 8051 had been established, the next iteration of the project was performed on a fully functional NCL 8051. As seen in the die micrograph pictured in Figure 10.6, the microcontroller was also laid out in a more uniform, industry-standard fashion.

10.3.2 Low temperature NCL circuit results

Initially, verifying correct logical functionality was prioritized above performing physical tests in low temperatures. In order to accomplish this task, the input vectors used to simulate the schematic design were converted into a format usable by a pattern generator. These inputs were sent to the DUT, the output was recorded by a logic analyzer, and the results were compared with the expected outcome from the schematic simulations by a custom script. Once the functionality of each component was verified logically, tests were performed to determine their minimum operating voltage and maximum operating frequency. The procedure to gather this data consisted of setting a constant input frequency and iteratively reducing the supply voltage while verifying the value of a single output pin at each interval. Then, the output of the DUT was saved to a text file in order to be verified after the voltage scaling process had been concluded. This labor-intensive procedure is the

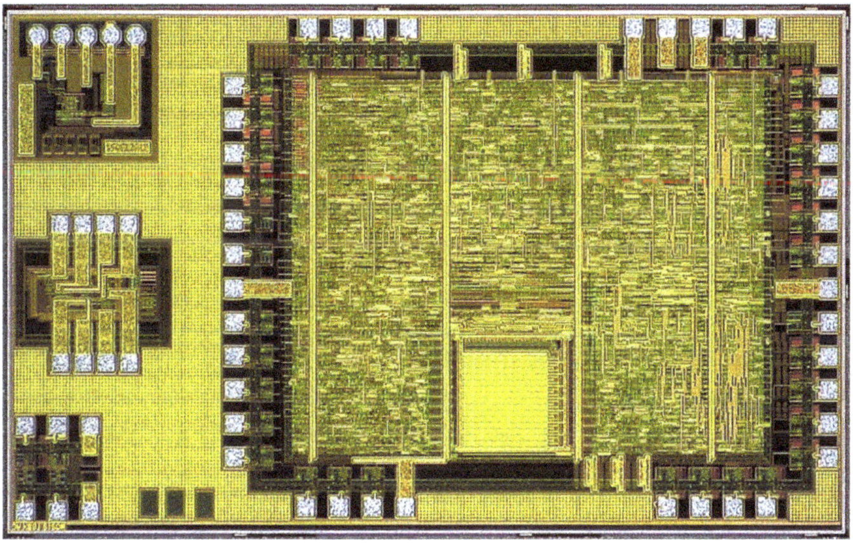

Figure 10.6 NCL 8051 microcontroller die micrograph

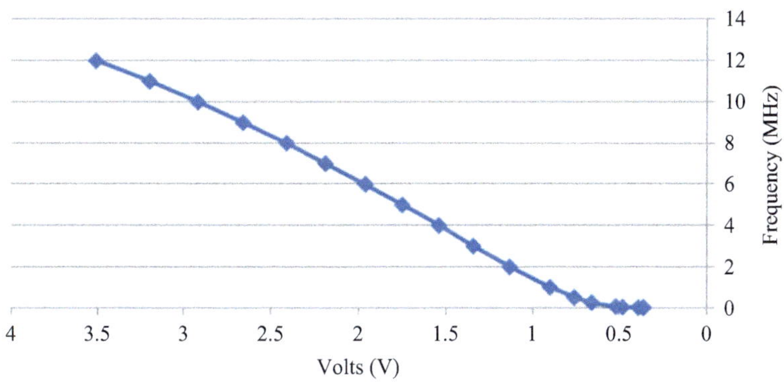

Figure 10.7 ALU minimum supply voltage versus maximum operating frequency

justification used for implementing the FPGA setup seen in the high-temperature project. Figure 10.7 displays the ALU results in the minimum supply voltage versus maximum operating frequency physical tests. As expected, there is a direct linear relationship between voltage level and performance.

Once all performance versus supply voltage tests were concluded, it was clear that the ALU would be the component in the design with the longest propagation delay. Taking this into account, the ALU performance can be directly correlated to the microcontroller's theoretical worst-case performance for a machine cycle. This concept was why it was chosen for the in-depth cryogenic physical testing process. The cryogenic test setup (seen in Figure 10.8) was primarily composed of a pattern generator, oscilloscope, and logic analyzer, as previously mentioned. Also, custom PCBs were leveraged for the voltage scaling control and interfacing from the DUT to the cryogenic enclosure. Lastly, a custom Janus cryostat test structure was used. This cryogenic tool uses long ribbon cables (approximately 4 feet), which may affect the signal to noise ratio during physical testing.

When performing tests, liquid helium is sent into the cryostat via a nozzle inside the test chamber. The helium immediately contacts the first of two temperature sensors, which is primarily used to control the temperature inside the chamber. The second sensor is located on the daughter board that the DUT is connected to, and the reading from this sensor is used for the data calculations. Cryogenic cross-temperature tests (2 K–297 K) were completed to verify correct functionality for all blocks in the design, and the ALU alone was subjected to the detailed supply voltage scaling and power consumption tests across the full temperature range. The results for the ALU minimum supply voltage across temperature, with respect to two different operating speeds (2.5 MHz and 10 MHz), are displayed in Figure 10.9. It is clear that the supply voltage required for correct logical behavior is reduced as temperature is decreased and operating at higher frequencies has a substantial impact on the required supply voltage.

As a more comprehensive experiment, the complete NCL 8051 microcontroller was redesigned and tested under a wide temperature swing between −180 °C and +125 °C using the test setup shown in Figure 10.10. The thermal chamber serves both as an electrically heated oven and a liquid nitrogen-cooled

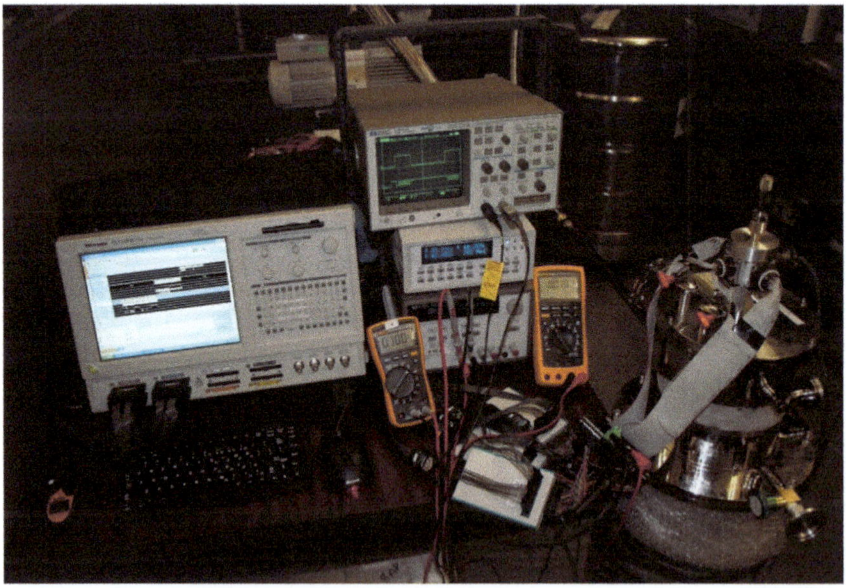

Figure 10.8 NCL ALU cryogenic test setup

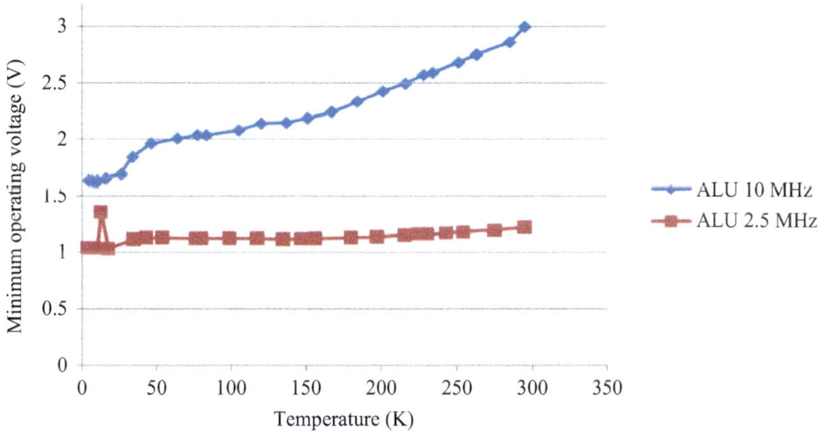

Figure 10.9 ALU minimum supply voltage across temperature

Figure 10.10 NCL 8051 temperature-swing test setup

cryogenic environment. All 255 instructions were fed into the microcontroller and were executed as the temperature was modified. The microcontroller successfully executed all instructions throughout the temperature swing, without any external adjustment or control. This is due to the delay insensitivity of NCL,

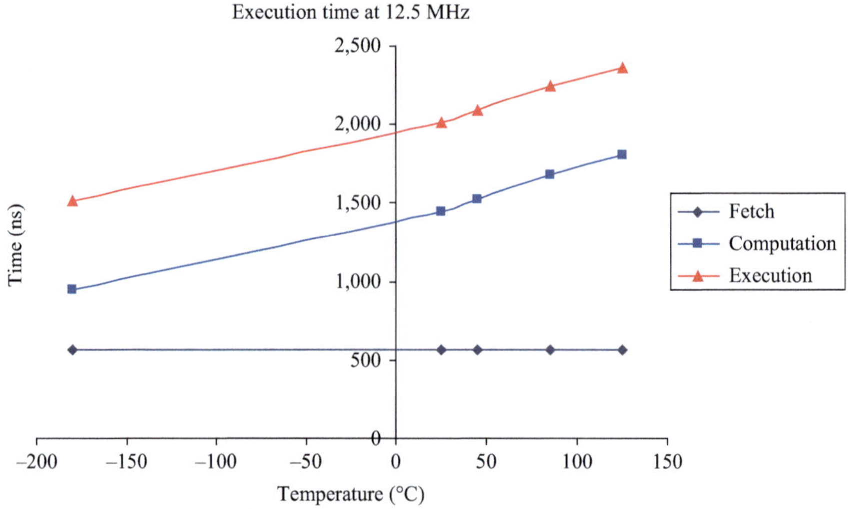

Figure 10.11   NCL 8051 temperature-swing test result

which guarantees proper circuit operation regardless of delay changes in the logic gates. Also, in this test setup, an Altera FPGA board was used for I/O control which allows for runtime asynchronous timing as discussed earlier. The execution time depicted in Figure 10.11 clearly shows that the microcontroller performs more complex operations at higher speeds while being subjected to colder temperatures [5].

10.4 Conclusion

This chapter presented successful physical testing results of multiple NCL circuit designs of varying size and complexity across a very large temperature range. For high-temperature applications, a SiC process developed by Raytheon was leveraged and exhibited circuits functioning at temperatures exceeding 500 °C. For low-temperature applications, the industry standard IBM 0.5 μm SiGe process was leveraged and exhibited circuits functioning as temperatures approached absolute zero. Through all these tests, the NCL circuits required no special considerations (due to environmental effects on the device level) to maintain correct operation across these wide temperature swings. In the same conditions, synchronous systems would require significant effort (either through complex logical design changes or physical setup considerations) in order to meet their timing constraints which always leads to a large amount of overhead incurred. These results have proven the flexibility and robustness advantage that asynchronous systems have over synchronous designs.

References

[1] Caley, L. "High temperature CMOS silicon carbide asynchronous circuit design." Ph.D. Dissertation, University of Arkansas, 2015

[2] Kuhns, N., Caley, L. Rahman, A., *et al.* "Complex high-temperature CMOS silicon carbide digital circuit designs." *IEEE Transactions on Device and Materials Reliability*, vol. 16, no. 2, pp. 105–111, 2016

[3] Hollosi, B. "8051-compliant asynchronous microcontroller core design, fabrication, and testing for extreme environment." Masters Theses, University of Arkansas, 2008

[4] Kuhns, N. "Power efficient high temperature asynchronous microcontroller design." Ph.D. Dissertation, University of Arkansas, 2017

[5] Hollosi, B., Di, J., Smith, S. C., Mantooth, H. A. "Delay-insensitive asynchronous circuits for operating under extreme temperatures." *2011 Government Microcircuit Applications & Critical Technology Conference (GOMACTech)*

Chapter 11

Asynchronous circuits for radiation hardness

John Brady[1]

The technological drive for smaller-node processes results in increased susceptibility to single-event effects (SEE) in integrated circuits, notably single-event upset (SEU), and single-event latch-up (SEL). An SEU may occur within an integrated circuit (IC) when an ionizing-particle strikes a node. The resulting change in charge within the node is referred to as a single-event transient (SET). If change effected by the SET causes an incorrect value to be latched within the circuit, an SEU has occurred. This incorrect data point propagates throughout the rest of the circuit, potentially corrupting other data, without an indication that the circuit is malfunctioning. SEL occurs when a short circuit is created between an IC's power and substrate. Even beyond potential corruption of the data within the IC, SEL causes unusually high current usage and may result in permanent damage to the IC if the power is not cycled.

Asynchronous circuits are inherently suitable for radiation-exposed environments due to their quasidelay insensitivity (QDI) and multirail logic systems. If an ionizing-radiation event is detected, the QDI property provides the ability to delay the current operation within the circuit until the effect has subsided. The dual-rail design provides additional support in this area because in many cases both rails must be affected in order for an SEU to occur. In addition to mitigating SEEs through asynchronous circuit-level architectures, radiation hardening techniques can be applied to transistor-level layout designs and circuit components, such as the DFF, for increased reliability.

11.1 Asynchronous architectures for mitigating SEE

While synchronous designs typically leverage triple-modular redundancy to detect errors or interruptions induced by SEEs, the dual-rail logic system of NCL, and a few modifications to the standard NCL pipeline architecture, enables dual-modular redundancy (DMR) to be an effective solution for mitigating SEUs. Figure 11.1

[1]Department of Computer Science and Computer Engineering, University of Arkansas, Fayetteville, AR, USA

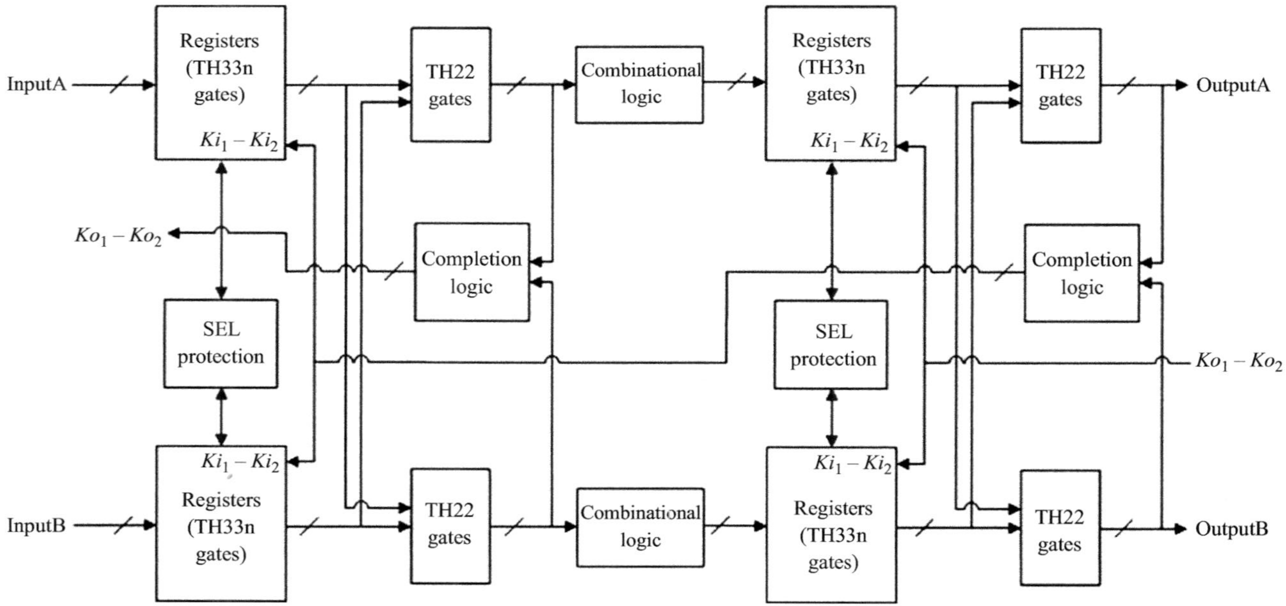

Figure 11.1 Single-bit SEU mitigation and SEL protection NCL architecture

illustrates the following required changes to the NCL pipeline in order to create an NCL architecture capable of mitigating a single-bit SEU or SEL [1]:

1. Add a duplication of the circuit.
2. Modify the registration logic to use TH33n gates instead of TH22n gates.
3. Add TH22 gates between every register output and the following stage's combinational input.
4. Place the completion logic at the output of the TH22 gates.
5. Insert an SEL protection component in between each stage and the corresponding V_{dd}.

Duplication of the original circuit allows the outputs from each stage to be compared by using two sets of TH22 gates. This ensures that as long as the data waves output from each register set do not match (which signals that an SET is occurring in either the first or second circuit), the corrupted data wave is unable to progress.

Once the SET has subsided, regardless of its location within the circuit, the output for each stage matches therefore allowing the new data wave to progress through both sets of TH22 gates. The completion logic then detects a complete data wave and sends a request for a new wave from the previous stage. Two sets of TH22 gates are required in order to generate the additional *Ko*.

The TH33n gates within the registration blocks are required in order to accommodate the second *Ko* per stage within the pipeline. This prevents a corrupt *Ko* from incorrectly requesting a new data wave or interrupting the operation of the current data wave.

In the case of this particular DMR-based NCL architecture, each stage only requires that the output data matches, meaning that the circuit does not need to be able to differentiate between incorrect and correct data during an SEE. This aspect of the architecture leverages the QDI feature of NCL and prevents the requirement of using triple-modular redundancy (TMR). The difference in requirements is due to the fact that a traditional synchronous design requires TMR in order to generate three votes to create a majority-voting system. When two out of the three total votes match, the system assumes that the third, nonmatching vote contains corrupted data due to an SEE and therefore propagates forward the value of the first two votes.

The SEL protection component operates by monitoring the amount of current usage per stage. During an SEL, the current demand begins to rise above predetermined normal operating standards and thereby prompts the SEL component to disable power to that particular stage. Once the power is re-enabled, the circuit will not deadlock but continues to operate normally, although any data previously within the effected stage will not be recoverable.

Using the 130 nm CMRF8SF 1.2 V process, this architecture is simulated at the transistor level through a 4 $\times$ 4-bit NCL multiplier [2]. The simulation results (Table 11.1) show a decrease of 1.31$\times$ in speed, an increase of 2.74$\times$ in area, and an increase of 2.79$\times$ in energy per operation.

Table 11.1 Architecture comparison data

Circuit	Transistors	T_{DD} (ns)	Energy per operation (pJ)
Original	1,695	7.2	1.05
Single-bit SEU and SEL resistant	4,646	9.4	2.93

11.1.1 NCL multibit SEU and data-retaining SEL architecture

The previous architecture can be improved to mitigate up to two simultaneous SEUs and prevent data loss during SEL recovery [3]. The following modifications are required to provide said functionality:

1. Replace TH33n gates within the registration logic with TH44n gates.
2. Generate a third *Ko* per stage in the pipeline.
3. Modify completion logic to add SEL data loss protection.

As seen in Figure 11.2, the TH33n gates in the registers are replaced with TH44n to accommodate the additional *Ko*. Additionally, the completion logic is modified to include the generation of a third *Ko* signal. Generating the third *Ko*, as opposed to using a copy of the first or second *Ko*, is necessary because it prevents one SEU from affecting multiple *Ko* signals. To accomplish this, the third *Ko* is created via both the first and second *Ko* signals. The third *Ko* results from a completion block that includes both inputs from a given stage's TH22 blocks.

The addition of the third *Ko* allows two of the *Ko* signals to be corrupted without likewise populating corrupted data. Because there are now three *Ko* signals, two of the signals may be corrupted without corrupting the overall request from that stage. This varies from a synchronous TMR voting system because the three *Ko* signals, although unique, do not represent votes. Because of this difference, the NCL architecture can sustain two incorrect values, whereas the traditional synchronous TMR system must have only one incorrect vote. The role these three *Ko* signals play is matching each other; meaning, the circuit will only progress once all SETs subside. Otherwise, the circuit concludes that the current value is incorrect and will not change the current request-for-NULL or request-for-DATA.

Once an SEL is detected by the SEL protection component, an individual stage is power cycled, effectively disconnecting the stage from V_{dd}. If the affected stage contains a NULL wave, typically data is not lost and the stage is able to be power cycled and reset to NULL without causing the rest of the circuit to malfunction. However, when an SEL affects a stage containing DATA, this DATA is lost and will create a gap within the data output of the circuit. In the case of a design containing feedback, the effect is worsened, considering that future data will also be affected by the initial lost data. For the architecture in Figure 11.2, data retention during an SEL is provided via temporal redundancy within the pipeline. The

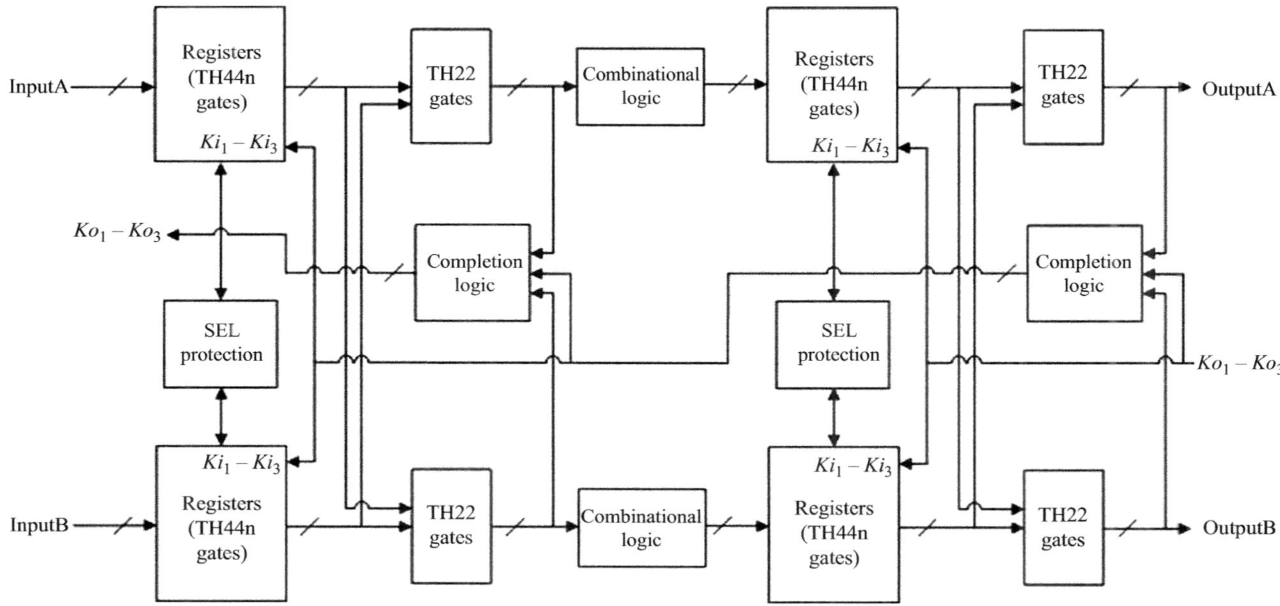

Figure 11.2 Multibit SEU NCL architecture

completion logic block for each stage within the pipeline contains a dependency on the two succeeding stages within the pipeline, instead of only the first succeeding stage. This dependency ensures that for any given DATA wave within the pipeline, an adjacent stage also contains a copy of the DATA wave. For this reason, any stage within the pipeline can suffer an SEL, recover from the loss of power, and reset to NULL without losing or corrupting the affected DATA wave.

11.2 Radiation hardened asynchronous NCL library and component design

In addition to circuit-level asynchronous architectures for improving radiation hardness, modifications to transistor-level gate libraries and circuit-level components can also provide improved SEE mitigation. One example of this is an asynchronous NCL library designed in the 90 nm IBM 9HP process that is devised to mitigate SEEs [4]. This library is built with standard threshold voltage transistors, and the gate layouts for the library are manipulated with multiple techniques to analyze their ability to improve the radiation hardness of the gate (see Section 11.3).

A layout of the TH22 gate is shown in Figure 11.3. The general layout of the gate (similar structure for all gates within the library) allows each transistor to be easily sectionalized which naturally allows the incorporation of guard rings. The PFETs are located in the top half of the layout while the NFETs are placed in the bottom half. The height for each half of the gate (lower and upper half) is based upon the maximum transistor width required to meet the gate's output timing requirements and drive strength. The general template for these layouts is designed

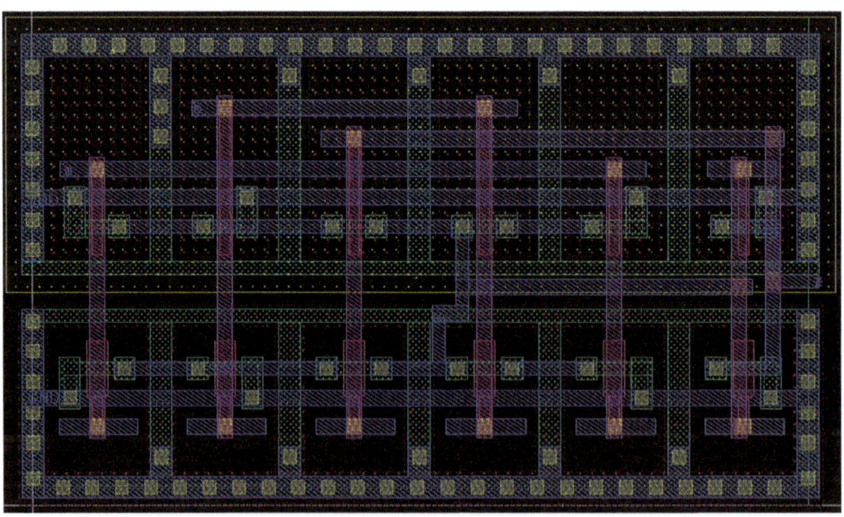

Figure 11.3 TH22 gate layout

to be modular such that adding transistors for higher transistor-count gates is accomplished by extending the layout horizontally.

Guard rings are included around each transistor within the layout. Guard rings isolate an SEE to an individual device, but they also increase the distance between each device thereby decreasing the probability of a multiple-event upset (MEU). Contacts are placed in all areas of the guard rings except where internal routing is inhibited. Due to the placement of guard rings, this prevents the continuous connection of the gate between a corresponding PFET and NFET. Instead, the second metal layer (the first metal layer is primarily utilized in intra-gate routing) is used to connect the corresponding transistors. A similar structure is applied to connect adjacent NFETs (and PFETs) with the first metal layer. Note that the increased NFET channel length is included for low-temperature reliability and not radiation hardness. This change has an effect on the radiation hardness of the gate (further discussed in Section 11.3).

In addition to modifying an NCL gate library to provide increased radiation hardness, circuit components such as the DFF can be modified to increase radiation hardness [4]. For a synchronous circuit, the clock signal controls the storage of new data within the DFF. Due to the absence of a clock within asynchronous circuits, the NULL and DATA states of dual-rail data signal must be used to control storage of new data. This is accomplished by using the output of the first latch within the DFF in combination with the incoming data signal. The output of the first latch, *Qint*, is required to ensure that the data has reached the input of the DFF's second latch before the asynchronous DFF is "clocked."

The asynchronous DFF can be radiation-hardened by replacing both latches with dual interlocked storage cell (DICE) latches [5]. A DICE latch mitigates SEUs by adding redundant nodes within the latch, meaning that there are two nodes within the latch that store the current value and two nodes that store the inverted current value. Spatial redundancy is leveraged, along with limited controllability from node to node, such that while an SET is able to change the value of one node, the latch is heavily resistant to corrupting the values of all four nodes. After an SET subsides, the three unaffected nodes are able to drive the affected node to its original valid state. Because of the DICE latch's strong resistance to state change, a strong driver is required to update the value within the latch.

The DICE-based DFF is shown in Figure 11.4. In addition to including the DICE latches, the required additional buffering is shown between the output of the first DICE latch (*Qint*) and the input of the second DICE latch. This is required because the first DICE latch lacks the drive strength required to change the state of the second DICE latch.

The logic for converting a synchronous DFF to an asynchronous DFF is shown in Figure 11.5. Signal *Input* represents the incoming dual-rail data signal input into the DFF. *Qint* is the output of the first latch, as seen in Figure 11.4. The inversion of $Input^0$ is required and inputting $Input^1$ instead will cause the logic to function incorrectly. In the case where *Input* has a value of NULL, the value of *Qint* becomes 0. The result is for both AND gates within the logic, the output is 0 which results in a *Clock* value of 0. When the value of *Input* changes to DATA0, the

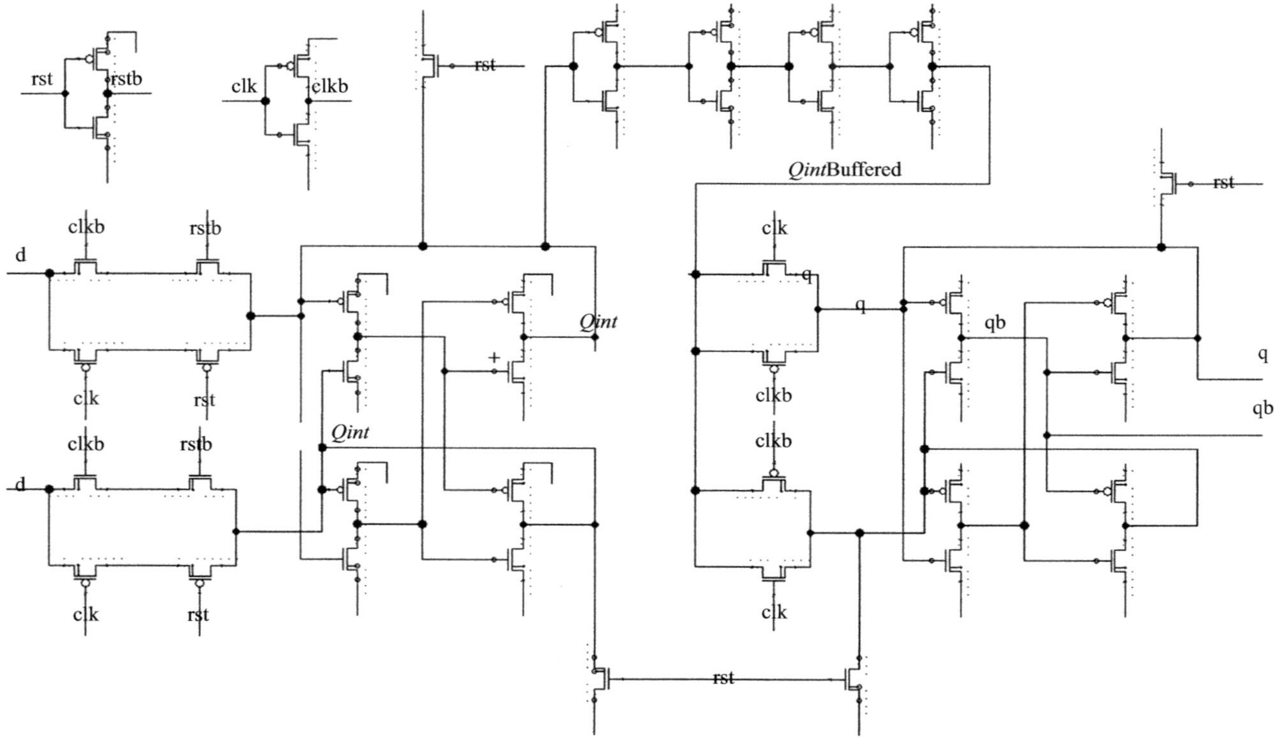

Figure 11.4 DICE-based DFF schematic [4]

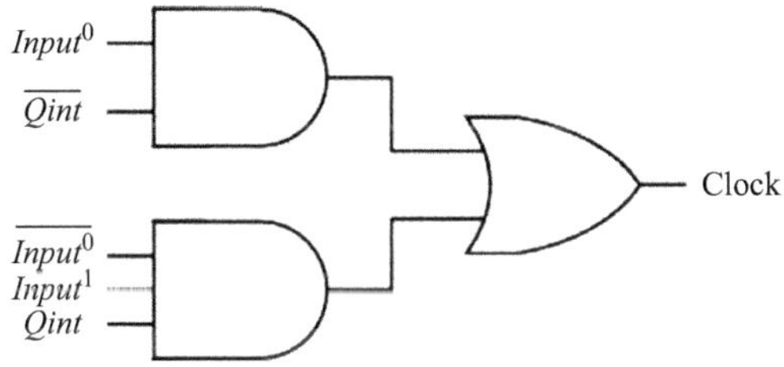

Figure 11.5 Asynchronous clocking logic

output of the bottom AND gate remains 0, but the top AND gate outputs a value of 1. This results in *Clock* becoming a 1 in which case the value within the DFF is updated to 0. When *Input* has a value of DATA1, the output of the top AND gate is 0, and the output of the bottom AND gate is 1. The two outputs, 1 and 0, are inputs to the OR gate resulting in a value of 1 for *Clock*; the data stored in the DFF is updated to a value of 1. As seen in the three cases, the asynchronous DFF only responds to inputs that are DATA0 or DATA1, meaning that the DFF will constantly be storing the value of the current or most recent DATA. The value stored will not be interrupted or change to 0 during NULL waves.

11.3 Analyzing radiation hardness

Each gate in the NCL library is simulated for radiation-hardness by generating a list of ionizing-radiation scenarios detailing the time and location of a particle strike event. The model selected for simulating particle strike events leverages the worst-case upset through a double-exponential function [6]. For each strike scenario, the strike-enabled model is simulated and the output is compared to the original gate simulation in order to determine whether an SEU occurred.

The schematic for each gate is analyzed using the netlist and a set of stimuli that includes each state of the gate. The number of simulations varies per gate due to the amount of possible states. As expected, the number of states generally increases for gates with a larger number of devices. Because of this, the number of simulations varies per gate meaning that the amount of upsets per simulation is a more appropriate value for comparison.

A particle-strike-simulation waveform for the TH24 is shown in Figure 11.6. Near 118 ns, transistor 0 is struck by a particle (100 LET) resulting in a single-event transient (SET). When simulating the TH24 for SEUs, the strike-event data from all gate transistor scenarios is included to determine when an SEU occurs.

The results for simulating the 90 nm IBM 9HP NCL library for strike events are shown in Table 11.2. It is important to see that the upset percentage is the most important statistic per gate, as the number of simulation passes is dependent on the gate. Overall, the nonhysteretic gates perform best. The TH22 and TH24 gate are the test cases for each of the tested techniques. Multiple techniques were used to improve the layouts for mitigating SEUs and MEUs.

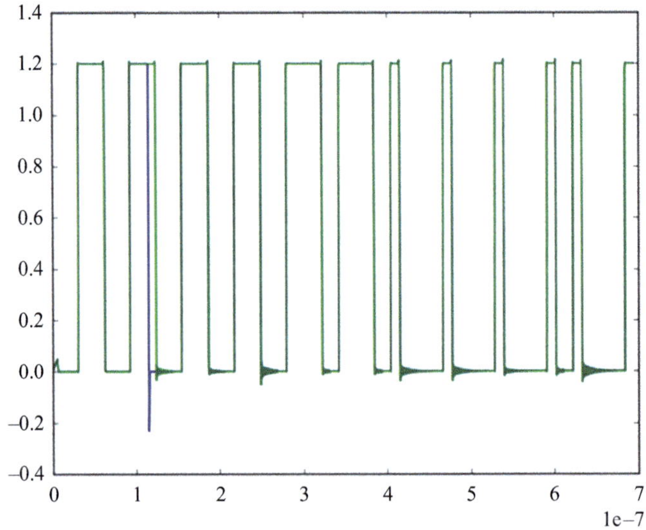

Figure 11.6 TH24 particle-strike simulation waveform

Table 11.2 SEUs per gate per LET

Gate	Passes	SEUs	SEU (%)	LET = 1	LET = 5	LET = 20
TH12b	1,728	0	0	0	0	0
TH13b	5,832	0	0	0	0	0
TH14	17,280	0	0	0	0	0
TH22n	12,195	93	0.76	27	33	33
TH24	339,309	459	0.14	141	153	165
TH24w22	29,945	55	0.18	15	19	21
TH33	17,226	102	0.59	30	36	36
TH33n	26,367	93	0.35	27	33	33
TH33w2	20,236	92	0.45	26	32	34
TH34	227,091	357	0.16	64	141	152
TH34w3	9,042	84	0.93	24	24	36
TH34w22	164,652	348	0.21	105	120	123
TH44	34,395	165	0.48	49	58	56
TH44w2	89,152	272	0.31	60	104	108
TH44w3	34,835	157	0.45	45	56	56
TH44w22	80,555	295	0.37	87	97	111
TH54w22	42,144	192	0.46	58	66	68
TH54w32	34,381	179	0.52	53	62	64
TH54w322	144,941	211	0.15	65	73	73

In addition to testing the NCL library, a set of modifications were applied to the TH22 gate in order to determine, without modifying the area of the gate, if the radiation hardness could be improved. The techniques along with the results are shown in Table 11.3.

For many gate layout modification techniques, the SEU response worsens, most notably when doubling the channel length. The main exception is when the overall channel area is increased, as seen in the last modification technique. Even though a 3% improvement is seen, this technique has a few drawbacks: it requires wider guard rings and therefore a larger gate area, and the power usage increases by a factor of 9 for the TH22 gate. Overall, these techniques either provide no additional protection or require a large overhead in terms of area and power.

Analysis of the effect of adding multiple fingers per transistor is included. Layout parasitics are included in a separate simulation to illustrate their effect. As you can see in Figure 11.7, adding multiple fingers and extracted parasitics increase

Table 11.3 SEUs per modification technique on the TH22 gate

Type of gate modification	Number of SEUs	Improvement (%)
Original gate	654	—
2× channel width	672	−3%
4× channel width	667	−2%
Maximum channel width	672	−3%
2× channel length	651	0%
4× channel length	1,700	−160%
Maximum channel width and 4× channel length	632	3%

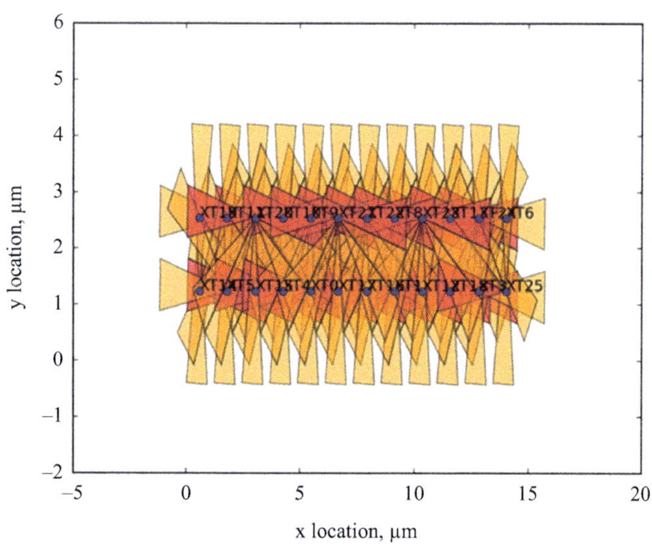

Figure 11.7 Multifinger simulation including device relative location

the radiation hardness for the TH22 gate. This, however, increases both the area and power usage. While the original TH22 gate's power consumption simulated at 14.63 nW, the multifingered TH22 gate's power consumption increased to 29.26 nW in addition to doubling the area of the gate layout.

In order to analyze the distribution of the gate area via multiple fingers, the simulation must include the relative location of each device within a gate. Figure 11.7 illustrates how this information is used by the simulator. Additionally, the parasitic capacitances and resistances have been added.

The results in Table 11.4 indicate that multiple fingers per device improve the overall gate radiation hardness, and the parasitic capacitances and resistances improve the gate's response. The parasitics are not included in the earlier simulation results (Table 11.2), but the improvement would apply to each of the standard NCL gate layouts.

Figure 11.8 is a waveform of the data stored in the asynchronous DICE-base DFF during six separate strike events of LET values from 1 to 100 MeV*cm^2/mg. Around 45 ns, the six strike events occur. As displayed, only four out of the six events affect SETs resulting in the data stored temporarily rising to 1. However, the data within the asynchronous DFF is not upset and the original value is restored once the SET subsides.

Table 11.4 Multifinger layout simulation results

TH22 test cases	Simulation passes	MEUs (%)
Original	6,862	0.729
Multiple fingers	13,746	0.567
Multiple fingers and parasitics	13,756	0.415

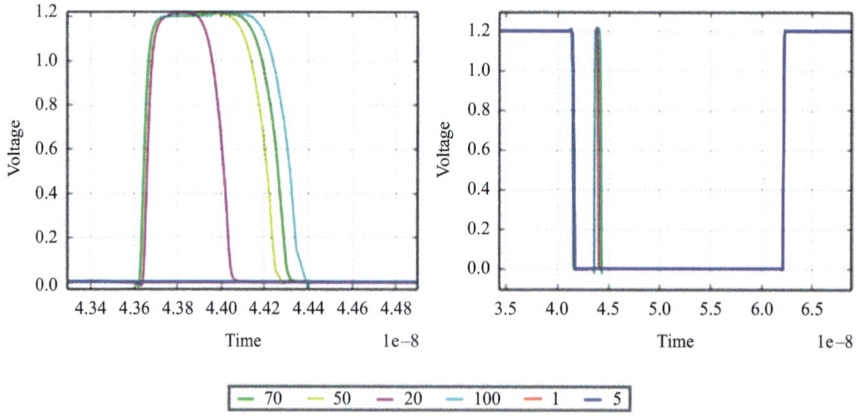

Figure 11.8 Asynchronous DICE-based DFF SET simulation

Table 11.5 SEUs per component

LET (MeV*cm^2/mg)	Asynchronous DFF	DICE-based asynchronous DFF
20	20	0
50	20	0
70	21	0
100	21	0

Table 11.5 shows that for the same LET values, the asynchronous DFF suffered up to 21 SEUs while the asynchronous DICE-based DFF simulations resulted in 0 SEUs.

References

[1] J. Di, "A framework on mitigating single event upset using delay-insensitive asynchronous circuits," *IEEE Region 5 Technical Conference*, April 2007.

[2] L. Zhou, S. C. Smith, and J. Di, "Radiation hardened NULL convention logic asynchronous design," *Journal of Low Power Electronics and Applications*, 2015.

[3] J. Brady, "Radiation-hardened delay-insensitive asynchronous circuits for multi-bit SEU mitigation and data-retaining SEL protection," University of Arkansas, May 2014.

[4] J. Brady, A. M. Francis, J. Holmes, J. Di, and H. A. Mantooth, "An asynchronous cell library for operation in wide-temperature and ionizing-radiation environments," *IEEE Aerospace Conference*, 2015.

[5] T. Calin, M. Nicolaidis, and R. Velazco, "Upset hardened design for submicron CMOS technology," *IEEE Transactions on Nuclear Science*, Vol. 43, No. 6, pp. 2874–2878, 1996.

[6] A. M. Francis, D. Dimitrov, J. Kauppilla, *et al.*, "Significance of strike model in circuit-level prediction of charge sharing upsets," *IEEE Transactions on Nuclear Science*, Vol. 56, No. 6, 2009.

Chapter 12

Dual rail asynchronous logic design methodologies for side channel attack mitigation

Jean Pierre T. Habimana[1] *and Jia Di*[1]

Side channel attacks (SCAs) remain a great threat to hardware security. In most CMOS circuitries, electrical behaviors are correlated to processed data which makes them vulnerable to SCAs. Dual-rail circuitries present an advantage in mitigating SCAs due to the inherited balance in data representation. NCL circuits present more stable power traces compared to industry standard synchronous counterparts; however, NCL circuits are still vulnerable to some SCAs due to the lack of balance in data propagation. In this chapter, the vulnerability of NCL circuits to SCAs is explained, and more secure dual-rail design methodologies are presented. Derived from NCL, dual-spacer dual-rail delay-insensitive logic or D^3L methodology produces crypto hardware with great resilience against SCAs. D^3L resilience, overheads associated with it as well as improved methodologies for overhead reduction are explained in this chapter.

12.1 Introduction

12.1.1 Side channel attacks

As technology advances, more personal and sensitive data are found on everyday use electronics such as cell phones, laptops, portable storages, and so on. The sensitive nature of the data stored or processed by these devices obliges certain security measures. In most cases, the data are encrypted using standards cryptographic algorithms such as DES, RSA, and AES. Although these algorithms are mathematically sophisticated and practically immune to brute force attacks, their electrical behaviors often leaked during data processing are highly correlated to processed data and can be used by attackers to deduct critical information. The collection of these electrical behaviors in various forms such as power consumption, execution time, electromagnetic emissions, and the statistical analysis required to reveal the critical data is what constitute a SCA.

[1]Department of Computer Science and Computer Engineering, University of Arkansas, Fayetteville, AR, USA

Most cryptographic hardware components are implemented with CMOS logic. Different input patterns processed by CMOS gates translate into different charging and discharging activities which result in a measurable current flow. This can be done with various tools and methods, and the current through a targeted circuitry can be determined with high precision. In the last two decades, different studies and real-life cases have shown that statistical analysis on leaked power information from a cryptographic chip can reveal its secret key [1–5].

This chapter focuses on power attacks performed on the advanced encryption standard (AES) adopted by NIST in 2001 [6]. It is shown that the encryption key used by an AES core can be successfully revealed by statistical analysis on leaked power information captured during an encryption process [1,3,4,7,8]. Some of the most known SCAs are differential power analysis (DPA) attack [1] and correlation power analysis (CPA) attack [2,3]. Many times over, attackers have succeeded breaking AES encryptions with both DPA and CPA attacks [1,3,8].

On the other hand, different countermeasures against SCAs have been proposed [9–14]. For instance, there are masking and randomization techniques. It has been shown that masking and randomizing some encryption variables increases the resilience against SCAs. An AES implementation that combines masking and randomization techniques is presented in [9]. It was demonstrated that masking all intermediate values makes first-order DPA impossible while randomization of operations achieves great resilience against second-order DPA. In [13], Montgomery ladder and scalar multiplications were adopted in AES S-BOX implementation to create resilience against simple power attack (SPA) and to intensify security against DPA. Moreover, fast elliptic curve multiplication was introduced to improve the performance of elliptic curve cryptography (ECC) hardware and their resilience against DPA attacks [14].

However, the stated countermeasures share the same data representation mechanism where logic 0 and logic 1 are represented by a connection to ground and power, respectively. Therefore, despite masking or randomization techniques being used, the correlation of data to circuit side channel information still exists, and it can be exploited by different attack models. In fact, for every aforementioned countermeasure technique, a more sophisticated or simply modified attack model has been developed to bypass or weaken the added resilience. For instance, zero-value point attack (ZPA) introduced in [15] is capable of breaking ECC cryptography despite randomization techniques being used. The masking of intermediate values proposed in [9] as a solution for typical DPA attacks is annihilated by different DPA models that do not depend on intermediate values. A toggle count model proposed in [2] bypasses intermediate values thus weakening masking techniques effectiveness. Furthermore, it has been proven that randomization-based encryptions are weakened by simply increasing the number of analyzed patterns [5], and with increasing technology, more and more processing power is available to attackers allowing them to process as many patterns as needed in a reasonable amount of time.

Table 12.1 NCL data representation

D0	D1	State
0	0	NULL
0	1	DATA*1*
1	0	DATA*0*
1	1	Not allowed

12.1.2 Dual-rail logic solution to SCAs

In contrast to most proposed SCAs mitigation techniques, dual-rail design methodologies present a distinguished potential for a better-balanced switching activity compared to synchronous single rail circuits. Using NULL convention logic (NCL) [16,17] as example, it is seen that logic 1 (DATA*1* in NCL representation) and logic 0 (DATA*0* in NCL representation) are both represented by one high rail and one low rail and only differ in the order in which the high and low rails are assigned as explained in Table 12.1.

As a result of its data representation scheme, most components of a dual-rail circuit are affected the same way by a logic 0 or a logic 1 input. Nevertheless, side channel information from NCL circuits can still be correlated to processed data, and it has been proven that NCL crypto hardwares are still vulnerable to SCAs [18,19].

Dual-spacer dual-rail delay-insensitive logic (D^3L) methodology [18] is an extension of the NCL methodology that balances the switching activities in a circuit and eliminates the correlation of power consumption to processed data. By adopting a dual-spacer scheme where an all-zeros spacer alternates with an all-ones spacer to serve as NULL states between DATA states, the switching activities between the two rails of every signal are completely balanced for every DATA/NULL cycle. This characteristic decouples data from leaked side channel information, and it is the cause for D^3L crypto hardware resilience against SCAs.

Although D^3L succeeded in mitigating tried SCAs such as CPA and timing attacks [18], D^3L circuits area, power consumption, and delay overheads remain big hindrances for its adoption on a bigger scale. Consequently, attempts have been made to improve the D^3L design methodology to reduce its overheads while keeping its SCAs mitigation capabilities. Applying techniques used to create the Multi-threshold NULL Convention Logic (MTNCL) [20] such as the use of high-V_T and low-V_T transistors for fast switching yet low leakage current circuits and the early completion detection technique, a multithreshold dual-spacer dual-rail delay-insensitive logic (MTD^3L) methodology was created [19]. An AES core implemented using this methodology withstood attempted SCAs attacks proving that MTD^3L methodology preserves D^3L SCA resilience. However, area, power consumption, and delay overheads reduction achieved by this methodology was modest [19], and on its turn, this MTD^3L version overheads still needed substantial reduction [21].

An improved MTD^{3}L version presented in [21] aimed to reduce area, power, and delay overheads to optimal levels. In addition to the use of multi-threshold transistors and early completion detection techniques, a new transistor level register cell was created allowing an easy generation of NULL states (all-zeros and all-ones spacers) and costless handling of the all-ones spacer. This addressed the main causes of overheads in D^{3}L circuits by eliminating the need for extra circuitries to handle two different types of spacers and the need for spacer generator and spacer filter registers [18].

This chapter is organized as followed. Section 12.2 discusses NCL: its added balance to power consumption compared to synchronous logic and its weakness to SCAs. Section 12.3 introduces D^{3}L methodology and explains its resilience to SCAs. In Section 12.4, MTD^{3}L is presented as an improved dual-rail methodology with D^{3}L SCAs mitigation capabilities, but with lower area, power, and delay overheads. In Section 12.5, NCL, D^{3}L, and MTD^{3}L methodologies are compared against their resilience to SCAs and their area, power, and delay overheads. Finally a conclusion is given in Section 12.6.

12.2 NCL SCAs mitigation capabilities and weaknesses

12.2.1 NCL balanced power consumption

The dual-rail data representation scheme allows more balanced switching activities in NCL circuits compared to their synchronous single rail counterparts. In synchronous circuits, a clock event triggers simultaneous switching activities emanating from registers and latches to combinational logic units. During this event, the number of switching nodes is determined by the data that is being latched out of registers and latches. As a result, the current flow (or power consumption) in synchronous single rail circuits is highly correlated to processed data.

In NCL circuits, an equivalent situation is the transition of a DATA wave after a NULL state. At this time, combinational logic units hold signals in NULL states waiting for DATA to be released by NCL registers. In contrast to synchronous single rail circuits, the number of nodes going high and low for some components can be determined without knowing the data coming from the registers. For example, for an 8-bit register output bus, it is known that eight rails will go high while other eight rails will remain low regardless of the data. Therefore, for these components, the switching activity does not depend on the data, but on the hardware.

The same scenario happens at the input side of the registers. In fact, in NCL, one-bit value is kept by two register cells, one register cell for Rail1 and another for Rail0. During the transition from NULL to DATA, the number of register cells changing their state is known regardless of the data; in fact, one-half of register cells will switch, and another half will not. This shows that NCL eliminates the correlation of data to switching activity during stages when current flow is the most critical.

12.2.2 NCL unbalanced combinational logic

Although NCL design methodology manages to keep the switching activity balanced during DATA and NULL cycle transitions around registers, a correlation between power consumption and processed data still exists. This correlation is due to some combinational logic implementation where NCL signal rails are not necessarily driving equivalent nodes. For instance, consider the implementation of the NCL AND function as depicted in Figure 12.1. It is shown that rail X0 and X1 and rail Y0 and Y1 of inputs X and Y are not driving the same capacitance loads. In fact, rail X0 and Y0 are only driving gate TH34W22 while rail X1 and Y1 drive both TH34W22 and TH22 gates. This explains that in NCL circuits, even though either Rail1 or Rail0 is asserted for every NULL/DATA cycle, the resulting current flow and power consumption is different and dependent to which rail is asserted. In other words, the current flow in NCL circuits still depends on processed data.

12.2.3 NCL SCA mitigation

NCL SCA mitigation capabilities were tested [18]. A CPA attack was performed on an NCL AES core, and it was determined that NCL is vulnerable to SCAs. Performing the attack on a single SUBBYTE, the correlation between energy consumption and the total number of switching transistors was high enough to confirm the encryption key guess [18]. The same experience shows that the dual-rail data representation has a considerable impact in reducing the correlation between data and side channel information. In fact, even though CPA attack successfully broke the NCL AES core, the correlation coefficient was considerably lower compared to a synchronous AES core attacked in the same situations. The attack succeeded with a 0.668 correlation coefficient for the synchronous AES core and a 0.428 correlation coefficient for the NCL AES core [18].

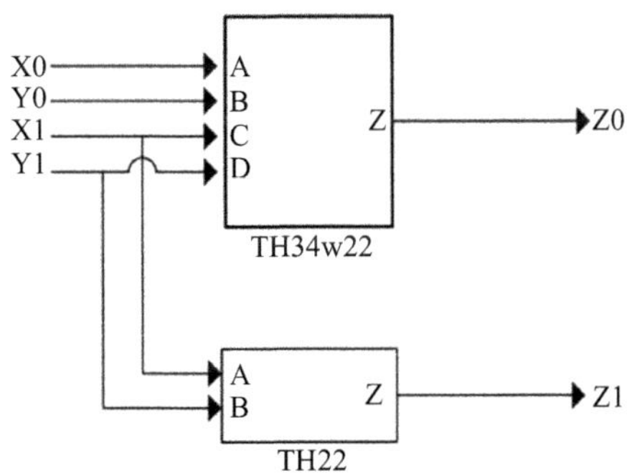

Figure 12.1 NCL input incomplete AND function

12.3 Dual-spacer dual-rail delay-insensitive logic (D^3L)

Dual-spacer dual-rail delay-insensitive logic or D^3L methodology was invented to further decouple the correlation between data and side channel information. Considering the improved power consumption balance achieved by NCL circuits and the known cause of SCAs weakness in NCL crypto hardware, D^3L methodology aimed to completely eliminate the correlation between data and side channel information by forcing both rails of every signal to switch once for every DATA/NULL transition (from NULL state to DATA state back to NULL state cycle).

12.3.1 Introducing an all-ones spacer

As explained in Section 12.2.2, SCAs weakness in NCL circuits comes from combinational logic functions where switching one rail does not necessarily draw the same current as switching the other rail. This problem would not exist if both rails switched in every cycle for any given data pattern. D^3L enforces this rail switching behavior by using two representations of a NULL state (referred to as "spacers" in D^3L context). As opposed to NCL, where two rails of a signal are to be asserted exclusively, D^3L signal rails are asserted simultaneously to form an All-ones spacer. Therefore, in D^3L context, a NULL state means one of the two states: all-zeros spacer, when both rails are logic 0 or all-ones spacer, when both rails are logic 1. In other words, D^3L signals can be in one of the four states: DATA*1*, DATA*0*, all-zeros spacer (*A0s*) or all-ones spacer (*A1s*) as detailed in Table 12.2.

With this data encoding scheme, a complete D^3L data transition cycle goes from *A0s* spacer to DATA then to *A1s* spacer. Alternatively, the cycle goes from *A1s* spacer to DATA then to *A0s* spacer. As a result, both rails of every signal switch for every data transition cycle as exemplified in Table 12.3.

It is illustrated by Table 12.3 that for every cycle, each rail switches once and that for three cycles as the data value changes three times, each rail switches three times. In contrast, for NCL circuits as shown in Table 12.4, for each cycle one rail switches twice while the other remains unchanged, and for three data cycles, for the first scenario Rail1 switches six times while Rail0 remains unchanged, and for the second scenario, Rail1 remains unchanged while Rail0 switches six times. It is concluded that in D^3L circuits, the switching activity within a cycle does not depend on processed data.

Table 12.2 D^3L data encoding

D0	D1	States
0	0	*A0s*
0	1	DATA*1*
1	0	DATA*0*
1	1	*A1s*

Table 12.3 Example of D³L balanced switching activity

	Initial state	Cycle 1		Cycle 2		Cycle 3		Total switching
Data value	NULL *A0s*	Logic 1	NULL *A1s*	Logic 1	NULL *A0s*	Logic 1	NULL *A1s*	3
Rail0	0	0	1	0	0	0	1	3
Rail1	0	1	1	1	0	1	1	3
Data value	NULL *A0s*	Logic 0	NULL *A1s*	Logic 0	NULL *A0s*	Logic 0	NULL *A1s*	3
Rail0	0	1	1	1	0	1	1	3
Rail1	0	0	1	0	0	0	1	3

Table 12.4 Example of NCL unbalanced switching activity

	Initial state	Cycle 1		Cycle 2		Cycle 3		Total switching
Data value	NULL	Logic 1	NULL	Logic 1	NULL	Logic 1	NULL	3
Rail0	0	0	0	0	0	0	0	0
Rail1	0	1	0	1	0	1	0	6
Data value	NULL	Logic 0	NULL	Logic 0	NULL	Logic 0	NULL	3
Rail0	0	1	0	1	0	1	0	6
Rail1	0	0	0	0	0	0	0	0

12.3.2 Adapting NCL register to the dual-spacer scheme

The use of all-ones spacer as a NULL state does not come without cost. Great modifications have to be implemented to allow the NCL register structure to be used with D³L data representation scheme.

12.3.2.1 D³L *ko* generation

The logic 1 in an *A1s* spacer poses a conflict in the way the *ko* signal is generated and its meaning to the rest of the circuit. Consider an NCL register with input *A* and output *Z*. When *Z* is DATA *ko* needs to be logic 0 to request a NULL cycle, and when *Z* is in a NULL state *ko* needs to be logic 1 to request DATA. A TH12b gate with *Z0* and *Z1* as inputs and *ko* as output complies perfectly with these requirements. Figure 12.2 details the composition of an NCL register and the *ko* generation logic.

However, with *A1s* spacers being used as NULL states, the TH12b gate can no longer correctly generate *ko* signal. *Z0* and *Z1* being both logic 1 results in *ko* being logic 0, which is a request for a NULL input while the register is still holding a NULL output (*A1s* spacer). According to the truth table presented by Table 12.5, *ko* values required to comply with the dual-rail asynchronous logic handshake

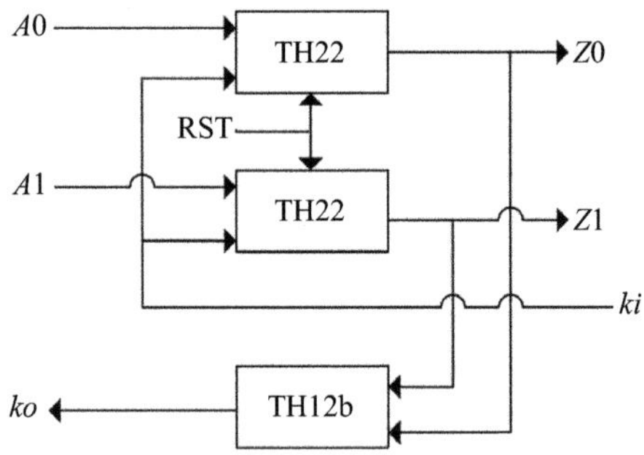

Figure 12.2 NCL register and ko generation logic

Table 12.5 ko signal relation to register output state

State	*Z0*	*Z1*	*ko*
A0s	0	0	1
DATA*1*	0	1	0
DATA*0*	1	0	0
A1s	1	1	1

protocol follow an XNOR function. Therefore, D^3L register replaces the NCL TH12b gate with a Boolean XNOR gate to generate *ko* signal.

12.3.2.2 D^3L *ki* generation

Another D^3L adjustment to accommodate the *A1s* spacer is the generation and interpretation of the *ki* signal. In NCL register as presented in Figure 12.2, the *ki* signal complements the TH22 gate holding capabilities to ensure that DATA and NULL waves are released when requested [17]. To elaborate, when a NULL input A is presented, both *A0* and *A1* are logic 0, outputs *Z0* and *Z1* remain unchanged until *ki* is logic 0. In the same way, when *Z0* and *Z1* are logic 0 (NULL output), and a DATA input is presented (either *A0* or *A1* is logic 1), *Z0* and *Z1* do not change until *ki* is logic 1 because a TH22 gate requires both inputs to be asserted for the output to be asserted and both inputs to be de-asserted for the output to be de-asserted [17].

This arrangement does not hold with an *A1s* NULL inputs. When the register output is DATA, *ki* remains logic 1 for as long as the next stage requires DATA. However, when an *A1s* NULL input is presented, the TH22 gate with logic 0 output would assert it to logic 1 given that both inputs would be logic 1. This problem is solved by adding more logic to the NCL base register. The architecture described in [18] uses a KI generator component that set the *ki* value depending on the next spacer to be transmitted by the register.

The KI generator uses a *previous spacer (ps)* signal to determine which spacer to expect. The *ps* signal holds logic 0 at reset until the register output becomes an

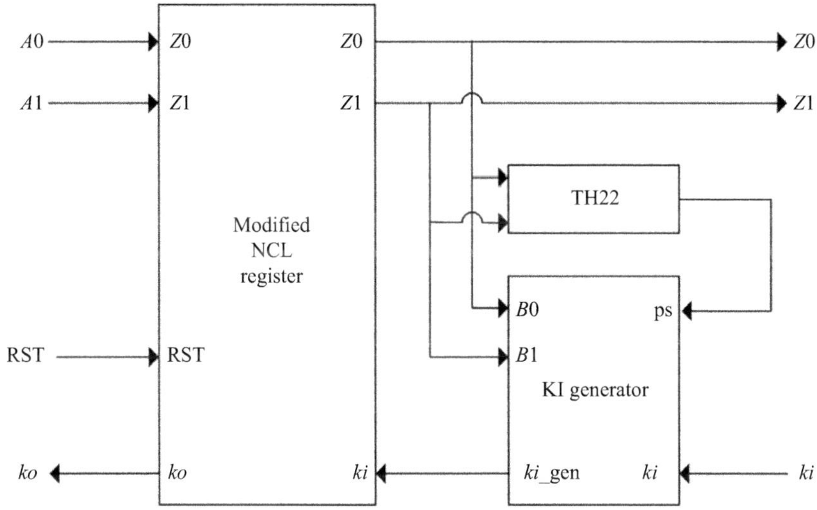

Figure 12.3 Complete D^3L register

A1s spacers, then it changes to logic 1 until the register output becomes an *A0s* again. The KI generator logic uses the *ps* signal along with *ki* as inputs and generates an output *ki_gen* signal to be used in lieu of the original *ki*. A D^3L register including the KI generator component and its inputs and internal connections is presented in Figure 12.3.

To comply with dual-rail handshaking protocol, *ki_gen* signal works as followed. First scenario, as a request for DATA *(rfd)* with register output being an *A0s* spacer, *ki_gen* signal will be logic 1 to allow TH22 gate to propagate logic 1 DATA input (one output rail rises to logic 1 while the other remain logic 0). The second scenario, DATA is requested with register output being an *A1s* spacer. *ki_gen* signal will be logic 0 to allow logic 0 to propagate (one output rail falls to logic 0 while the other remain logic 1). The third scenario, register holds a DATA output and a NULL is requested with *ps* being logic 1. *ki_gen* signal will be logic 0 to allow an *A0s* spacer to propagate (output rail with logic 1 falls, and the other rail remains logic 0 forming an *A0s* spacer). The final scenario, NULL is requested with *ps* being logic 0. *ki_gen* signal will be logic 1 to allow an *A1s* spacer to propagate (output rail with logic 0 rises, and the other rail remains logic 1 forming an *A1s* spacer).

It is critical to notice that more inputs are required to distinguish when *ki_gen* is logic 1 as a request for DATA from when it is logic 1 as a request for an *A1s* spacer. Table 12.6 summarizes the relationship between *Reset*, input *A*, output *Z*, *ki_gen*, and *ps*.

The adaptation of NCL gates to D^3L data representation scheme is more complex than simply allowing the all-ones spacer state. For a complete explanation

Table 12.6 D³L register DATA transition mechanism

Time	0	1	2	3	4	5	6	7	8	9
Reset	1	0[1]	0	0	0	0	0	0	0	0
A0	0	0	0	1	1	1	1	1	0	0
A1	0	0	1	1	1	1	0	0	0	0
Z0	0	0	0	0	**1	1	1	1	***0	0
Z1	0	0	*[2]1	1	1	1	*0	0	0	0
ki	1	1	1	1	0	*1	1	1	*0	0
ko	1	1	**0	0	***1	1	**0	0	****1	1
ki_gen	1	1	**0	0	*1	1	0**1	1	**0	0
ps	0	0	0	0	***1	1	1	1	****0	0

[1]Table entry in red color represents the main event in the cycle.
[2]The star (*) symbol marks a delay. An entry with * occurs following entries with no *, and entries with ** occur after entries with * and so on.

on KI generation logic and implementation as well as the complete picture of a D³L register further reading on D³L methodology is encouraged.

12.3.2.3 D³L filter register

In some situations, the D³L spacer alternation scheme does not work properly. For example, without further improvement, a three register ring would fall into a deadlock state after reset. D³L DATA/spacer transitions rely on correct DATA/NULL input streams. However, for a three-ring register, all outputs are forced to *A0s* spacer after reset. The inability for a D³L register as described in Section 12.3.2 to propagate an *A0s* spacer when an *A1s* is expected would stall the circuit indefinitely [18]. This problem is addressed by adopting a special filter register capable of changing an *A1s* spacer to an *A0s* spacer and vice versa depending on the previous spacer and the input spacer. For instance, if the previous and the input spacers are both an *A1s* spacer, the filter register changes the *A1s* spacer input into an *A0s* spacer output [18].

12.3.2.4 D³L spacer generator register

Another scenario that could be problematic during D³L DATA/spacer transitions is when an alternating input spacer is not available. For example, a component X might need to perform many cycles before its input stream changes because its previous stage does not have to perform as many cycles. A spacer generator register is used to provide appropriate spacers as requested by component X [18].

12.3.3 D³L resilience to side channel attacks

The dual-rail data representation scheme along with balanced switching activities between signal rails address the correlation between side channel information and processed data problem. With both rails of every D³L signal switching just once for every DATA/spacer cycle, the switching activity in D³L circuits is not correlated to processed data. Therefore, statistical analysis on side channel information such as power consumption or energy consumption does not yield any detail about protected data.

An AES core implemented and tested in [18] shows that CPA attack fails to break D^3L crypto hardware. With dual-rail data representation, the hamming distance in D^3L circuitries is always the same for any state change. This makes the original CPA attack [3] nonapplicable. However, a modified version specialized for dual-rail circuits was used [18]. A program was written to compute the number of switching transistors for the execution of a given input pattern using a guessed encryption key. Another program compares the measured energy consumption to the computed number of switching transistors and assign a correlation coefficient to each encryption key guess. The highest correlation coefficient should align with the correct key guess [2,3,18].

Results presented in [18] show that the performed CPA attack failed to break a D^3L AES core even attacking one SUBYTE at a time. It is reported that the maximum correlation coefficient obtained was 0.354, which was the lowest amongst the tested methodologies and it did not correspond to the right encryption key guess.

12.4 Multi-threshold dual-spacer dual-rail delay-insensitive logic (MTD^3L)

12.4.1 The first MTD^3L version

Multithresholds transistors used in MTNCL methodology [20] along with techniques such as early completion detection and sleep-able gates saw MTNCL circuits perform better than their NCL counterparts for most criteria. Linder, Di, and Smith applied these techniques on the D^3L methodology to reduce D^3L circuits overhead and created the MTD^3L methodology [19]. In this methodology, the sleep mechanism used eliminates the need for the NCL-X completion circuitries required by D^3L combination logic functions. Similar to MTNCL gates, MTD^3L low-V_T transistors allow MTD^3L gates to switch faster than their D^3L counterparts while high-V_T transistors keep the leakage current to a minimum [19].

This design methodology was tested for its resilience against SCAs and its overhead compared to synchronous, NCL, and D^3L methodologies. For this comparison, four AES cores were designed using the four stated methodologies in the same environment. Cadence tools were used with the IBM 8RF-DM 130 nm process. It was proved that MTD^3L methodology attained success in energy, speed, and area overhead reduction compared to D^3L methodology. For example, for the tested AES cores, energy consumption for a full encryption was reduced from 6.012 nJ to 3.84 nJ, and the core area was reduced from 6.27 to 3.37 mm^2 [19]. For resilience against SCAs, a modified CPA attack was performed and it only succeeded on the synchronous AES core [19]. Correlation coefficients and attack results for the four AES cores are presented in Table 12.7.

Table 12.7 SCA results on synchronous, NCL, D^3L, and MTD^3L AES cores

Design methodology	Correlation coefficient	Key guess Success/Failure
Synchronous	0.872	Success
NCL	0.207	Failure
D^3L	0.376	Failure
MTD^3L	0.353	Failure

Table 12.8 NCL and MTD^3L energy consumption by component

Component	NCL energy (nJ)	MTD^3L energy (nJ)
Registers	0.252	1.836
Logic	1.512	1.26
Buffering	0.444	0.744
Total	2.208	3.84

Although, this MTD^3L methodology achieved significant overhead reduction in area and energy consumption compared to D^3L methodology, MTD^3L circuits overheads were still very high compared to NCL and synchronous counterparts. Table 12.8 presents NCL and MTD^3L AES core energy consumption by component. It can be observed that the biggest difference between NCL and MTD^3L energy consumption comes from registration where MTD^3L registration costs 1.836 nJ compared to 0.252 nJ used by NCL registration. Despite a roughly 700% difference, this observation did not come as a surprise. In fact, the main causes of overhead in D^3L registers, namely KI generator components, spacer filter registers and spacer generator registers were still part of the MTD^3L registration.

Considering registration overhead in MTD^3L circuits, an improved MTD^3L methodology aiming to create a new and simplified registration scheme was suggested [21]. Section 12.4.2 presents the improved MTD^3L methodology explaining a reinvented transistor-level register cell and mechanisms used to replace KI generator components, spacer filter registers, and spacer generator registers. For the remainder of this chapter, the MTD^3L methodology by Linder, Di and Smith will be referred to as MTD^3L_v1 while the improved MTD^3L methodology is referred to as simply MTD^3L.

12.4.2 Reinvented MTD^3L design methodology

12.4.2.1 Approach

D^3L and MTD^3L_v1 registration circuitries are based on the NCL register structure to which KI generator, spacer generator, and spacer filter functionalities are added to handle the dual-spacer mechanism. However, these functionalities were added at the gate level, which did not allow much flexibility and optimization. To solve the registration overhead problem, the improved MTD^{3}L version adopts a new register cell designed at the transistor-level to handle the dual-spacer mechanism within itself. *ki* generator components, spacer generator, and spacer filter registers are not needed with this register cell. Furthermore, this register cell is simple and small in terms of transistor count; in fact, it is smaller than its NCL counterpart. The register cell transistor-level implementation is shown in Figure 12.4.

12.4.2.2 Spacer generator registers elimination

Sleep signals and register cell output relationship

Spacer generator registers were replaced by a sleep-able register capable of generating *A1s* and *A0s* spacers by respectively asserting the *sleep-to-one (s1)* and *sleep-to-zero (s0)* signals. Given the fact that each rail of a dual-rail signal connects to its own register cell, sleeping both register cells at the same time gives either an *A1s* when the *sleep-to-one* signal is asserted or an *A0s* when the *sleep-to-zero* signal is asserted.

The register cell in this methodology is implemented in a way that the output is logic 1 when the *sleep-to-one* signal is asserted and logic 0 when the *sleep-to-zero* signal is asserted. However, for a simplified transistor level implementation, the inverted *sleep-to-one* (*not-sleep-to-one*) signal, denoted by *ns1* is used instead of *sleep-to-one (s1)*. Nonetheless, the *ns1* signal is used such that the output is logic 1 when *sleep-to-one* is asserted or when *ns1* is logic 0. Figure 12.4 highlights the sleep signals and their function in the register cell implementation.

Mutual exclusivity between *s0* and *s1* is crucial to this architecture. Both sleep signals are set to logic 0 for a DATA state, and for the NULL state either *s0* or *s1* is asserted. A sleep signal controlling circuit is used to generate *s0* and *s1 (ns1)* signals. The gate level implementation of this circuitry is presented in Figure 12.5.

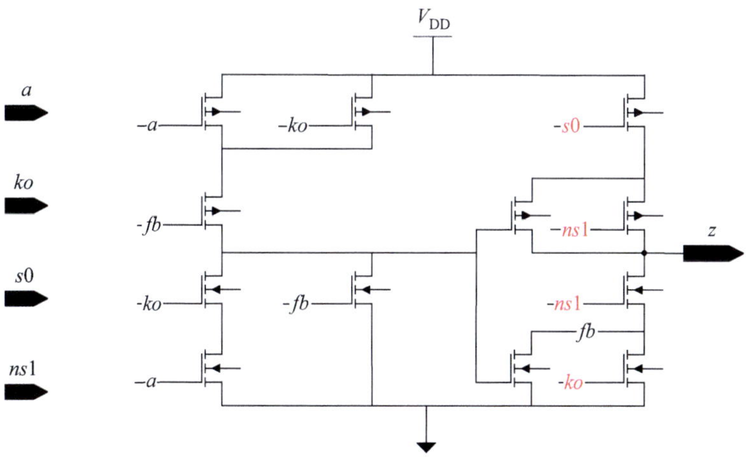

Figure 12.4 Reinvented transistor-level MTD^{3}L register cell

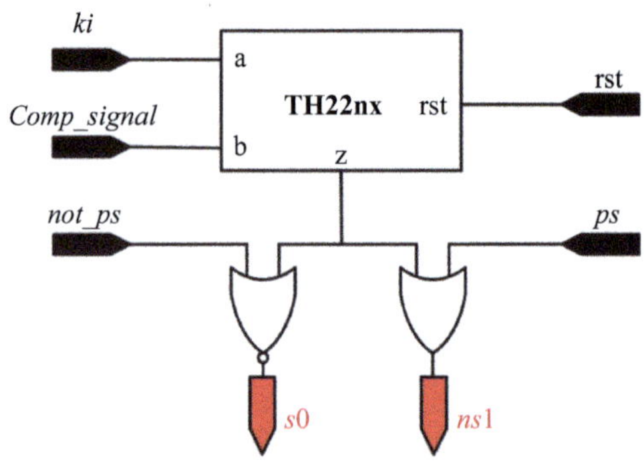

Figure 12.5 *MTD³L sleep signal controlling unit*

Table 12.9 *Previous spacer relationship to sleep signals and generated spacer*

ps	*s0*	*s1*	Generated spacer
1	1	0	All-zeros spacer
0	0	1	All-ones spacer

With overhead reduction in mind, the sleep signal controlling logic is a simple circuitry, and only one instance is needed for the whole stage registration.

The previous spacer (*ps*) signal is used in the same way as in D^{3}L methodology to determine which sleep signal to assert. When *ps* is logic 1 (the previous spacer was an *A1s* spacer), *s0* is asserted so that the register cell generates an *A0s* spacer and when *ps* is logic 0 (the previous spacer was an *A0s* spacer), *s1* is asserted so that the register cell generates an *A1s* spacer. Table 12.9 summarizes the relationship between the previous spacer (*ps*) and the generated spacer.

12.4.2.3 Register cell transistor-level implementation

Compared to other dual-rail methodologies such as NCL or MTNCL, MTD^{3}L methodology must overcome the challenge of distinguishing a logic 1 of a DATA state from a logic 1 of an *A1s* spacer. The main objective of this MTD^{3}L methodology is to overcome this challenge without added cost in area, power consumption, and delay.

The new register cell implementation takes advantage of the relationship between *ko* and sleep signals. In the dual-rail handshaking protocol, *ko* being logic 1 means that the output of the register is in a NULL state, and the register is ready for DATA input, while *ko* being logic 0 means that the output is in a DATA state, and the register is ready for a NULL input. Also, the output of the register is in a NULL state when either *s0* or *s1* is asserted. Therefore, when *ko* is logic 1, the output of the register is controlled by the sleep signals, and the value of the register

Table 12.10 Relationship between ko signal and register output

ko	*s0*	*s1*	**Register outputs**
1	1	0	All-zeros spacer
1	0	1	All-ones spacer
0	0	0	DATA

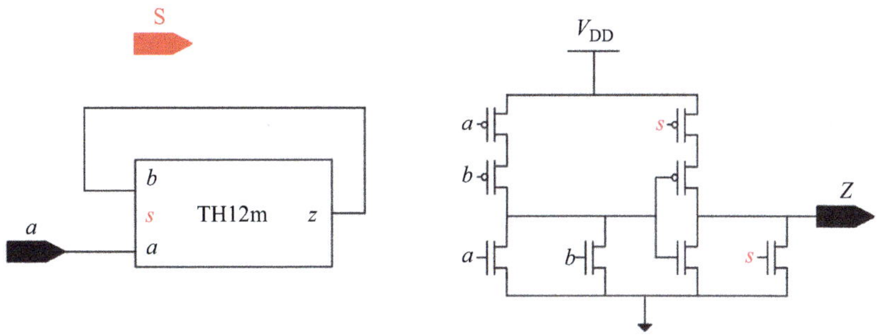

Figure 12.6 MTNCL OR-like register cell structure with a feedback input

input should not matter. The relationship of *ko* signal to register output state is summarized in Table 12.10. This relationship was used to incorporate *ko* in the transistor-level design of the MTD^{3}L register cell by following these three steps.

Step 1: Adapting an OR gate-like structure with a feedback path

This is the same structure used for MTNCL register in the form of a TH12m gate. The sleep signal (*s*), when asserted, forces the gate output to logic 0, and the feedback path forces the output to remain asserted once it has been asserted. This functionality is explained by the transistor-level schematic in Figure 12.6.

Since MTD^{3}L cells need two sleep signal inputs, the first step is to add an *ns1* input to the TH12m structure as shown in Figure 12.4. With two sleep signal inputs, this structure handles the spacer generation process, but it is not capable of stopping logic 1 of an *A1s* spacer input from overriding the output DATA state. In other words, a logic 0 output would be overridden by an incoming *A1s* spacer before the next stage of the pipeline requests for a NULL state.

Step 2: Using ko to stop A1s logic 1 input from overriding logic 0 output before NULL is requested

As presented in Table 12.10, when the output of the register is in a DATA state, *ko* is logic 0. Moreover, placing *ko* in series with the register input *a* cancels out the effect of input *a* being logic 1 when *ko* is logic 0. In other words, for input *a* to pull the output to logic 1, both *a* and *ko* need to be logic 1 simultaneously, and this cannot happen with the register output being DATA. *ko* role in register cell implementation is highlighted by the schematic in Figure 12.7.

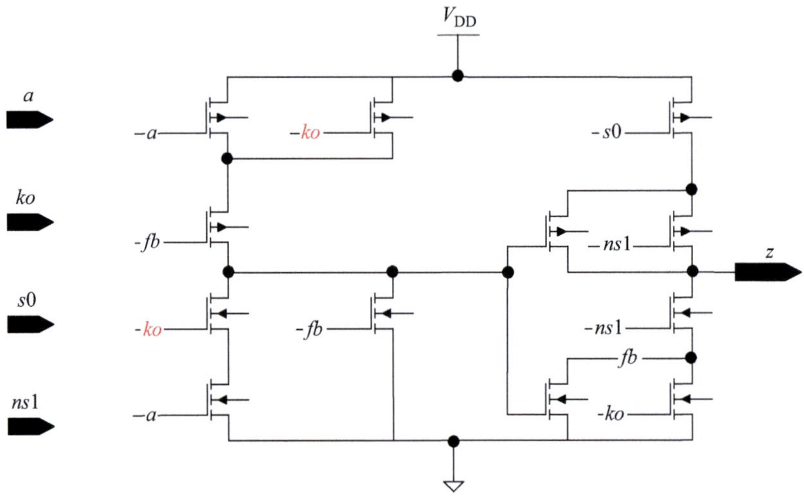

Figure 12.7 MTD³L register cell transistor-level highlighting ko function

Step 3: Controlled timing between sleep signal (s0) and ko falling edges

The MTD³L register cell as presented in Figure 12.7 generates the spacers itself, and it correctly stops *A1s* spacer inputs from overriding DATA outputs. However, by observing the succession of events, it can be seen that a hold time violation might occur following an *A0s* spacer output. In fact, using early completion detection, the propagation delay between *ko* and *s0* allows by default both signals to be de-asserted simultaneously. Given that logic 1 input is not latched unless *ko* is logic 1, the completion logic unit that generates both *ko* and sleep signals has to ensure that *ko* is held at logic 1 for an appropriate time after *s0* falls. The delay between *s0* and *ko* falling edges needs to be just enough for the output to rise, and once it has risen the feedback path takes over. At this point when *ko* falls, the output is not affected. It has been determined that one buffer delay between *s0* and *ko* is an appropriate time for both speed and leakage purposes.

Combining the three steps results in a simple MTD³L register cell capable of generating and filtering its own spacers thus eliminating the need for spacer generator and spacer filter registers used in D³L and MTD³L_v1 circuits. Moreover, the completion logic unit that generates sleep signals and *ko* uses the *ki* input in its traditional state as defined by the dual-rail handshaking protocol, so KI generator components used in D³L and MTD³L_v1 circuits are not needed. Figure 12.8 shows the complete MTD³L register and Figure 12.9 shows the overall MTD³L architecture including registers, completion logic and combinational logic units as well as the handshaking signal connections.

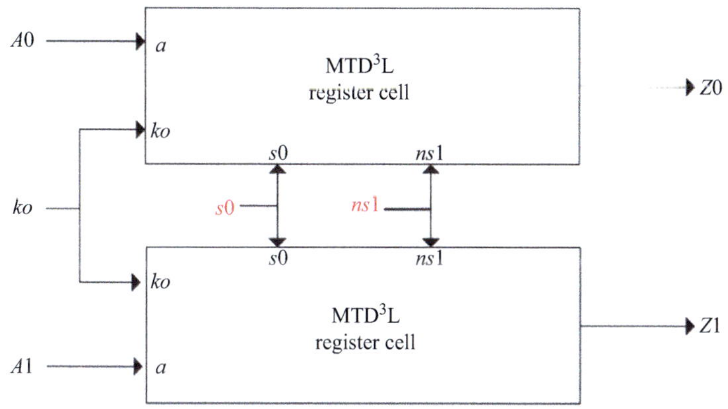

Figure 12.8 Complete MTD^{3}L dual-rail register

12.4.2.4 MTD^{3}L simulation and results

The reinvented MTD^{3}L methodology was tested for its resilience against SCAs and its area, energy and delay overheads. To attain a fair comparison to other dual-rail design methodologies discussed in this chapter, an AES core was designed using the same CAD tools with the same IBM 130 nm process [18,19,21]. The improved MTD^{3}L methodology achieved very significant area, energy, and delay overhead reductions compared to D^{3}L and MTD^{3}L_v1 methodologies. Table 12.11 presents NCL, MTD^{3}L_v1, and MTD^{3}L delay and energy consumption for one encryption round. The improved MTD^{3}L AES design presents the best numbers for both delay and energy consumption. In Table 12.12, delay, energy, and registration transistor count are compared. The registration scheme adopted in the new MTD^{3}L methodology is the main reason for a 353.9% improvement in energy consumption; in fact, registration energy consumption was improved by 443.5%, going from 1.836 nJ used by MTD^{3}L_v1 registration to 0.414 nJ used by the improved MTD^{3}L registration.

12.4.2.5 Side channel attacks resilience

DPA and CPA attacks were used to test the improved MTD^{3}L AES core resilience against SCAs. The best attack conditions for DPA and CPA were attempted. Statistical analysis was performed on schematic simulation information.

CPA attack

As explained in Section 12.3.3, the original CPA attack using the correlation between hamming distances from input patterns and guessed key and circuit energy consumption [3,22] is not applicable to dual-rail crypto hardware. The dual-rail data representation forces hamming distances to be the same across different input patterns [18]. Moreover, the modified CPA model using the number of switching transistors for a given input pattern as used on the D^{3}L and MTD^{3}L_v1 AES cores [18,19] is not applicable to the improved MTD^{3}L methodology. A byproduct of

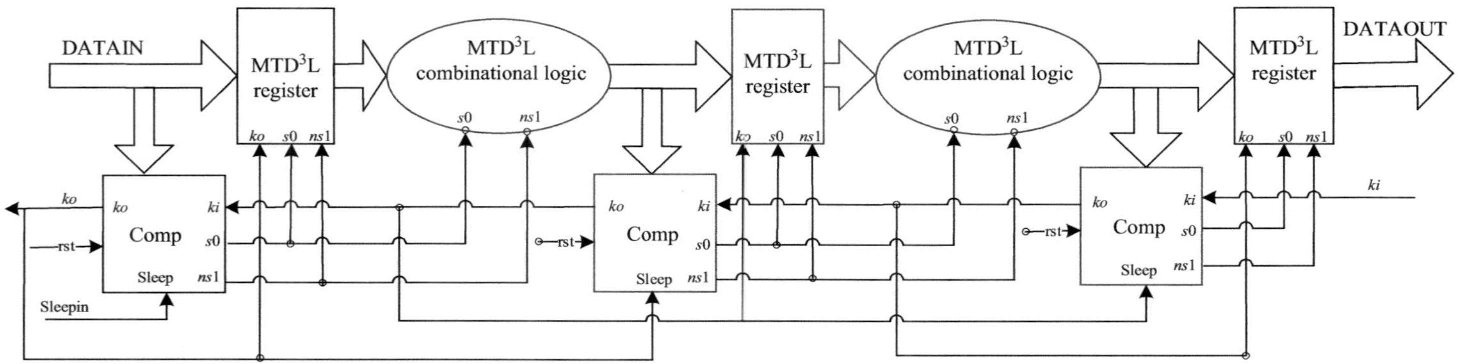

Figure 12.9 Improved overall MTD^3L architecture

Table 12.11 Delay and energy consumption between NCL, MTD³L_v1, and MTD³L AES cores

Design	Delay (ns)	Energy (nJ)
New MTD³L	85	1.085
MTD³L_v1	330	3.84
NCL	462	2.208

Table 12.12 Delay, energy, and size improvement between MTD³L_v1 and improved MTD³L

Design component	MTD³L_old	New MTD³L	Improvement
Delay (ns)	330	84.9	388.7%
Registration energy (nJ)	1.836	0.414	443.5%
Total energy (nJ)	3.84	1.085	353.9%
Registration transistors	242,364	44,344	546.6%

eliminating the KI generator, spacer generator, and spacer filter functions is that remaining components of the MTD³L architecture are implemented with only dual-spacer dual-rail gates whose switching activities does not depend on processed data. MTNCL and synchronous gates used in completion logic units for early completion detection and handshaking signals undergo the same switching activities every cycle regardless of input data. As a result, the program used to compute the number of switching transistors for attacks on NCL and D³L AES cores gave the same number for all possible input patterns, which was the expected outcome.

A CPA attack needs a characteristic that can be predicted depending on guessed key to compare to the measured energy consumption in order to compute a correlation coefficient and determine the correct key guess [3,22]. With hamming distances and the number of switching transistors being the same for all input patterns, it was concluded that a CPA attack on an MTD³L AES core is not applicable.

In addition to CPA attack not being applicable, the energy consumption uniformity across different input patterns proved further that MTD³L methodology balances the switching activity and decouples energy consumption from processed data. Energy values collected for all 256 possible inputs to an S-BOX circuit for one SUBBYTE operation present a standard deviation of 1.773E−14 joules or 0.0365% the average SUBBYTE energy consumption.

DPA attack

A successful DPA attack on a 128-bit key AES core requires a great number of power traces, elevated processing capabilities, and a long processing time. To reduce the number of required power traces, the attack focused on one S-BOX at a time. In this case, if successful, the attack would recover one byte of the key. It is assumed that if one byte can be recovered, the whole key can be recovered by

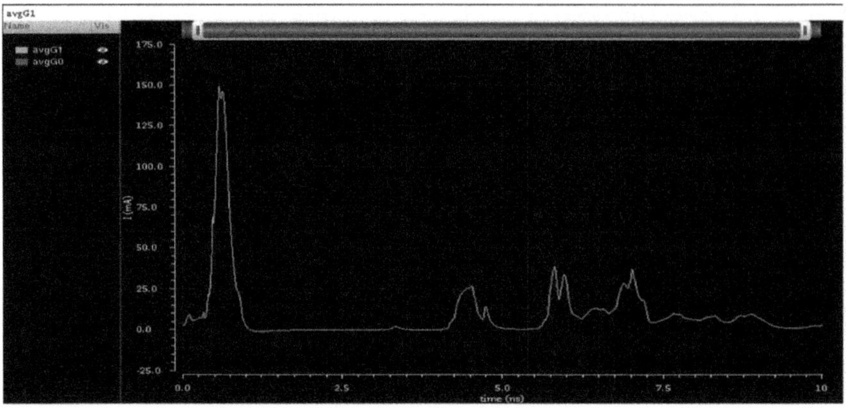

Figure 12.10 Grouping G0 and grouping G1 average power consumption traces

repeating the same process. This technique reduces the number of key guesses from 2^{128} required for a whole 128-bit key attack to 2^8 guesses required for a one-byte attack. Each key guess is used with multiple plain texts to achieve a practical amount of power traces for a statistical analysis. Moreover, the first-round attack was chosen over the last round attack to allow shorter simulation times. A first-round attack targets the first AddRoundKey and the first round SUBBYTE operations [1,2]; therefore, only the corresponding power trace portions were required for the attack, which allowed each simulation to be stopped after the first encryption round. Finally, the actual key and the exact time when the key addition and SUBBYTE operations are performed are known; therefore, the attack conditions allowed for this study were much more favorable than usual DPA attacks conditions.

Statistical analysis performed on more than 4,000 power traces proved that there was no sign of correlation detected. Observed power traces corresponding to different input patterns showed very close similarities with spikes corresponding to executed operations (steps) rather than processed data. These similarities proved again that power consumption in MTD^{3}L circuits depends on the operation being executed, but totally independent of the processed data.

Statistical analysis for the attack was based on a software model representation of the S-BOX circuitry that helped separate power traces into two groupings. Grouping *G0* consists of traces corresponding to inputs that should cause the circuit to draw less current at the points of interest, while *G1* consists of power traces corresponding to higher power consuming inputs. For a successful attack, it is anticipated that the difference in power consumption at those points of interest keeps amplifying as more traces are grouped together. However, *avgG1* trace (trace representing the average of power traces in grouping *G1*) and *avgG0* trace (trace representing the average of power traces in grouping *G0*) as depicted in Figure 12.10 show no detectable differences. In fact, *avgG1* trace (green) covers *avgG0* trace (red) almost completely. In a successful DPA attack, at the point of interest, the first AddRoundKey stage for example, *avgG1* would have spiked

Table 12.13 SCAs mitigation capabilities across different design methodologies

Design methodology	CPA attack	DPA attack
Synchronous	Success	Not tested
NCL	Success	Not tested
D^3L	Failure	Not tested
MTD^3L_v1	Failure	Not tested
MTD^3L	Not applicable	Failure

Table 12.14 Delay and energy consumption across different design methodologies

Design methodology	Energy consumption (nJ)	Delay (ns)
Synchronous	1.356	153
NCL	2.208	462
D^3L	6.012	325
MTD^3L_v1	3.84	330
MTD^3L	1.085	85

higher separating itself from *avgG0* [1]. The two average traces covering each other in their entirety is another proof of lack of correlation between power consumption and processed data. Therefore, it is concluded that DPA attack would not break an MTD^3L AES core.

12.5 Results

Different methodologies used for crypto hardware design are discussed in this chapter. AES cores designed using these different methodologies were tested for SCAs mitigation capabilities and for their area, power (energy) consumption and delay overheads. Table 12.13 summarizes discussed methodology as it pertains to their resilience against SCAs.

D^3L, MTD^3L_v1, and MTD^3L AES cores have all proved great resilience against SCAs but power consumption, area, and delay overheads in D^3L and to a lesser degree, MTD^3L_v1 circuits remained great concerns. Table 12.14 presents energy consumption and delays associated with the execution of one data encryption by an AES core designed in each one of the discussed methodologies.

All discussed dual-spacer dual-rail design methodologies namely D^3L, MTD^3L_v1, and MTD^3L present crypto hardware with great resilience against SCAs. Respective researches on these methodologies have proven that AES cores implemented using dual-spacer dual-rail methodologies have withstood performed SCAs. The improved MTD^3L methodology presents an advantage over other methodologies because in addition to its resilience against SCAs, it presents the lowest energy and delay overheads besting even the synchronous methodology.

12.6 Conclusion

Data security becomes more and more critical as more sensitive data are stored and processed by everyday use personal devices. Encryption hardware, though highly safe against brute force attacks are vulnerable to side channel attacks. Moreover, advance in technology comes with advances in side channel attacks. Various countermeasures against side channel attacks have only stood strong for a limited period of time and eventually failed the test of time. Dual-rail-based methodologies such as NCL have the potential to decouple processed data from side channel information which would present a permanent solution to side channel attacks.

In this chapter, a dual-spacer dual-rail-based methodology capable of producing side channel attack resilient crypto hardware was presented. An AES core implemented using the D^3L methodology presented resilience against a modified CPA attack. Moreover, two other methodologies were derived from D^3L in pursuit of SCAs resilient crypto hardware with minimal overheads. The last effort in this path led to an improved MTD^3L methodology that achieved both great resilience against SCAs and very low overheads. In fact, an AES core designed in this methodology presented the best delay and energy consumption over all other discussed methodologies including the synchronous AES core. Moreover, a DPA attack performed on this AES core showed no sign of weakness and failed to break it. As a result of its completely balanced switching activities, the MTD^3L methodology also rends a CPA attack nonapplicable as hamming distances and switching activities remain the same for all possible input patterns. It is concluded that even though all three dual-spacer dual-rail-based methodologies are resilient against known SCAs, only the MTD^3L methodology achieves a completely balanced switching activity and produces the best circuits in terms of area, energy consumption and delay.

References

[1] P. Kocher, J. Jaffe, and B. Jun, "Differential power analysis," Proc. Adv. Cryptogr., Ser. LNCS, vol. 1666, pp. 388–397, 1999.

[2] A. Moradi, O. Mischke, and T. Eisenbarth, *Correlation-Enhanced Power Analysis Collision Attack*, Springer, Berlin, Heidelberg, 2010, pp. 125–139.

[3] E. Brier, C. Clavier, and F. Olivier, "Correlation power analysis with a leakage model," Cryptogr. Hardw. Embed. Syst., vol. 3156, pp. 16–29, 2004.

[4] T. S. Messerges, E. A. Dabbish, and R. H. Sloan, "Investigations of power analysis attacks on smartcards," in *USENIX Workshop on Smartcard Technology*, Chicago, IL, USA, May 10–11, 1999, pp. 151–161.

[5] P. Bottinelli and J. W. Bos, "Computational aspects of correlation power analysis," J. Cryptogr. Eng., pp. 1–15, 2016.

[6] S. Heron, "Advanced encryption standard (AES)," Netw. Secur., vol. 2009, no. 12, pp. 8–12, 2009.

[7] S. Aumonier, "Generalized correlation power analysis," in *Proceedings of the Ecrypt Workshop Tools for Cryptanalysis*, 2007.

[8] P. Kocher, J. Jaffe, B. Jun, and P. Rohatgi, "Introduction to differential power analysis," J. Cryptogr. Eng., vol. 1, no. 1, pp. 5–27, 2011.

[9] C. Herbst, E. Oswald, and S. Mangard, "An AES smart card implementation resistant to power analysis attacks," in Lecture Notes in Computer Science (including subseries Lecture Notes in Artificial Intelligence and Lecture Notes in Bioinformatics), 2006, vol. 3989, pp. 239–252.
[10] S. Nikova, V. Rijmen, and M. Schläffer, "Secure hardware implementation of nonlinear functions in the presence of glitches," J. Cryptol., vol. 24, no. 2, pp. 292–321, 2011.
[11] E. Oswald, S. Mangard, N. Pramstaller, and V. Rijmen, "A side-channel analysis resistant description of the AES S-box," Fast Softw. Encryption, pp. 413–423, 2005.
[12] D. Canright and L. Batina, "A very compact 'perfectly masked' S-box for AES," in *Applied Cryptography and Network Security*, Springer, Berlin, Heidelberg, 2008, pp. 446–459.
[13] J.-H. Ye, S.-H. Huang, and M.-D. Shieh, "An efficient countermeasure against power attacks for ECC over GF(p)," in 2014 *IEEE International Symposium on Circuits and Systems (ISCAS)*, 2014, pp. 814–817.
[14] T. Izu and T. Takagi, "A fast parallel elliptic curve multiplication resistant against side channel attacks," in Lecture Notes in Computer Science (including subseries Lecture Notes in Artificial Intelligence and Lecture Notes in Bioinformatics), 2002, vol. 2274, pp. 280–296.
[15] T. Akishita and T. Takagi, "Zero-value point attacks on elliptic curve cryptosystem," Inf. Secur., vol. 1, pp. 218–233, 2003.
[16] S. C. Smith and J. Di, "Designing asynchronous circuits using NULL convention logic (NCL)," Synth. Lect. Digit. Circuits Syst., vol. 4, no. 1, pp. 1–96, 2009.
[17] K. M. Fant and S. A. Brandt, "NULL convention logic: a complete and consistent logic for asynchronous digital circuit synthesis," in *Proceedings of International Conference on Application Specific Systems, Architectures and Processors: ASAP '96*, 1996, pp. 261–273.
[18] W. Cilio, M. Linder, C. Porter, J. Di, D. R. Thompson, and S. C. Smith, "Mitigating power- and timing-based side-channel attacks using dual-spacer dual-rail delay-insensitive asynchronous logic," Microelectr. J., vol. 44, no. 3, pp. 258–269, 2013.
[19] M. Linder, J. Di, and S. Smith, "Multi-threshold dual-spacer dual-rail delay-insensitive logic (MTD3L): a low overhead secure IC design methodology," J. Low Power Electron. Appl., vol. 3, no. 4, pp. 300–336, 2013.
[20] J. Di and S. C. Smith, "Ultra-low power multi-threshold asynchronous circuit design," U.S. Patent 7,977,972 B2, July 12, 2011.
[21] J. P. T. Habimana, F. Sabado, and J. Di, "Multi-threshold dual-spacer dual-rail delay-insensitive logic: an improved IC design methodology for side channel attack mitigation," in *2016 IEEE International Symposium on Circuits and Systems (ISCAS)*, Montreal, QC, 2016, pp. 750–753.
[22] C. Clavier, J. S. Coron, and N. Dabbous, "Differential power analysis in the presence of hardware countermeasures," In *Proc. CHES 2000*, pp. 252–263, 2001.

Chapter 13

Using asynchronous clock distribution networks for timing SFQ circuits

Ramy N. Tadros[1] and Peter A. Beerel[1]

Single Flux Quantum (SFQ) technology has the potential to meet the booming demands for lower power consumption and higher operation speeds in the electronics industry and future exascale supercomputing systems. Nevertheless, the promised benefits of three orders of magnitude lower power at an order of magnitude higher performance have yet to be attained. In particular, variability and scalability have been long-term obstacles for the technology to advance, compete, and replace silicon CMOS. These issues have been confounded with an absence of an established tool flow, a main trump of CMOS digital design.

In this chapter, we discuss the use of asynchronous clock distribution networks (ACDNs) to provide the timing for SFQ circuits. In particular, we review the *hierarchical chains of homogeneous clover-leaves clocking*, or $(HC)^2$ LC [1], a self-adaptive clocking technique designed to be resilient in such uncertain environments. $(HC)^2$ LC inherits its robustness from its asynchronous nature that adapts to the spatially correlated cell delays, trading-off reasonable area, and power overheads for higher reliability and improved scalability.

13.1 Introduction

Whilst the present has big data and supercomputers, the future must analyze even bigger data and therefore requires even more powerful computers. Nevertheless, the power consumption of a modernly designed exascale computing platform is in the range of dozens of megawatts [2]. Following the most optimistic of assumptions, future supercomputers would require an amount of power similar to what is generated by a small power plant [3], hence the need for low-power and high-performance processors.

[1]Ming Hsieh Department of Electrical and Computer Engineering, University of Southern California (USC), Los Angeles, CA, USA

13.1.1 Why superconductive?

Achieving these projections is challenging because the VLSI industry is approaching the physical limit of semiconductor scaling. Many in the VLSI community are seeking the future of More-of-Moore in beyond-CMOS devices [4]. With a theoretical potential of one order of magnitude higher speed with up to three orders of magnitude lower power in the case of non-resistive bias networks [3], superconductive electronics—and SFQ in particular—has been conveyed as the definite near future back in the late 1980s [5], despite the cryocooling overhead [3].

Nevertheless, this promise has never been attained. Among many suspects, variability and scalability seem to play a significant impeding role [6–10]. First, regarding scalability, CAD tools and their development and evolution throughout the decades of technology advancements have enabled CMOS technology to thrive. The superconductive electronics community lacks such an established flow [8–10]. Second, regarding variability, SFQ has high uncertainties [6,7]: (i) global process variations [6], (ii) local mismatches due to fabrication limitations [7], (iii) RLC parasitics [11], (iv) bias distribution mismatches [12], (v) flux trapped during superconductive transitioning [13], (vi) thermal fluctuations [6], and (vii) local resistors heating [14]. This forced a 1 THz device to function at a substantially lower clock frequency of 20 GHz.

13.1.2 Timing is a challenge

In such an uncertain environment, the operation at high frequencies remains very challenging [6,15]. While some researchers assert that balanced tree zero-skew clocking is not sufficiently robust [5,16], radical asynchronous solutions are often deemed too expensive [17,18]. It is true that asynchronous solution seems very natural to SFQ as explained by [5], but high-performance objectives lured researchers away from pure asynchronous solutions. Unfortunately, hybrid approaches are too custom to be generalized [15,19–21].

13.1.3 Asynchronous clock distribution networks

In [22], we suggested the use of an ACDN [1] in order to provide the SFQ clocking of a circular shift register (CSR), the simplest form of an algorithmic loop that can be used to study timing [23]. For a 32-gate CSR, the proposed technique achieved up to a 93% yield improvement at the same cycle time compared to zero-skew tree clocking. Nevertheless, the work of [22] did not address how to extend the proposed clocking structure to more generic and complex pipelines than a basic CSR loop. In particular, its cycle time was $O(N_{\text{gates}})$, which is highly impractical for large-scale VLSI designs.

Our previous work in [1,24] extended [22] and proposed a robust and self-adaptive clocking technique for generic and complex pipelines, the cycle time of which is independent of the total number of gates. In this chapter, we discuss the $(\text{HC})^2$ LC which inherits its robustness from (1) the spatial correlation of various sources of variations [6,7,11,12,14,25] and (2) the timing robustness of traditional counterflow clocking [5].

13.1.4 Chapter overview

The chapter is structured as follows. Section 13.2 provides some background about SFQ technology, the timing fundamentals of synchronous systems, and clock distribution networks (CDNs) in SFQ. Then, Section 13.3 explains the theoretical aspects of ACDNs based on marked graph (MG) theory. After that, Section 13.4 discusses the architecture of the $(HC)^2$ LC scheme, its timing properties, and how to construct it, and reviews some preliminary results. Finally, Section 13.5 is a discussion.

13.2 Background

First, this section explains the basics of SFQ technology and circuits, points to its differences with CMOS, and discusses the various sources of uncertainties. Then, it provides some fundamentals about timing, synchronization, and CDNs. Additionally, it discusses the CDN design and challenges in SFQ in particular.

13.2.1 SFQ technology

Superconductivity [26,27] is the zero electrical resistance and the expulsion of magnetic field in certain materials when cooled below a characteristic critical temperature. It is not the idealization of perfect conductivity based on classical physics, but rather a quantum mechanical phenomenon.

Both [26] and [27] explain the quantum phenomenon of superconductivity and the tunneling properties of the main basis of superconductive electronics—the Josephson Junction (JJ). In short, a JJ is formed by two superconductors separated by a small discontinuity, and it possesses interesting IV characteristics that permits its use to build logic gates [5]. In the first paper of the IEEE transactions on applied superconductivity, Likharev and Semenov [5] summarized the fundamental properties of SFQ technology, where overdamped JJs are connected in a way such that binary information is presented in short quantized pulses—called fluxons, or SFQ pulses—instead of voltage dc levels. An overdamped JJ could leap producing a fluxon in as little as 0.5 ps back in the 1980s.

The main differences between SFQ and traditional CMOS [5] can be summarized as follows.

Representation of bits. SFQ circuits follow the "*SFQ basic convention*". During a certain period of time, the arrival of the SFQ pulse is identified as logic high, and the absence of a pulse during that period is identified as logic low.

Gates-level pipelining. Intrinsically, every logic cell requires a clock signal to operate; traditional combinational logic cells cannot be implemented. This modifies the structure designers traditionally use to abstract a sequential netlist where they use two registers separated by a cloud of combinational logic. In SFQ, there is no combinational cloud and there are only clocked gates. This results in gate-level pipelines, which is challenging because of

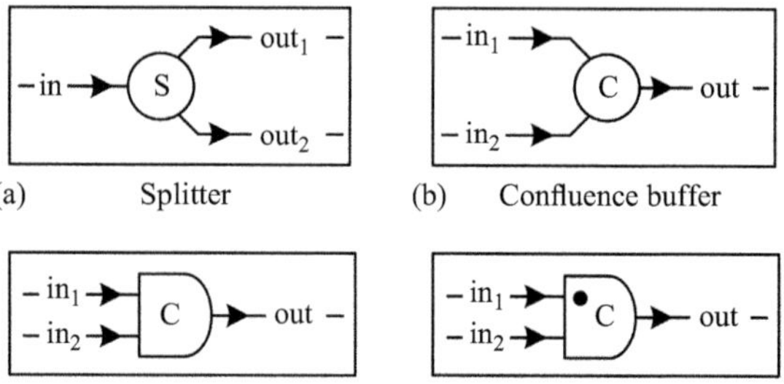

(a) Splitter (b) Confluence buffer

(c) Coincidence junction (d) Primed coincidence junction

Figure 13.1 Some SFQ basic cells symbols [1]

(a) pipeline starving [28,29] in which the majority of the pipeline is empty of data, reflecting a low effective throughput, (b) having relatively many more clock sinks than traditional CMOS circuits, aggravating the challenges in CDN design, and (c) increased sensitivity to setup and hold times because they represent a larger fraction of the clock period as there is no significant combinational logic delay.

Interconnects. Simple connecting wires do not exist in the conventional sense. Interconnections in SFQ ICs are either a passive transmission line (PTL) which has to be matched, or an active transmission line which is known as a Josephson transmission line (JTL) [6].

Destructive read-out (DRO). In SFQ logic gates, the evaluation of a logic cell using a clock pulse is a destructive process. The gates are thereby called DRO cells. CMOS registers are fundamentally on the contrasting side.

Fan-out. The fan-out of SFQ cells is strictly one. This necessitates the use of a tree of splitters (see Figure 13.1(a)) to obtain a higher fan-out.

Besides the synchronous logic cells, SFQ uses several other cells to perform nonlogic functions. These cells are asynchronous by nature [5] in that they do not rely on a clock. Figure 13.1 shows the symbols of some basic asynchronous SFQ cells, first, the *splitter*; Figure 13.1(a) introduces the splitter cell which solves the fan-out issue and uses current amplification in order to generate two separate pulses copying the pulse on its input port as soon as it occurs. Second, Figure 13.1(b) shows the *confluence buffer* (CF) which is a merger that generates an output pulse whenever a pulse is detected at either of its input ports. If the two input pulses occur within a certain setup time, only one pulse is generated. That is what distinguishes it from a combinational OR gate. Third, the *coincidence junction* (or C-junction) is depicted in Figure 13.1(c) and is similar in functionality to the Muller C-element [16]. An output pulse is spawned if and only if both input ports sense an incoming pulse. Its functionality follows a finite state machine (FSM) as described in [5]: if two pulses arrive sequentially on one input while the other input received none, the second pulse does not change the state and has basically no effect. Its primed conjugate is illustrated in Figure 13.1(d), where using a tweak in the bias current distribution circuit results in the junction being initialized in a different state of the FSM. This means that the first pulse on in_2 would result in an output pulse, and it shall behave as an un-primed C-junction for all subsequent pulses.

As mentioned in Section 13.1.1, SFQ technology exhibits a high level of variability. We hereby recount the known sources of variations.

Global process variations. Manufacturing-induced variations have a significant impact on the physical and timing parameters in SFQ [6,7].

Local mismatches. Due to photolithography diffraction limits, manufacturing tolerances, and layout inaccuracies and asymmetries, local uncertainties should be considered [6,7,25].

RLC parasitics. Although designers tend to consider the lumped-element impedance values as precise, many parasitics could affect the practical values of those impedances in addition to their mutual interactions [11].

Bias distribution network. Mismatches in the bias current alter the gates operation and modify their timing parameters [12].

Flux trapping. During cooling, devices experience a transitional partially superconductive phase. When a "normal" region is surrounded by superconductive regions, some quanta of flux can reside and become trapped, altering the critical current of the associated JJs [13,30].

Thermal fluctuations. Cooling is imperfect leading to thermal fluctuations that can lead to storage and decision errors [6,31].

Quantum fluctuations. Variations in transport properties of Niobium can lead to variations in the critical current of associated JJs [6].

Local resistors heating. Both RSFQ and ERSFQ/eSFQ [3] use overdamped junctions where a resistor is shunted to every JJ [5], while RSFQ uses resistors for the bias distribution networks as well. In addition to other resistors used in various gates or components. All these resistors cause thermal noise which affects the timing parameters of the nearby gates [14].

Jitter accumulation. As the number of JJs in a path increases, the amount of jitter on the gates constituting that path accumulates resulting in an additional source of timing and delays variation.

13.2.2 Timing fundamentals

Yield failures can emerge from either fabrication malfunctions, flux trapping during cooling, dc biasing, or timing violations. This chapter is solely concerned about the latter. Towards that end, this subsection provides background on timing and clocking.

A pair of registers are sequentially adjacent if only combinatorial logic (no sequential elements) exists between the two registers [32]. Figure 13.2 shows two sequentially adjacent cells in a synchronous system [16]. Note that we use Δ to denote the maximum delay value, and δ to denote the minimum (contamination [33]) delay value. The difference between an RSFQ system and a CMOS system — depicted in Figure 13.2(a) and (b), respectively—is that in CMOS, there is a combinatorial logic block between the register cells; while in SFQ, the logic cells are

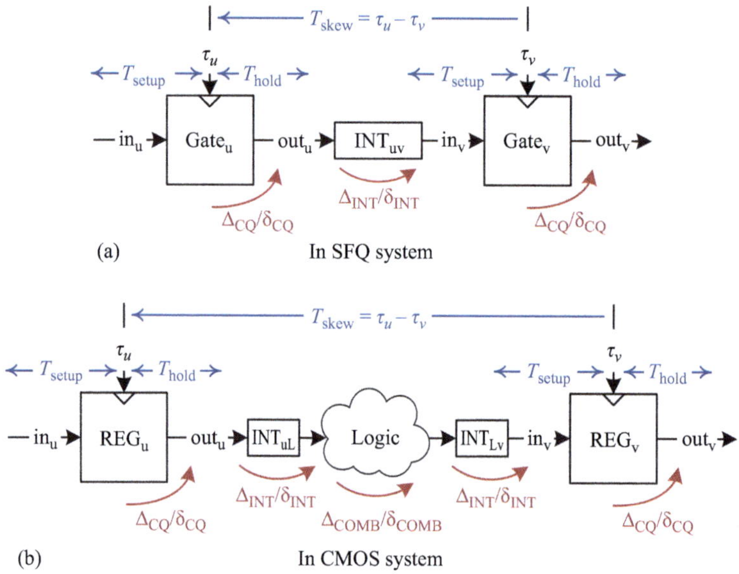

Figure 13.2 Two sequentially adjacent gates in a synchronous system [1]. τ_x represents the clock input of the xth gate. Data and clock paths parameters are depicted as red (curved) and blue (straight), respectively. INT stands for the interconnect between the gates; REG and COMB stand for register and combinational logic, respectively

themselves the registers, and there is no more logic in between the cells—only the interconnects delay takes place between them. This is because all regular SFQ logic gates are clocked.

The main timing conditions that characterize conventional synchronous clocking [32] are (1) steady-state periodicity, where an SFQ pulse occurs on the clock input every time period T_{sys}, and (2) well-determined skew value, where the difference of the arrival time of the clock between any two clock sinks is well-defined. These conditions can be formalized for any clock sink, $Gate_n$, as follows [1]:

$$\tau_n^{(i)} = T_{up} + T_{skew_n} + i \cdot T_{sys}, \quad \forall i \geq i_{up} \tag{13.1}$$

where i represents the ith occurrence of the clock signal. Here, T_{up} and i_{up} are the warming-up period and occurrence index, respectively, which are constants that represent the transient phase of the clocking system before entering the periodic steady state [1,34,35]; T_{skew_n} is the known skew at $Gate_n$ with respect to a common reference.

In case of a system with feedback—which is the case of most sequential systems—the clock skew is said to be *conserved* [32]. In other words, the clock skew in a feedback path between any two gates, G_u and G_v, is related to the clock skew of the forward path by the following relationship:

$$T_{\text{skew}_{uv}} = -T_{\text{skew}_{vu}} \tag{13.2}$$

These concepts are crucial for the system to avoid timing violations and are usually checked through some form of static timing analysis (STA) [33]. In particular, in order to respect setup time on an arbitrary data path from a $Gate_u$ to a $Gate_v$ as depicted in Figure 13.2(a), the system should satisfy the *setup constraint* as follows [16]:

$$T_{\text{sys}} \geq \Delta_{\text{DATA}} + T_{\text{setup}} + T_{\text{skew}} \tag{13.3}$$

where $\Delta_{\text{DATA}} = \Delta_{\text{CQ}} + \Delta_{\text{INT}}$ is the maximum delay from the clock pulse at τ_u to the data pulse arrival at in_v. T_{setup} is the setup time of $Gate_v$. T_{skew} is the clock skew between the two gates. Second, in order to respect hold time on that data path, the system should satisfy the *hold constraint* as follows [16]:

$$T_{\text{skew}} + \delta_{\text{DATA}} \geq T_{\text{hold}} \tag{13.4}$$

Here, $\delta_{\text{DATA}} = \delta_{\text{CQ}} + \delta_{\text{INT}}$ is the minimum delay that a clock pulse at τ_u can take before its associated data pulse reaches in_v. T_{hold} is the hold time of $Gate_v$.

The value of T_{skew} is known as the systematic clock skew, which is the difference in the arrival time of the clocking signal at various clock sinks [33]. However, when we take the previously discussed uncertainties into consideration, this value cannot be considered as fixed. Instead, it should be dealt with as a random variable. Additionally, other high-frequency environmental variations such as noise or any irregularities in the clock source result in *clock jitter* [33]. Similar to clock skew, clock jitter stiffens the timing constraints and forces the system design to accommodate for larger margins, and hence limits performance [16,32,33].

As defined in [32], the CDN is the network designed to generate the clock signal waveform and deliver it to each register for synchronization. Conventionally, the common CDN structure is based on *equipotential* clocking, where the entire network is considered a surface which must be brought to a specific voltage at each half of the clock cycle.

Even in CMOS, the design of CDN is not a trivial task, especially for high frequency. This is due to the following:

- Reverse scaling [36] where interconnects do not scale with the same rate as devices. When the channel length decreases, the gate delay decreases, but the interconnect delay increases. This keeps the *insertion delays*—which is the delay from the clock source to the clock sinks—at the same value while the clock cycle decreases. The insertion delay has to be minimized because the probability density functions (pdfs) of the skew values depend on the absolute value of the delay. A larger insertion delay results in more skew uncertainties [37].

- At high frequencies, the clock waveform strays away from the ideal step response due to parasitics, and Elmore delay [38]—which is commonly used to model the interconnects delay for design—becomes less accurate [32].
- At high frequencies, on-chip inductance and lossy transmission line behavior starts to manifest, which increases the delay uncertainty.

There are several CDN techniques in CMOS such as grids [33,37], H-trees [33], serpentines [37], and spines [33]. Practically, a hybrid approach is often used for high-performance chips [37]. Three levels of distribution are thus used [39]: (1) global CDN—symmetrical buffered trees are used, shielding is done for all the wires, and either static or active deskewing is incorporated; (2) regional CDN—a spine or a grid is commonly used; (3) local CDN—each sink of the regional network is considered as a local clock buffer and then a balance tree is used for the timing of the local region.

13.2.3 Clocking in SFQ

As previously mentioned, designing a CDN for CMOS clocking at high frequency is very challenging. In SFQ, the challenges are even harder because of the following:

- More variations.
- Higher sensitivity to variations in skew, setup, and hold times because there is no significant combinational logic delay.
- Extremely high number of clock sinks due to gate-level pipelining.
- It is not clear how to perform deskewing in SFQ. In particular, the mechanisms needed to monitor the clock skew and adjust programmable delay lines accordingly have not been well studied.
- Due to the quantum based transmission, grids seem physically incompatible with SFQ pulses.

Consequently, CDNs and timing of SFQ at high frequencies is a challenging problem. The authors in [5] explained the two basic clocking techniques for SFQ, which are concurrent flow clocking that is suited for high performance, and counterflow clocking offering a more robust solution. However, the inevitable existence of *algorithmic loops* in computing hinders the use of such techniques due to the conservation of skew as stated in (13.2). These loops aggravate the clocking complexity. Several attempts have been made to build a processor in SFQ. Both FLUX [19] and CORE [15] used variants of the globally asynchronous locally synchronous (GALS) approach. TIPPY [20] used the data-driven self-timed technique of [40], and the processor in [21] used the asynchronous technique of [41]. All of those processors relied on significant manual and custom optimizations, and none of them used a fully synchronous CDN. Even for linear pipelines, the work of [42,43] suggested the use of asynchronous wave-pipelined structure for an 8-bit ALU and a 16-bit sparse tree adder. This suggests our assertion that clocking algorithmic loops in SFQ is still, as stated in 1996 [16], an unresolved problem.

The reasons of the inability to use a zero-skew balanced tree on a very large-scale SFQ chip can be summarized as follows:

- The uncertainty of physical design due to fabrication in addition to parameters fluctuations [12,44] is inconsistent with a zero-skew scheme that requires an exhaustive knowledge of every parameter in every branch.
- A large-scale inductive clock network carrying a very high speed wave is susceptible to various electromagnetic effects. Especially that no grid or adaptive de-skewing buffers [39] can be naturally implemented in SFQ.
- On-chip temperature fluctuations intensify clock jitter and skew [31,44].
- Flux trapping [13] affects the current bias distribution network [45], and hence alters any tree-designed symmetry.

13.3 Asynchronous clock distribution networks

In this section, we explain the fundamentals of ACDNs [1]. First, background on some aspects of MG theory is provided to establish a basis of the results. After that, the ACDN theory is discussed.

13.3.1 MG theory

Petri nets (PNs) are a graphical and mathematical modeling tool [46]. They are constituted of *places* (depicted as circles and denoted as p) that hold markings or tokens (depicted as black dots inside a place) which represent the conditions leading to this place were met, and *transitions* (depicted as bars and denoted as t) which represent actions or events. Arcs are used as interconnections between the places and the transitions to signify causal dependence. PNs can be timed by assigning a certain execution time value to either places or transitions. In case of assigning them to transitions, if all the input places to a transition have tokens in them, that transition is said to be enabled, and after its execution time, the transition is said to *fire*, signaling the completion of that event. Once fired, one token per input place is removed and one token is added to every output place. That is known as *the firing rule*.

MGs are a subclass of PNs where each place has exactly one input transition and exactly one output transition [46]. For the purposes of this chapter, some additional related definitions are needed. An MG is said to be *live* if it is always possible to fire any transition—i.e. it models deadlock-free operation. An MG is said to be *safe* if the number of tokens in each place does not exceed one. A *source* is a transition that is initially enabled. A *directed path* is a sequence of alternating places and transitions. The *directed path delay* is the sum of the execution times of the transitions forming that path. A *directed circuit* is a directed path that starts and ends at the same transition with every other node being distinct. A *critical circuit* is any directed circuit whose delay divided by the total number of tokens in it equals the maximum of such values across all directed circuits, T_{sys}.

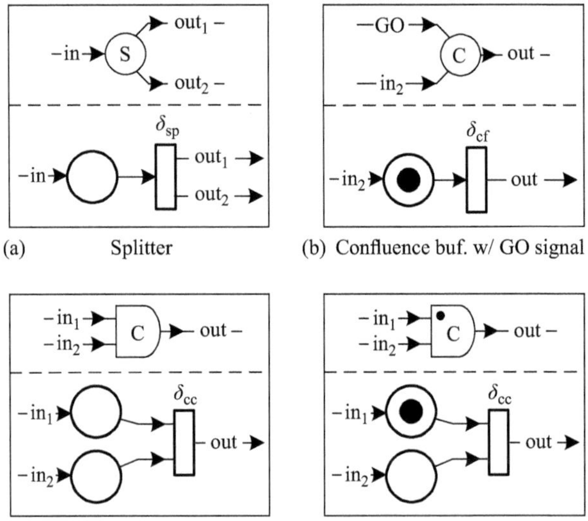

(a) Splitter (b) Confluence buf. w/ GO signal

(c) Coincidence junction (d) Primed C-junction

Figure 13.3 Basic SFQ cells and their MG models [1]

Given their properties, timed MGs are well-suited to model the timing of circuits that exhibit no choice and are periodic. Any gate can be modeled as a transition with execution time associated with it equal to the gate delay. For SFQ, any interconnect can be modeled as an input place connected to a transition. Figure 13.3 shows some basic asynchronous SFQ cells and their corresponding MG models. The splitter which is the cell needed to solve the SFQ single fan-out issue [5] is shown in Figure 13.3(a). A CF cannot be modeled as an MG due to its peculiar characteristics. However, the combination of the CF and a start-up GO signal as the one employed in Figure 13.7(a) can be modeled as an MG as illustrated in Figure 13.3(b), assuming that this GO signal is the first event to happen in the system and that it happens only once, which is the case in Section 13.4.3. The coincidence junction (C-junction) [47] is depicted in Figure 13.3(c), and its primed counterpart is illustrated in Figure 13.3(d).

13.3.2 ACDN theory

In this section, we explain the fundamentals of ACDNs. In particular, we discuss the timing characteristics of the timing signals arriving at the clock sinks, and the constraints that an asynchronous system must satisfy in order to achieve these characteristics.

13.3.2.1 Synchronous clock sinks

If we model the CDN as a timed MG, then we can consider all the transitions to be either clock sinks or other parts of the CDN, both of which have to satisfy the synchronous property discussed in Section 13.2.2 and formalized in (13.1). From [1], we recite the following:

Theorem 13.1: *In a live and safe MG, if (1) there is a single token in every directed circuit, (2) the net has a single source t_s, and (3) this source belongs to a critical circuit, then for any transition t_n in the graph, at any $i \geq 0$,*

$$\tau_n^{(i)} = \left(\max_{h \in \hat{\mathscr{H}}(t_s, t_n, 0)} \Delta(h) \right) + i \cdot T_{\text{sys}} \tag{13.5}$$

where $\tau_n^{(i)}$ *is the ith firing of the transition* t_n, $\hat{\mathscr{H}}(t_s, t_n, 0)$ *is the set of all zero-tokens directed paths from* t_s *to* t_n, *and* $\Delta(h)$ *is the delay of the directed path h.*

13.3.2.2 Uncertainty case

The third condition of Theorem 13.1 makes a strong assumption about the source transition. It requires knowledge of the exact delay of every directed circuit in the system to ensure it holds. Meanwhile, the main motivation in the work discussed in this chapter is the high level of uncertainty. Fortunately, Theorem 13.2 in [1] relaxes this restriction which we recite as follows:

Theorem 13.2: *In a live and safe MG, if (1) there is a single token in every directed circuit, (2) the net has a single source t_s, and (3) this source does not belong to any critical circuit,*

$$\begin{aligned} &\textit{then } \exists t_n \in C_n \in \mathscr{C}(t_s), \ \textit{and } \exists t_u \in \mathscr{J}, \ t_u \neq t_s, \\ &\qquad \textit{such that } \tau_s^{(i)} + \max_{h \in \hat{\mathscr{H}}(t_s, t_n, 0)} (\Delta(h)) < \\ &\qquad \tau_u^{(i - M(h_{u \to n}))} + \Delta(h_{u \to n}), \ \forall i \geq i_{\text{up}} \end{aligned} \tag{13.6}$$

where t_n shares a directed circuit with t_s, t_u is some other transition in the net. $h_{u \to n}$ is a path from t_u to t_n, and $M(h_{u \to n})$ is the number of tokens along this path.

This theorem means that if the source is not part of a critical circuit, it is always possible to probe a certain *neighboring* transition to find out the value of T_{sys}. In addition, if it is possible to modify the delay of one of the source circuits accordingly, forcing it to be critical, the system will then follow the synchronous property in (13.1) and (13.5) henceforth. This is discussed in detail in Sections 13.4.3 and 13.4.4.

13.4 Hierarchical chains of homogeneous clover-leaves clocking

This section reviews the $(\text{HC})^2$ LC approach [1,24]. It is a robust and self-adaptive clocking technique for generic and complex pipelines. The clocking network inherits its robustness from (1) the spatial correlation of various sources of variations

[6,7,11,12,14,25] and (2) the timing robustness of traditional counterflow clocking [5]. While trading-off a reasonable area and power overheads, it obtains higher reliability and improves scalability.

First, Section 13.4.1 describes the main body of the clocking hierarchy. Then, Sections 13.4.2 and 13.4.3 discuss the architecture of the bottom and top clocking levels, respectively, which permit the initiation and termination of the hierarchical structure. After that, Section 13.4.4 discusses the theoretical aspects of the architecture and Section 13.4.5 discusses its timing properties. Finally, Section 13.4.6 explains the architecture evaluation flow and provides some preliminary results.

13.4.1 Hierarchical chains

First, we explain the architectural foundation, which is the Hierarchical Chain's Link (HCL), illustrated in Figure 13.4(a). An HCL has a single input and a single output. Assuming a periodic signal on CLK_{in} with period T_{in}, the output signal on CLK_{out} shall have a period of

$$T_{\text{HCL}} = \max(T_{\text{in}}, \delta_{\text{ov}} + T_{\text{core}}) \tag{13.7}$$

where δ_{ov} is the HCL overhead delay and T_{core} is the propagation delay of the blackboxed *core*.

Note that since this core has one input and one output, it also can be an HCL. Using HCL as the building block of a fractal hierarchical architecture is based on this property. Now we connect such links together to form a chain of HCLs as shown in Figure 13.5. Once again, note that a chain of HCLs can be considered as an HCL. Similarly, assuming an existing T_{in}, the output period can be written as follows:

$$T_{\text{chain}} = \max(T_{\text{in}}, (C+1) \cdot \delta_{\text{ov}}, \max_i T_{\text{HCL}_i}) \tag{13.8}$$

where C is the number of links in the chain, and $\max_i$ is over the ith link in the chain.

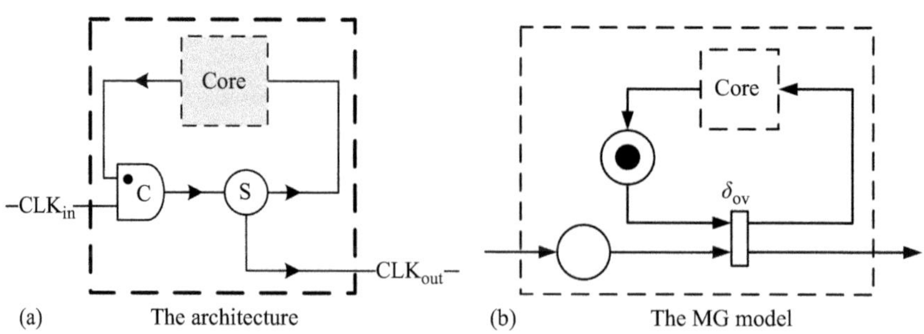

Figure 13.4 The hierarchical chain's link [1]

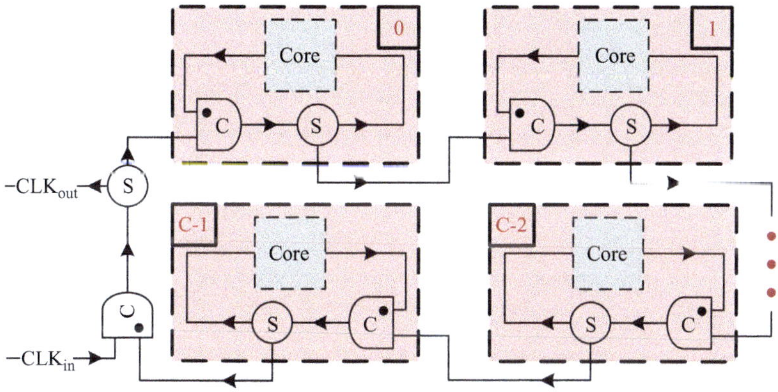

Figure 13.5 A chain of C HCLs [1]

If we assume that we have a large number of *cores* with some delay, T_{core_i}, then linking each core to the HCL circuitry of Figure 13.4(a), and chaining them, building the hierarchy upwards using the HCL chain of Figure 13.5, we end up with a hierarchical chain that has a top HCL. Hence, with the top HCL's CLK_{in} having a period of T_{in}, the period of the signal at the input of every *j*th bottom level core, T_{core_j}, shall be

$$T_{\text{core}_j} = \max(T_{\text{in}}, (C_{\max} + 1) \cdot \delta_{\text{ov}}, \delta_{\text{ov}} + \max_i \delta_{\text{core}_i}) \tag{13.9}$$

where $C_{\max}$ is the maximum number of links in an HCL chain.

13.4.2 Bottom level

This section describes how the logic cells are clocked at the bottom level of the hierarchy. In a similar way to the hybrid clover of [22], $(\text{HC})^2$ LC bottom clocking level uses only counterflow clocking as shown in Figure 13.6. The gates per clover are divided into L gates per leaf, where the clover has N leaves. First, CLK_{in} is fed to a splitters tree to obtain one input per leaf. Within one leaf, we use counterflow clocking [5,16] (see Section 13.2.1), where a sequence of L splitters is used to clock L gates sequentially in an opposite direction to the data flow. This naturally provides robustness to hold constraints at the cost of more stringent setup constraints. After that, we use a C-junctions tree to collect the leaves outputs producing a single CLK_{out} signaling the clocking of all the gates.

It is worth mentioning that the homogeneity of the flow (i.e. all leaves use counterflow) is not expected to be perfect for an arbitrary pipeline—i.e. the data path cannot always be consistently matched to counterflow clocking. For instance, a connection can exist across two distinct clovers or HCLs and the clocks associated with those connections would thus not follow the counterflow pattern. However, there will exist a certain gates-to-clock-sinks assignment that makes the homogeneity as close to perfect as possible. In fact, different orders

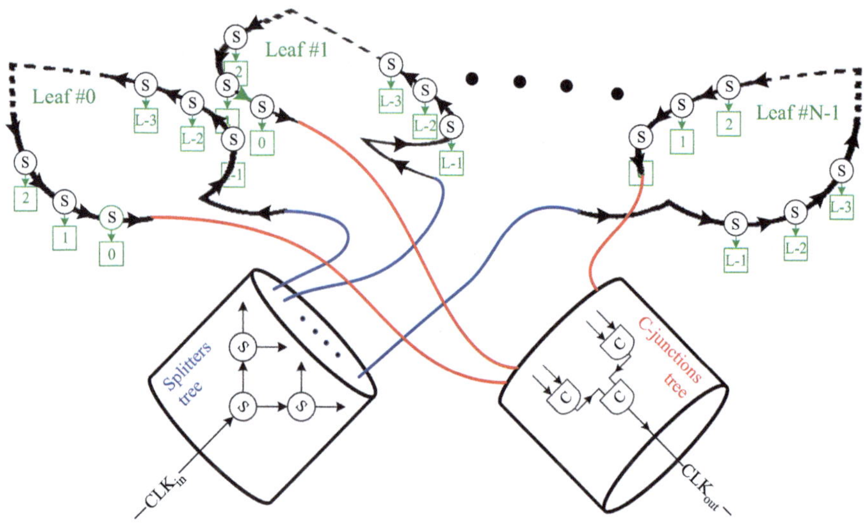

Figure 13.6 A homogeneous clover with N leaves and L gates per leaf [1]

can be created that optimize either the homogeneity, minimize the hold/setup buffers overhead [24], maximize the performance, or strike a balance between these various objectives.

Based on the diagram of Figure 13.6, T_{core}, or the *clover* delay, δ_{clover}, which is the delay from CLK_{in} to CLK_{out} can be written as

$$\delta_{\text{clover}} = [\log_2 N] \cdot \delta_{\text{ov}} + L_{\max} \cdot \delta_{\text{sp}} \tag{13.10}$$

where N is the number of leaves per clover, and $L_{\max}$ is the maximum number of gates per leaf belonging to that clover.

13.4.3 Top loop

This section explains how the highest hierarchy HCL, simply referred to as the *top loop*, is managed. Figure 13.7(a) illustrates the block diagram of the loop. (HC)^2 LC needs a single *GO* pulse to initiate the clocking system. This signal is coupled to the loop using a confluence buffer (see Figure 13.3(b)). After the initial firing, when no more pulses are generated on the *GO* port, the buffer shall function as a mere JTL. This results in an oscillating loop which explains the omission of a clock source. Notice that a dotted C-junction and a splitter are used to couple the rest of the hierarchy similar to the HCL structure. The output signal CLK_{out} could be used to probe the system from off-chip for either testing or communication, and it has a period of

$$\begin{aligned} T_{\text{top}} &= \max(T_{L1}, T_{L2}) \\ &= \max\left(\delta_{\text{ov}} + \delta_{\text{sp}} + T_{\text{hrcl}},\ \delta_{\text{cf}} + 2\delta_{\text{sp}} + \delta_{\text{ov}} + \delta_{\text{ctrl}}\right) \end{aligned} \tag{13.11}$$

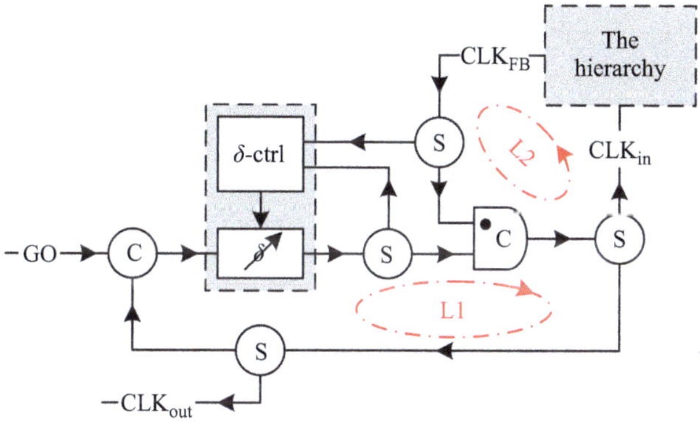

(a) The architecture

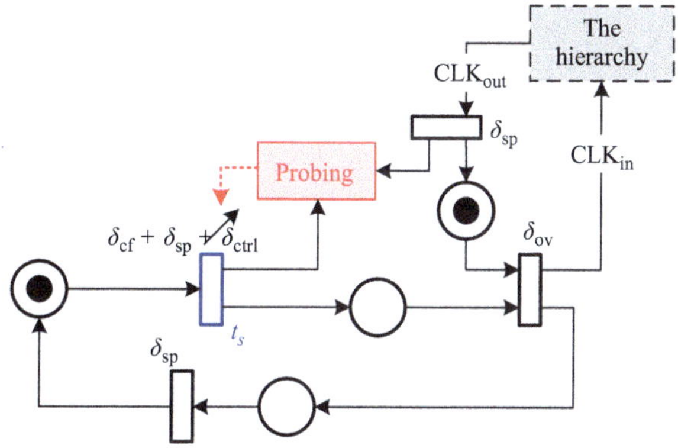

(b) The MG model

Figure 13.7 The top loop structure [1]

where T_{L1}, T_{L2}, and T_{hrcl} are, respectively, the periods of the loop $L1$, the loop $L2$, and the highest-level HCL which is depicted as the hierarchy box in Figure 13.7(a).

The circuit design of the shaded structure with variable control delay is left for future work, in addition to the loop stability analysis in the case of allowing the programmability to either increase or decrease the delay.

For our preliminary results, a behavioral model is used to verify the functionality of the proposed technique. It acts as a programmable delay line with delay δ_{ctrl}. Its value depends on two inputs: its own output, and on CLK_{FB} which is the CLK_{out} of the highest-level HCL. This ensures that the delay across the loop $L1$ is always longer than $L2$. In the resulting structure, the fact that $L1$ is longer than $L2$ would guarantee that the critical path of each clock path goes through the C-junction in the top loop of Figure 13.7(a). This is discussed in detail in the next section.

13.4.4 $(HC)^2$ LC theory

This section connects the structure of the $(HC)^2$ LC to the ACDN timing theory in Section 13.3.2. In particular, it shows that Theorems 13.1 and 13.2 (see (13.5) and (13.6)) can be applied to the clocking technique.

First, the $(HC)^2$ LC circuitry can be seen as a CDN which can be modeled as a timed MG as discussed in Section 13.3.1. Specifically, the MG model of the HCL structure and the top loop are illustrated in Figures 13.4(b) and 13.7(b). Second, if we prove that $(HC)^2$ LC satisfies the constraints of those theorems, then it can be used as an ACDN and hence used for the timing of a synchronous SFQ chip. The constraints are indeed satisfied as it is argued in the following:

1. *Live and safe MG.* Based on Theorem 6 in [46], which states that the number of tokens in a directed circuit does not change with firing, we can deduce that every HCL can be abstractly modeled as one place and one transition in series. Also, the bottom level (which does not contain any loops and thus has no tokens) can be abstractly modeled as one place and one transition, i.e. like a single gate with a single delay value. If we do this reduction recursively, replace every HCL with one place and one transition in series, and climb up the hierarchy, the top loop model becomes an alternating sequence of three places and three transitions that has a single token in it. It is clear that such a simple loop is a live and safe MG.
2. *A single token per circuit.* Similarly, every HCL has one single-token-directed circuit, and since no nodes are allowed to be repeated in a circuit by definition, we can abstract the HCL model and climb up the hierarchy recursively to show that no directed circuits can exist with more or less than one token.
3. *A single source.* The bottom level has no initial tokens, and the HCL structure does not contain a source. Therefore, the source created by the GO signal (see Figure 13.3(b)) is the only source, and the top loop is the only directed circuit that contains a source.
4. *Does the source belong to a critical circuit?* Assuming a high level of uncertainty—which is the case in SFQ as discussed in Section 13.2.1—the answer to this constraint can be either positive or negative. In case it is true and the source does indeed belong to a critical circuit—i.e. the top loop is the slowest loop—then Theorem 13.1 (see (13.5)) can be applied and the clocking technique works immediately upon start up. On the other hand, if the top loop is not critical, then Theorem 13.2 (see (13.6)) can be applied. In that case, let t_n in the theorem be the dotted C-junction in Figure 13.7(a). Thus, by probing its inputs, after i_{up} firings, the ith pulse on the un-primed input will occur before the $(i-1)$th pulse on the primed input, which means that the top loop needs to get slower. Then, slowing down progressively, the inequality will flip at some point, and the top loop will become critical. From this point onward, the circuit will behave following (13.5), which means it will satisfy the synchronous property of (13.1) and it can henceforth be used as an ACDN.

13.4.5 Cycle time and clock skew

Based on the discussions in Sections 13.2.2 and 13.3, all hierarchical chains share the same cycle time, T_{sys}. Substituting from (1.7) to (1.11), T_{sys} can be defined as

$$T_{\text{sys}} = \max \begin{cases} (\lceil \log_2 N_{\text{max}} \rceil + 1) \cdot \delta_{\text{ov}} + L_{\text{max}} \cdot \delta_{\text{sp}} \\ (C_{\text{max}} + 1) \cdot \delta_{\text{ov}} \\ \left(C_{\text{top}} + 1\right) \cdot \delta_{\text{ov}} + \delta_{\text{sp}} \\ \delta_{\text{cf}} + 2\delta_{\text{sp}} + \delta_{\text{ov}} + \delta_{\text{ctrl}} \end{cases} \tag{13.12}$$

where C_{max} is the largest chain length and C_{top} is the length of the highest level HCL (see the shaded box in Figure 13.7(a)).

However, this is not enough to fully determine the timing of the chip and perform STA. The value of T_{skew_n} in (13.1), which is found to be $\max_{h \in \mathscr{H}(t_s, t_n, 0)} \Delta(h)$ in (13.5), has to be well defined. Fortunately, in $(\text{HC})^2$ LC, there exists only one zero-token path from t_s to any transition t_n.

We can, therefore, conclude that the skew for any clock sink, i.e. gate n, has two components:

$$T_{\text{skew}_n} = \Delta_{\text{hrcl}}(n) + \Delta_{\text{local}}(n) \tag{13.13}$$

Here, $\Delta_{\text{hrcl}}(n)$ is the hierarchical delay that evaluates to the delay of the path $h_{s \to clv(n)}$ from t_s to the input of the clover (bottom level) that contains gate n. This clover is the core of one of the lowest level HCLs (see Figure 13.4(a)). $\Delta_{\text{local}}(n)$, on the other hand, is the delay of the local distribution within the clover (see Figure 13.6), i.e. the delay from the clover's input to the output of the splitter feeding this clock sink n.

We can further decompose the hierarchical delay. For H hierarchy levels where the 0th level is the lowest level HCL and the top loop is the $(H+1)$th level, and with $R(h, n)$ returning the rank (or order) of the HCL's core at the hth level that belongs to the path $h_{s \to clv(n)}$,

$$\Delta_{\text{hrcl}}(n) = \delta_{\text{cf}} + \delta_{\text{sp}} + \delta_{\text{ctrl}} + \delta_{\text{ov}} \cdot \left(H + 1 + \sum_{h=0}^{H} R(h, n) \right) \tag{13.14}$$

Note that $R(h, n)$ is in reversed order because of the counterflow clocking.

We can also more precisely define the local delay within a clover. If $N_{\text{sp}}(n)$ is the number of splitters from the input of the clover containing the gate n to the input of the specific leaf containing it, and $L_{\text{sp}}(n)$ is the number of splitters within that leaf leading to the gate n, then

$$\Delta_{\text{local}}(n) = \delta_{\text{sp}} \cdot \left(N_{\text{sp}}(n) + L_{\text{sp}}(n)\right) \tag{13.15}$$

Note that $N_{\text{sp}}(n)$ would equal $\log_2 N$ in case N is a power of 2, and that $L_{\text{sp}}(n)$ is also in reversed order because of the counterflow.

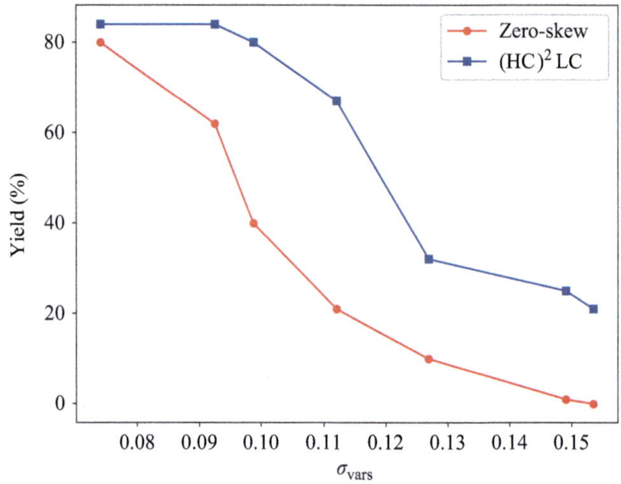

Figure 13.8 Monte-Carlo results for the c880 benchmark [48]

13.4.6 Comparison to conventional CDN

We evaluate (HC)^2 LC by comparing it to the baseline design, a conventional CDN based on a zero-skew clock tree. We leverage an evaluation flow [1] to implement these two CDNs ((HC)^2 LC and zero-skew tree) on SFQ netlists and test it on combinational ISCAS'85 benchmark circuits [48]. We obtain preliminary results from which Figure 13.8 shows a sample. In particular, this result was obtained performing the following steps:

1. Choose the CDN and configure the corresponding parameters. In case of (HC)^2 LC, the following are determined: leaf length, number of leaves per clover, and number of links per chain.
2. Use and modify the Berkeley open source synthesis tool, ABC [49], to
   - levelize the combinational graph for gate-level pipelining;
   - build the clock network as an asynchronous circuit without connecting it to the logic gates;
   - generate a gates to clock-sinks assignment map. It is worth mentioning that in the case of (HC)^2 LC, an arbitrary mapping is used. This assignment map could be optimized to minimize the hold/setup fixes overhead (described in Step 2) but this optimization is left as future work;
   - check for potential setup violations and add latency to fix them. For every data connection between a gate clocked at time τ_i to one at τ_j, the number of inserted flops equals

$$N_{\text{FF}} = \max\left(0, \left\lceil \frac{\tau_i - \tau_j + \Delta_{\text{DATA}} + T_{\text{setup}} - T_{\text{sys}}}{T_{\text{sys}} - \Delta_{\text{FF}} - T_{\text{setup}}} \right\rceil\right) \tag{13.16}$$

     where Δ_{FF} is the maximum C-Q delay of a flop;

- re-levelize the graph. Note that the flops added in both this step and the previous one are locally clocked from the associated clock sink;
- check for potential hold violations and add buffers to fix them as follows:

$$N_{\text{buf}} = \max\left(0, \left\lceil \frac{T_{\text{hold}} - \tau_i + \tau_j - \delta_{\text{DATA}}}{\delta_{\text{buf}}} \right\rceil\right) \tag{13.17}$$

where N_{buf} is the number of buffers, and δ_{buf} is the delay of one hold buffer.

3. Using an in-house script, modify the graph converting it to behavioral SystemVerilog SFQ interfaces in which they instantiate the cell models described in [22].
4. Generate random input combinations for each benchmark.
5. Run dynamic co-simulation to compare the functionality to the Verilog benchmark and use the built-in SystemVerilog timing checks, as in [22].
6. Using a grid-based variations model that exhibits the spatial correlation among many sources of variations [50–52], run Monte-Carlo generating different delay values for every single run.
7. A run is considered to pass if it reports a perfectly successful functional operation based on the co-sim outputs comparison, and zero timing violations reported. Otherwise, it is a fail.

Each point in Figure 13.8 is the yield obtained over 100 Monte-Carlo runs that simulate the c880 benchmark from the ISCAS'85 benchmark circuits [48]. The x-axis is the standard deviation of the variations applied hierarchically using a recursively fractal grid-based model, while the y-axis is the yield as previously defined. In case of zero-skew tree, T_{sys}, the average cycle time is an assigned value that is fixed among all the runs. On the other hand, it is run-based for $(\text{HC})^2$ LC since it follows the slowest loop whose delay value changes due to the variations of the gates delays. Regarding the implementation of the zero-skew tree, the values shown in the figure are based on a perfectly balanced tree with difference of depth strictly equal to zero. Moreover, no clock source jitter is applied, which favors the zero-skew tree.

The preliminary results in Figure 13.8 show that, on average, $(\text{HC})^2$ LC with an un-optimized gates-to-clock-sinks assignment has a yield improvement of 80.65% over ideal and jitter-free zero-skew trees. This improvement comes in the expense of an area overhead of 31.59% and a cycle time overhead of 5.49%. Our future work is focused on improving the gate-to-clock-sinks assignment which will further improve these benefits.

13.5 Discussion

The main motivation behind the work explained in this chapter is the low yield reported in many SFQ papers [6,7,19,21,25,40,43,53–57]. Though some chip failures are due fabrication and/or other issues, many failures were reported to be functional. We believe that due to unanticipated levels of uncertainty and other effects, timing violations and clock distribution may be the root of many of those failures. Consequently, we hypothesize that using the CMOS conventional zero-skew balanced trees is not the best fit for clocking large-scale digital SFQ chips.

$(HC)^2$ LC provides a higher level of tolerance to variations because of two key properties:

1. *Self-adaptive.* The top loop self-adapts to the worst-case delays of any lower-level loop. Moreover, the top loop increases its own intrinsic delay accordingly so that the clock skew at every clock sink can be exactly defined (see Section 13.4.4); that is, the relative time separation between (1) the clock edge reaching an arbitrary gate and (2) the reference net (CLK_{in} in Figure 13.7(a)) can be determined independent of which lower-level loop has the worst-case delay. Together, this provides a stable timing reference that can be used to ensure setup, and hold times are met across the entire circuit. This is in sharp contrast to synchronous clock trees where fixing a setup/hold delay on one portion of the clock path (using tunable delay lines as in [58,59]) can cause setup/hold problems in other sections of the tree.
2. *Spatial correlation of variations in a counterflow scheme.* Here we emphasize that the high resilience towards hold violations comes from the use of counterflow clocking at the bottom level and in ranking all the HCLs and clovers. Additionally, we assert that in such system with mostly local timing constraints, the spatial correlation between neighboring clock and data circuitry makes the proposed architecture more robust against setup and holds violations than zero-skew clocking. This is because of the built-in correlation between the dataflow and clock flow, and between the hold/setup buffers/flops and the clock skew path that would have potentially caused a violation. This concept was analyzed for CMOS logic in [60], where they used the data and clock paths together to improve the correlation among clock sinks to reduce the sensitivity of clock skew [39].

We believe that this timing scheme is more natural to SFQ, which makes it a more convenient solution to SFQ than zero-skew trees despite the performance and power overheads. It is true that zero-skew clock trees are simple and more intuitive; however, after decades of attempts, they failed to deliver when faced with the variability and scalability challenges of SFQ. On the other hand, $(HC)^2$ LC has the potential to offer the functionality and the reliability of a concrete and more realistic solution.

In summary, given a large level of uncertainty, the main advantage of this approach is that the clock follows the circuit's variability and makes less assumptions about the cells delay than zero-skew clock trees. The $(HC)^2$ LC clock path—in its own peculiar way—self-adapts to the data path and forces the whole distribution network to follow.

References

[1] Tadros RN, Beerel PA. A robust and self-adaptive clocking technique for SFQ circuits. IEEE Transactions on Applied Superconductivity. 2018;28(7): 1–11.

[2] Reed DA, Dongarra J. Exascale computing and big data. Communications of the ACM. 2015;58(7):56–68.

[3] Mukhanov OA. Energy-efficient single flux quantum technology. IEEE Transactions on Applied Superconductivity. 2011;21(3):760–769.

[4] ITRS. International Technology Roadmap for Semiconductors 2.0: Beyond CMOS; 2015.

[5] Likharev K, Semenov V. RSFQ logic/memory family: A new Josephson-junction technology for sub-terahertz-clock-frequency digital systems. IEEE Transactions on Applied Superconductivity. 1991;50(1):3–28.

[6] Bunyk P, Likharev K, Zinoviev D. RSFQ technology: Physics and devices. International Journal of High Speed Electronics and Systems. 2001;11(01): 257–305.

[7] Vernik IV, Herr QP, Gaij K, *et al.* Experimental investigation of local timing parameter variations in RSFQ circuits. IEEE Transactions on Applied Superconductivity. 1999;9(2):4341–4344.

[8] Gaj K, Herr QP, Adler V, *et al.* Tools for the computer-aided design of multigigahertz superconducting digital circuits. IEEE Transactions on Applied Superconductivity. 1999;9(1):18–38.

[9] Fourie CJ, Volkmann MH. Status of superconductor electronic circuit design software. IEEE Transactions on Applied Superconductivity. 2013;23(3): 1300205–1300205.

[10] IRDS. International Roadmap for Devices and Systems (IRDS): 2017 Edition, Beyond CMOS; 2017.

[11] Fourie CJ, Perold WJ, Gerber HR. Complete Monte Carlo model description of lumped-element RSFQ logic circuits. IEEE Transactions on Applied Superconductivity. 2005;15(2):384–387.

[12] Gaj K, Herr Q, Feldman M. Parameter variations and synchronization of RSFQ circuits. In: Conference Series-Institute of Physics. vol. 148. IOP Publishing Ltd; 1995. pp. 1733–1736.

[13] Polyakov Y, Narayana S, Semenov VK. Flux trapping in superconducting circuits. IEEE Transactions on Applied Superconductivity. 2007;17(2): 520–525.

[14] Çelik ME, Bozbey A. A statistical approach to delay, jitter and timing of signals of RSFQ wiring cells and clocked gates. IEEE Transactions on Applied Superconductivity. 2013;23(3):1701305–1701305.

[15] Ando Y, Sato R, Tanaka M, *et al.* Design and demonstration of an 8-bit bit-serial RSFQ microprocessor: CORE e4. IEEE Transactions on Applied Superconductivity. 2016;26(5):1–5.

[16] Gaj K, Friedman EG, Feldman MJ. Timing of multi-gigahertz rapid single flux quantum digital circuits. Journal of VLSI Signal Processing Systems for Signal, Image and Video Technology. 1997;16(2–3):247–276.

[17] Kameda Y, Polonsky S, Maezawa M, *et al.* Primitive-level pipelining method on delay-insensitive model for RSFQ pulse-driven logic. In: Advanced Research in Asynchronous Circuits and Systems, 1998. Proceedings. 1998 Fourth International Symposium on. IEEE; 1998. pp. 262–273.

[18] Ito M, Kawasaki K, Yoshikawa N, *et al.* 20 GHz operation of bit-serial handshaking systems using asynchronous SFQ logic circuits. IEEE Transactions on Applied Superconductivity. 2005;15(2):255–258.

[19] Dorojevets M, Bunyk P, Zinoviev D. FLUX chip: Design of a 20-GHz 16-bit ultrapipelined RSFQ processor prototype based on 1.75-μm LTS technology. IEEE Transactions on Applied Superconductivity. 2001;11(1):326–332.

[20] Yoshikawa N, Matsuzaki F, Nakajima N, *et al.* Design and component test of a tiny processor based on the SFQ technology. IEEE Transactions on Applied Superconductivity. 2003;13(2):441–445.

[21] Gerber HR, Fourie CJ, Perold WJ, *et al.* Design of an asynchronous microprocessor using RSFQ-AT. IEEE Transactions on Applied Superconductivity. 2007;17(2):490–493.

[22] Tadros RN, Beerel PA. A robust and tree-free hybrid clocking technique for RSFQ Circuits—CSR application. In: 2017 16th International Superconductive Electronics Conference (ISEC); 2017. pp. 1–4.

[23] Mancini CA, Vukovic N, Herr AM, *et al.* RSFQ circular shift registers. IEEE Transactions on Applied Superconductivity. 1997;7(2):2832–2835.

[24] Tadros RN, Beerel PA. A robust and self-adaptive clocking technique for RSFQ circuits—The architecture. In: 2018 IEEE International Symposium on Circuits and Systems (ISCAS); 2018. pp. 1–5.

[25] MIT Lincoln Laboratory. MIT-LL 10 kA/cm^2 SFQ Fabrication Process: SFQ5ee Design Rules; 2015. Version 1.2.

[26] Barone A, Paterno G. Physics and Applications of the Josephson Effect. vol. 1. Wiley Online Library; 1982.

[27] Gheewala T. The Josephson technology. Proceedings of the IEEE. 1982;70(1): 26–34.

[28] Sprangle E, Carmean D. Increasing processor performance by implementing deeper pipelines. In: ACM SIGARCH Computer Architecture News. vol. 30. IEEE Computer Society; 2002. pp. 25–34.

[29] Beerel PA, Ozdag RO, Ferretti M. A Designer's Guide to Asynchronous VLSI. Cambridge: Cambridge University Press; 2010.

[30] Ebert B, Ortlepp T, Uhlmann FH. Experimental study of the effect of flux trapping on the operation of RSFQ circuits. IEEE Transactions on Applied Superconductivity. 2009;19(3):607–610.

[31] Malakhov A, Pankratov A. Influence of thermal fluctuations on time characteristics of a single Josephson element with high damping exact solution. Physica C: Superconductivity. 1996;269(1):46–54.

[32] Friedman EG. Clock distribution networks in synchronous digital integrated circuits. Proceedings of the IEEE. 2001;89(5):665–692.

[33] Weste N, Harris D. CMOS VLSI Design: A Circuits and Systems Perspective. Boston, MA: Addison Wesley Publishing Company Incorporated; 2011.

[34] Hulgaard H, Burns SM, Amon T, *et al.* An algorithm for exact bounds on the time separation of events in concurrent systems. IEEE Transactions on Computers. 1995;44(11):1306–1317.

[35] Hua W, Manohar R. Exact timing analysis for asynchronous systems. IEEE Transactions on Computer-Aided Design of Integrated Circuits and Systems. 2017;37(1):203–216.

[36] Havemann RH, Hutchby JA. High-performance interconnects: An integration overview. Proceedings of the IEEE. 2001;89(5):586–601.

[37] Restle PJ, Deutsch A. Designing the best clock distribution network. In: VLSI Circuits, 1998. Digest of Technical Papers. 1998 Symposium on. IEEE; 1998. pp. 2–5.

[38] Elmore WC. The transient response of damped linear networks with particular regard to wideband amplifiers. Journal of Applied Physics. 1948;19(1):55–63.

[39] Guthaus MR, Wilke G, Reis R. Revisiting automated physical synthesis of high-performance clock networks. ACM Transactions on Design Automation of Electronic Systems (TODAES). 2013;18(2):31.

[40] Deng ZJ, Yoshikawa N, Whiteley SR, *et al.* Data-driven self-timed RSFQ digital integrated circuit and system. IEEE transactions on applied superconductivity. 1997;7(2):3634–3637.

[41] Müller L, Gerber H, Fourie C. Review and comparison of RSFQ asynchronous methodologies. In: Journal of Physics: Conference Series. vol. 97. IOP Publishing; 2008. p. 012109.

[42] Filippov T, Dorojevets M, Sahu A, *et al.* 8-bit asynchronous wave-pipelined RSFQ arithmetic-logic unit. IEEE Transactions on Applied Superconductivity. 2011;21(3):847–851.

[43] Dorojevets M, Ayala CL, Yoshikawa N, *et al.* 16-bit wave-pipelined sparse-tree RSFQ adder. IEEE Transactions on Applied Superconductivity. 2013; 23(3):1700605–1700605.

[44] Pankratov AL, Spagnolo B. Suppression of timing errors in short overdamped Josephson junctions. Physical Review Letters. 2004;93(17):177001.

[45] Mukhanov OA. Rapid single flux quantum (RSFQ) shift register family. IEEE Transactions on Applied Superconductivity. 1993;3(1):2578–2581.

[46] Murata T. Petri nets: Properties, analysis and applications. Proceedings of the IEEE. 1989;77(4):541–580.

[47] Mukhanov O, Rylov S, Semonov V, *et al.* RSFQ logic arithmetic. IEEE Transactions on Magnetics. 1989;25(2):857–860.

[48] Bryan D. The ISCAS'85 benchmark circuits and netlist format. North Carolina State University. 1985;25.

[49] Mishchenko A, *et al.* ABC: A system for sequential synthesis and verification. URL https://people.eecs.berkeley.edu/~alanmi/abc/. 2007.

[50] Nassif SR. Design for variability in DSM technologies [deep submicron technologies]. In: Quality Electronic Design, 2000. ISQED 2000. Proceedings. IEEE 2000 First International Symposium on. IEEE; 2000. pp. 451–454.
[51] Xiong J, Zolotov V, He L. Robust extraction of spatial correlation. IEEE Transactions on Computer-Aided Design of Integrated Circuits and Systems. 2007;26(4):619–631.
[52] Agarwal A, Blaauw D, Zolotov V, *et al.* Statistical delay computation considering spatial correlations. In: Proceedings of the 2003 Asia and South Pacific Design Automation Conference. ACM; 2003. pp. 271–276.
[53] Vernik I, Kaplan S, Volkmann M, *et al.* Design and test of asynchronous eSFQ circuits. Superconductor Science and Technology. 2014;27(4):044030.
[54] Volkmann MH, Vernik IV, Mukhanov OA. Wave-pipelined eSFQ circuits. IEEE Transactions on Applied Superconductivity. 2015;25(3):1–5.
[55] Kirichenko AF, Vernik IV, Vivalda JA, *et al.* ERSFQ 8-bit parallel adders as a process benchmark. IEEE Transactions on Applied Superconductivity. 2015;25(3):1–5.
[56] Sakashita Y, Yamanashi Y, Yoshikawa N. High-speed operation of an SFQ butterfly processing circuit for FFT processors using the 10 kA/cm^2 Nb process. IEEE Transactions on Applied Superconductivity. 2015;25(3):1–5.
[57] Narama T, Yamanashi Y, Takeuchi N, *et al.* Demonstration of 10k gate-scale adiabatic-quantum-flux-parametron circuits. In: Superconductive Electronics Conference (ISEC), 2015 15th International. IEEE; 2015. pp. 1–3.
[58] Tam S, Rusu S, Desai UN, *et al.* Clock generation and distribution for the first IA-64 microprocessor. IEEE Journal of Solid-State Circuits. 2000;35(11): 1545–1552.
[59] Geannopoulos G, Dai X. An adaptive digital deskewing circuit for clock distribution networks. In: Solid-State Circuits Conference, 1998. Digest of Technical Papers. 1998 IEEE International. IEEE; 1998. pp. 400–401.
[60] Guthaus MR, Sylvester D, Brown RB. Clock tree synthesis with data-path sensitivity matching. In: Proceedings of the 2008 Asia and South Pacific Design Automation Conference. IEEE Computer Society Press; 2008. pp. 498–503.

Chapter 14
Uncle—Unified NCL Environment—an NCL design tool

Ryan A. Taylor[1] *and Robert B. Reese*[2]

Uncle (Unified NULL Convention Logic Environment) is a tool for creating NULL Convention Logic (NCL) designs, which can be downloaded for free.[*] This chapter discusses Uncle internals and a detailed walk-through of an example design.

14.1 Overview

Uncle consists of Python scripts, binary executables written in the C programming language, Verilog files for library elements, and various technology files. The Uncle tool flow is shown in Figure 14.1.

An Uncle design begins with an register transfer level (RTL) Verilog *specification*. The term specification is purposefully used to indicate that this RTL generally cannot be simulated, unlike RTL in a clocked system. A Verilog netlist that can be simulated is generated after several transformations within the tool flow.

Uncle supports two distinct design styles, *data-driven* and *control-driven* (hence, the *Unified* designation). The two styles are distinguished by how NCL data flow is controlled within the design. In both styles, combinational logic gates are converted to their NCL dual-rail equivalents.

In a data-driven design, only the control network is the acknowledgment (ack) network; data moves through the network as it arrives and is acknowledged. Figure 14.2 shows a data-driven finite state machine, in which all D flip-flops that implement the state in the original netlist have been converted to three dual-rail half-latches. The outer half-latches (L1/L2 drlatn) have their data t/f rails reset to NULL, while the inner latch (L2) is either reset to DATA1 or DATA0 (the inner latch must be reset to DATA for the system to cycle).

In a control-driven design, the control network and data registers are separate from the NCL combinational logic that implements the data path logic. The control network uses Balsa-style control elements to selectively read/write data registers,

[1]Department of Electrical and Computer Engineering, The University of Alabama, Tuscaloosa, AL, USA
[2]Department of Electrical and Computer Engineering, Mississippi State University, Starkville, MS, USA

[*]Download at https://sites.google.com/site/asynctools/

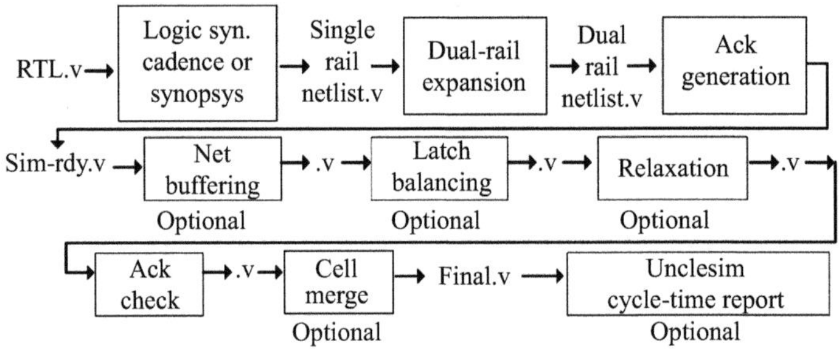

Figure 14.1 Uncle tool flow

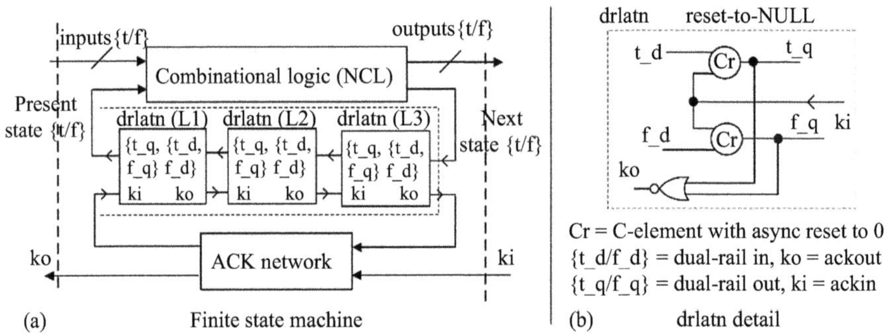

Figure 14.2 Finite state machine

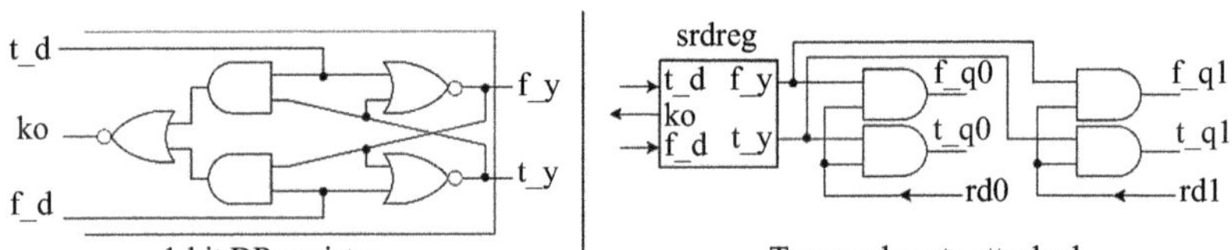

Figure 14.3 Dual-rail register based on an SR-latch

which are implemented as dual-rail registers based on a set-reset (SR) latch. Figure 14.3 gives the implementation for a 1-bit dual-rail (DR) register and illustrates how two read ports can be attached to it. The *rd0*/*rd1* signals are used for reading the network and would be driven by Balsa-style control elements.

An example of two Balsa control elements, the S-element and the T-element, are given in Figure 14.4. These are typically connected in chains, with the *Ia* output connected to the *Ir* input of the next element of the chain. The *Or* output is used to start a data path action by connecting it to the read port of a register. The *Oa* input is the ack that the data path action has completed. The T-element offers more concurrency by asserting the *Ia* output before the *Oa* input has returned low.

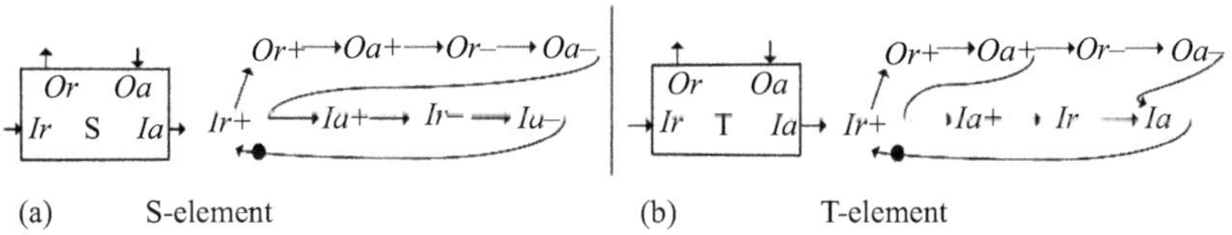

Figure 14.4 Balsa handshaking elements

Other elements can be added to form loops (loop-while) or choice (if-else) state machines.

The following guidelines are offered for determining which design style to use:

- The data-driven style is the best choice in terms of performance for linear pipelines. For transistor count/energy, the better choice (data-driven/control-driven) is design dependent.
- The data-driven style is generally the best choice in terms of performance for a block that has feedback (e.g. accumulators and finite state machines) if ALL registers/ALL ports are read/written each compute cycle. This assumes that the block is performance-optimized using the automated delay balancing tool available in the tool flow. If minimal energy/transistor count is required, then the control-driven style is generally better.
- The control-driven style is the better choice in terms of transistor count/energy for blocks that have registers with conditional reads/writes, and/or ports with conditional activity. It can also be better in performance than the data-driven implementation, but it depends on the block.

Design styles can be mixed in a design, but each Verilog module within a design should adhere to one design style.

14.2 Flow details

This section gives the details of each step of the Uncle tool flow.

14.2.1 RTL specification to single-rail netlist

The first step in the flow uses a logic synthesis tool (Synopsys and Cadence are supported) to transform the RTL specification into a single-rail gate netlist. For combinational logic, the synthesis library contains two-input AND/OR/XOR gates, an inverter, and a couple of complex cells (mux2, full-adder) that have well-known, efficient NCL implementations. For sequential logic, data flip-flops (DFFs) and latch variants are provided. There are also many specialized cells in the library that are used for implementing control-driven designs. These gates are manually specified in the RTL specification using parameterized Verilog modules that implement common functionality in a control-driven design (e.g. demuxes and merges). The output of this synthesis step is a gate-level netlist comprised of gates from this synthesis library.

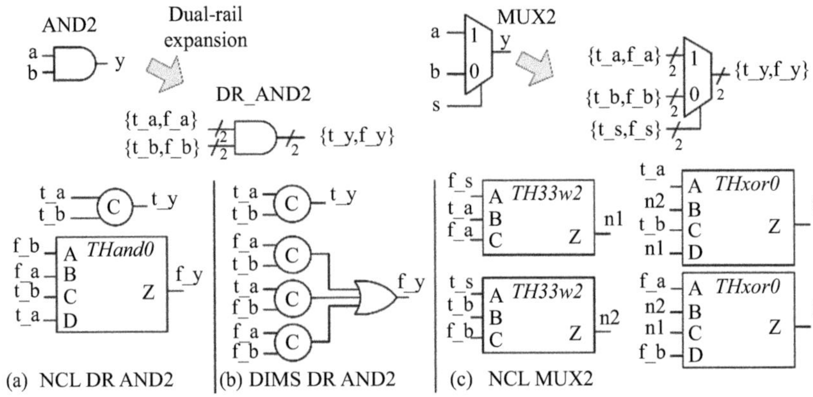

Figure 14.5 NCL AND2 and MUX2 implementations

14.2.2 Single-rail netlist to dual-rail netlist

The second step expands gates in the single-rail netlist into their dual-rail equivalents. Figure 14.5 shows an NCL AND2 function, along with its delay insensitive minterm synthesis (DIMS) equivalent [1] for comparison purposes. Note that dual-rail NCL logic functions require fewer transistors and are faster than their DIMS counterparts (e.g. 31 vs. 56 transistors for a static dual-rail AND2 function). The NCL implementation of a dual-rail 2-input multiplexer (MUX2) is also given in Figure 14.5.

In a control-driven design, there are both single-rail and dual-rail signals based on the gate elements that are used.

14.2.3 Ack network generation

After dual-rail expansion, the ack networks comprised of C-element trees are automatically built. Both merged ack networks (to reduce gate count) and non-merged ack networks (to promote performance via bit-level pipelining) are supported via a configuration file option. By using special demux gates in the RTL specification file, Uncle also recognizes when acks from demux targets must be OR'ed to support wave front steering [2], since only one demux target will provide an active ack signal in this case. The netlist produced after the ack generation step is a complete implementation and is simulation ready.

14.2.4 Net buffering, latch balancing (optional steps)

The example cell library in the Uncle distribution has timing data from a 65-nm process for output transition time and propagation delay represented in industry-standard nonlinear delay model (NLDM) lookup tables. One use of this timing information is to perform buffering of heavily loaded nets to meet a minimum transition time specified via a configuration option.

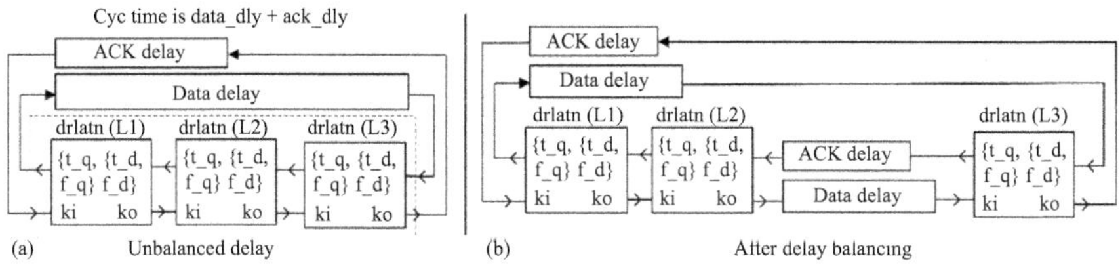

Figure 14.6 Latch balancing to reduce cycle time

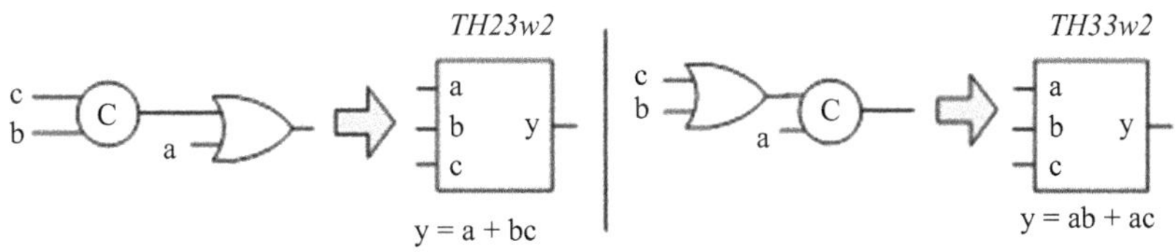

Figure 14.7 Example of cell merging

Another use of the timing data is to perform latch delay balancing in data-driven designs to reduce cycle time. Figure 14.6 shows how this works in an finite state machine (FSM) data-driven design. The cycle time of the network in Figure 14.6(a) is the data delay plus the backward delay through the network. By pushing some of the logic between three half-latches that were expanded from a D-flip-flop specification, the overall cycle time is reduced, as seen in Figure 14.6(b). Latch balancing also works with data-driven linear pipelines. In all cases, only existing latch stages are moved; new latch stages are not added.

14.2.5 Relaxation, ack checking, cell merging, and cycle time reporting

Relaxation [3] is an optional optimization that searches for redundant paths between a set of primary inputs and a primary output in a combinational netlist. *Eager* dual-rail functions that have reduced transistor count are placed on the redundant paths, with all primary inputs having at least one path to a primary output that go through all non-eager (i.e. input-complete) dual-rail functions. This is an area-driven optimization that is applied to combinational logic blocks in the netlist.

The ack checking stage reverse engineers the ack network and checks it for correctness. This stage is in the tool pipeline as an error check when developing new approaches for ack network generation.

The optional cell merging step is an area optimization in which adjacent gates with no fan-out are merged into more complex gates. Figure 14.7 shows an example of 2 of 15 different types of cell merges that Uncle supports.

Before doing a full Verilog simulation, the final gate-level netlist can be simulated using the internal Uncle simulator (*unclesim*). This tool uses the timing characterization data previously discussed, and is useful for reporting the average cycle time of the netlist using either randomly generated input vectors or user-supplied input vectors. The simulator also reports error/unusual conditions such as failure to cycle (error), X values after reset, simultaneous assertion of dual-rail nets (error), and orphans/glitches (warning). An orphan/glitch warning means that a net transition that fanned out to at least one NCL gate did not cause a corresponding transition in at least one of the fan-out gates. In general, the dual-rail expansion methodology used in Uncle does not cause gate orphans in combinational logic, and the ack generation strives not to generate gate orphans. Long chains of gate orphans may cause timing problems in NCL. It is possible that something the designer has done using demux or merge gates can cause orphans. Orphans [4] are an unusual condition and should be checked by the designer.

14.3 Example—a 16-bit GCD circuit

The greatest common divisor (GCD) of a set of numbers is the largest number that can divide all numbers in the set while leaving a remainder of zero in each case. An interesting example to analyze for this software system is Euclid's successive subtraction algorithm, a solution to solve for the GCD of two numbers.

The simple implementation of this algorithm can be seen in the algorithmic state machine (ASM) chart of Figure 14.8. By glancing at this ASM, the reader might notice that only one state exists, a state which simply serves as a *reset* state. This is slightly misleading, as the Euclidean successive subtraction algorithm is typically designed with three states: one to accept input, a second to calculate the result, and a third to output said result to the external requestor. The ASM in Figure 14.8 is shown only to explain the calculations made in the second stage as described here.

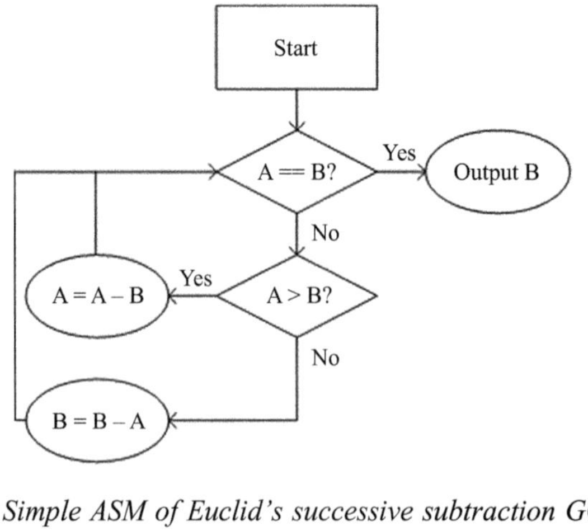

Figure 14.8 Simple ASM of Euclid's successive subtraction GCD algorithm

Assuming two inputs are provided to the algorithm, a loop is entered which will only terminate upon both numbers being equal. If both numbers are equal, then this number is the resultant GCD of the system. Note that it is possible for this number to be the value of the original inputs. If the numbers are not equivalent, then the smaller of the two is subtracted from the larger, with this result replacing the larger, and then equivalency is checked again. Eventually, this will lead to both numbers becoming equivalent and finding the GCD, even if that means that the GCD of the two inputs is one.

14.3.1 Synchronous implementation

To implement this system as a synchronous hardware circuit, let us begin with the design of the control and data circuitry. Three states should exist, like the description stated earlier. The first state should simply capture a pair of inputs. The second state should continually calculate the difference of the larger and smaller numbers, exiting only when the numbers are equal. The third state should output the result to the receiving entity.

A set of registers is needed that will store the values of the present and next states. Also included in the control path is a comparator, to determine when to exit the second state and to determine which of the numbers is the larger and which is the smaller during the calculations in the second state.

The data path is straightforward. A pair of subtraction modules can be used to perform the calculations in the second state. The rest of the data path consists only of multiplexors controlled by signals from the control path.

14.3.2 Data-driven NCL implementation

Likewise, the design of the data-driven NCL implementation of the GCD circuit will also be built around the same finite state machine, consisting of the same three states. The data path is largely the same, having the same subtractor modules and including multiplexors to determine the path of the current values. The major difference between the synchronous and data-driven implementations comes when analyzing the way that data moves into and out of the circuit, and the way that said data flows through the circuit.

To allow a DATA wave to pass through the system, there must be a method for allowing the circuit to accept data only when it is prepared for a new DATA wave. The Uncle system implements this control as a module called a read port. The read port serves to supply data to a system when notified that the system is ready for data. Physically, this is implemented by a single control line, *read_control* in Figure 14.9. When the *read_control* signal is asserted, the read port module

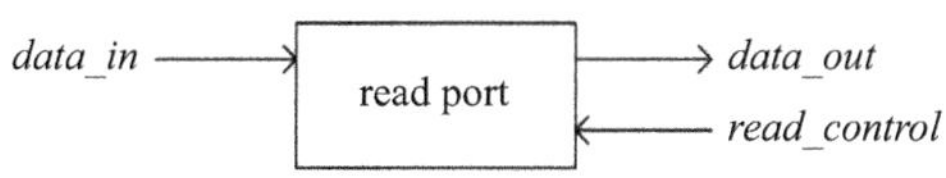

Figure 14.9 Basic read port module

supplies data from *data_in* to the *data_out* signal. Otherwise, the *data_out* signal will be provided with dummy data.

The inner workings of the read port module are shown in Figure 14.10. The input from the supplying system provides data to the initial D-latch, which provides this data unchanged to a module called a *demux_half1_noack* at the RTL level. This is not a mappable module, as it currently has no library definition in either the Cadence or Synopsys libraries. During synthesis, it will remain unchanged. In this case, this is a flag to the mapper to handle the control signals in a specific way. On the other side of the module, the dummy data, implemented by a constant zero signal, is supplied to the partner *demux_half0_noack* module. The outputs of both modules are then routed to a merge gate, whose output gets applied to the *data_out* output signal. In this way, the Uncle system can gate input into a new system conditionally using only a control signal from the original system. In a data-driven NCL design, the ability to move data into the system conditionally in this fashion is crucial to proper operation of the DATA and NULL waves.

Figure 14.11 shows the inner workings of the portion of the circuit that is identified by the *demux_half0_noack*, and *demux_half1_noack* modules. As can be seen, the implementation of these modules is a flag to Uncle to mix the read and ack control signals appropriately. The data modules that receive the dummy and *data_in* values are implemented with two different modules, considering the need to supply them with two types of data. Since the *read_control* signal's value will determine which of these elements supplies data to the merge network, the rails

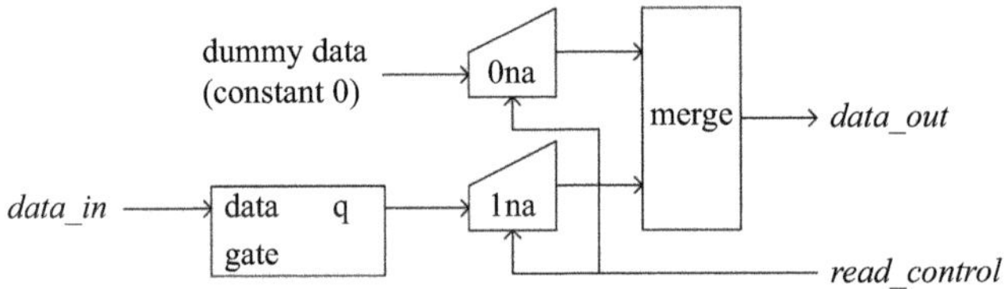

Figure 14.10 Implementing the read port module

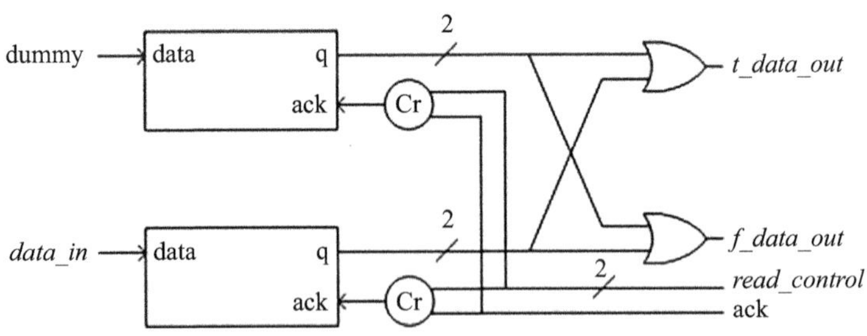

Figure 14.11 Implementing the demux_half0_noack and the demux_half1_noack control modules

are split between the two, and the acknowledge signal is sent to both. If the *read_control* signal is DATA1 and *ack* is asserted, the *data_in* value will pass along to the rest of the system. On the other hand, if the *read_control* signal is DATA0 and *ack* is asserted, the dummy data will be provided. This, along with the network in Figure 14.10, completely implements the read port module.

This read port module can be incorporated into the GCD design to effectively port data into the system. However, we also need a module that will be able to conditionally port data out of the data-driven NCL system. To this end, Uncle supplies a write port module that will allow for this operation and control.

The write port module in Uncle serves to control writes to an output port of an NCL system. As the DATA wave passes through the circuit, it will eventually need to pass out of the system through an output port to another entity in the larger system. The write port controls this interface. The basic black-box representation can be seen in Figure 14.12.

Figure 14.13 shows the RTL representation of the write port. As can be clearly seen, the write port is simply a two-output demultiplexor with one of the outputs disconnected from the network. The basic multiplexor operates by copying the input data signal, *data_in* in this case, to one of the outputs depending on the value of the control signal, in this case *write_control*. If *write_control* is DATA1, the *data_in* signal is copied to the *data_out* signal. Otherwise, if *write_control* is DATA0, the *data_in* signal is copied to the "0" output of the demultiplexor, which in the case of the write port implementation, remains disconnected to prevent the output from being written to the output port. Instead of writing the output in this latter case, the demultiplexor is said to *consume* the data. This behavior is further explained in the gate-level implementation of the write port shown in Figure 14.14.

As can be seen in Figure 14.14, if the *write_control* signal is DATA0, the dual-rail *data_in* signal is combined through an NOR gate (to invert and combine) and added to the acknowledge network by the Uncle toolset. This successfully routes

Figure 14.12 Basic write port module

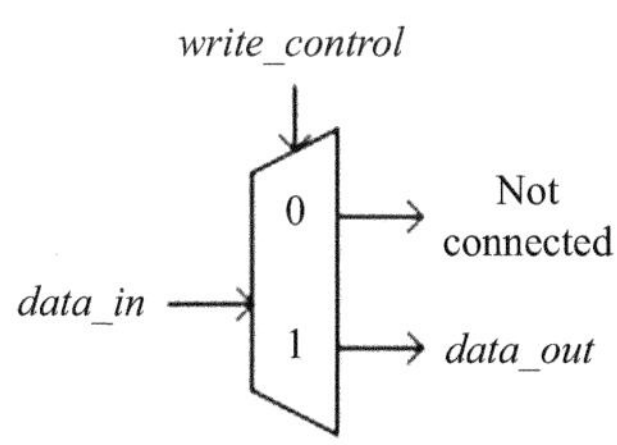

Figure 14.13 Implementing the write port module

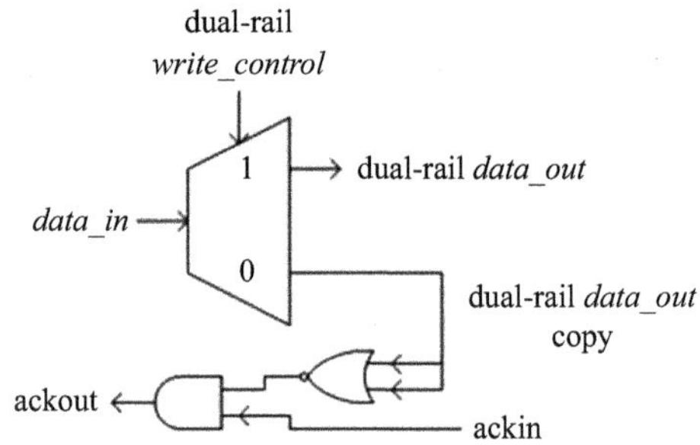

Figure 14.14 Gate-level implementation of the write port module

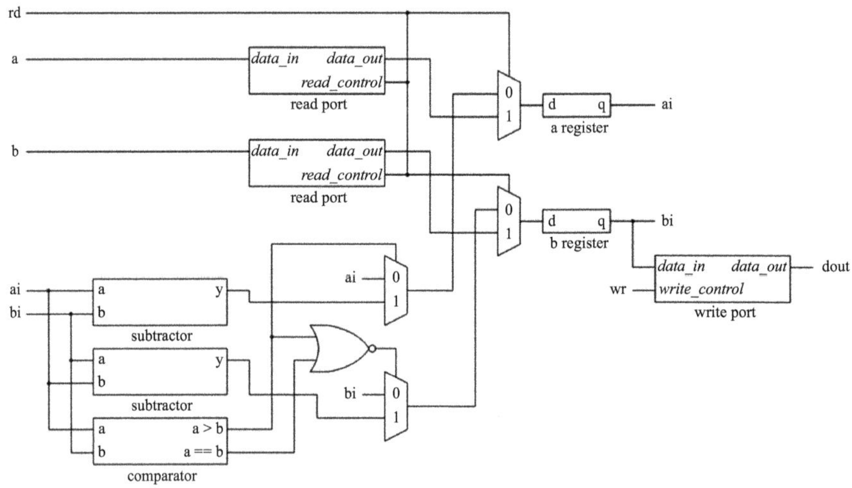

Figure 14.15 Data-driven NCL implementation of the Euclidean GCD circuit

the DATA wave back into the control network, effectively "consuming" and acknowledging the data.

Using these two new modules, the read port and the write port, along with some other common elements, we can implement the GCD circuit as described earlier. The data arrives at the read ports and is gated in by the *rd* signal in the first state. The generation of this signal is not shown as it is straightforward from the discussion earlier in this chapter. Each signal is then immediately stored into a register. The system now moves into the second state. This state is identified by the deassertion of both the *rd* and the *wr* signals. When both signals are low, the circuit will calculate the successive subtraction algorithm using the rest of the elements. The comparator in the bottom left of Figure 14.15 calculates whether the values are equivalent or, if not, which is greater. If equivalent, the FSM moves onto the third state and asserts the *wr* signal to port the output. Otherwise, the smaller of *a* or *b* is

```
// assign a_lt_b = ai < bi;
// assign a_eq_b = ai == bi;
compare_16 compmod (.a(ai), .b(bi), .agb(a_gt_b), .aeqb(a_eq_b));
assign a_lt_b = ~(a_gt_b | a_eq_b);

wire [15:0] sub_a_b, sub_b_a;
subripple_n #(.WIDTH(16)) subcompa (.s(sub_a_b), .a(ai), .b(bi));
subripple_n #(.WIDTH(16)) subcompb (.s(sub_b_a), .a(bi), .b(ai));

// assign ai_val = a_gt_b ? sub_a_b : ai;
// assign bi_val = a_lt_b ? sub_b_a : bi;
mux2_n #(.WIDTH(16)) u_mux0 (.y(ai_val), .a(sub_a_b), .b(ai), .s(a_gt_b));
mux2_n #(.WIDTH(16)) u_mux1 (.y(bi_val), .a(sub_b_a), .b(bi), .s(a_lt_b));

wire [15:0] ai_rd, bi_rd;
mux2_n #(.WIDTH(16)) u_mux2 (.y(ai_rd), .a(a_in), .b(ai_val), .s(rd));
mux2_n #(.WIDTH(16)) u_mux3 (.y(bi_rd), .a(b_in), .b(bi_val), .s(rd));
```

Figure 14.16 Verilog RTL for the data-driven NCL GCD data path

subtracted from the larger and stored in the appropriate register. This continues until the comparator moves the system into the third state.

The implementation of such a circuit using Uncle is not complex. Although multiple elements in the circuit of Figure 14.15 could be implemented in standard Verilog by using simple assignment statements and arithmetic operators, the proper method when using Uncle is to use supplied parameterized modules to ensure that the specified and expected architecture for each entity is used. Figure 14.16 shows an example of how the data path of this system might be implemented in the Verilog RTL code.

The code for the control path that controls the *rd* and *wr* signals of Figure 14.15 is not shown here, as it is identical to that of the synchronous system.

14.3.3 Control-driven NCL implementation

As discussed earlier in this chapter, the overarching difference between the data-driven and control-driven NCL design styles is that the registers in the control-driven design style are made to be conditionally read and written. In the data-driven design style, all registers are read unconditionally and written unconditionally during each DATA wave. Opposed to this paradigm, control-driven registers are read only when the data in question is needed and written only when necessary. Therefore, the control logic of these systems tends to be a bit more complex, but area and power savings are typical when compared with comparable traditional data-driven NCL systems.

Because of this fine-grained control over which registers are read and written, these systems must currently be designed independently of the synchronous version. This gives responsibility to the designer to make decisions that will benefit this aspect of the design style.

In this method of design, there are relatively more states because of the need to have fine-grained control over the system. However, there are typically fewer actions required in each of those states. This framework leads to longer processing times, on average, but fewer transitions and, thus, less power dissipated.

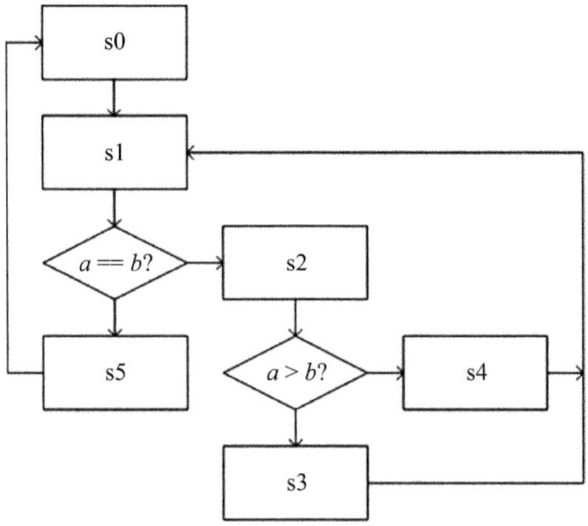

Figure 14.17 Control-driven NCL GCD circuit control ASM

For example, consider the ASM chart in Figure 14.17. In s0, the communication with the external input port takes place. Unconditionally, we move to s1, which computes the control flags coming from the comparator (the data-driven equivalent of this is shown in the bottom left of Figure 14.15). Once these values have been computed and stored, the system decides if we should immediately exit. If the "equivalent" comparator flag is true, then we enter s5, which then outputs the result on the output port and unconditionally loops back to s0. If the flag is false, we move to s2, where we now read the other output of the comparator to determine which value is larger. Depending on which is larger, we enter s3 or s4 to perform the subtraction and update of the appropriate *a* or *b* value, and then unconditionally loop back to s1.

Figure 14.18 displays the data path associated with the control-driven NCL implementation of the GCD circuit discussed here. A slight variant of the read port from Figures 14.9 to 14.11 is used in control-driven systems because of the slight change in control. Because the user is responsible for setting up the control path, he or she is also responsible for connecting the control signal. In this case, s0 reads these values from the supplying system and stores them in their appropriate registers, which the help of a pair of merge elements.

After these values have been stored, s1 reads both values from said registers and uses them to compute the comparator flags, shown in the bottom right of Figure 14.18. Another "half-state" not shown in the diagram of Figure 14.17 is the check for equivalency. We must split these transactions because the *aneb* register must first be written, and then it is read. An additional element in the control path called a while loop is used to set up the loop of states one through five and check for termination iteratively. This element is not shown here, but supplies the *rd_aneb* signal to the data path.

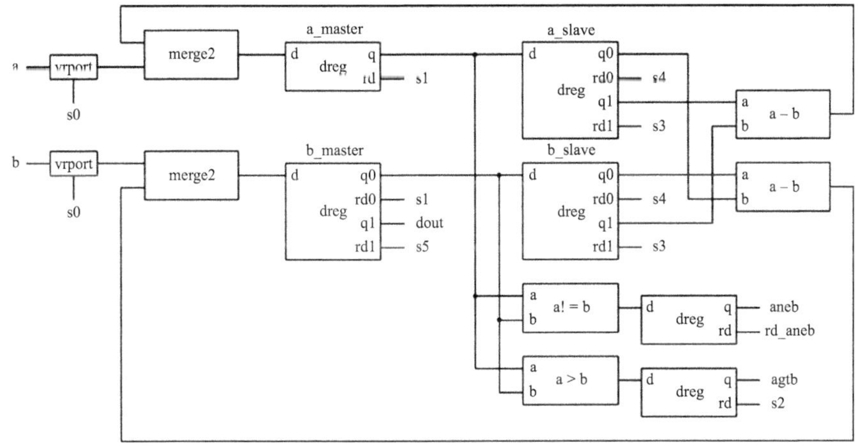

Figure 14.18 Control-driven NCL GCD circuit data path

Once the *aneb* signal is read, it determines which of s5 or s2 to enter. If s5, an additional port on the *b* register is read to be sent to the output write port (not shown in Figure 14.18). This is another concept that is unique to the control-driven design style. Registers may have multiple ports, as discussed earlier in this chapter.

If the *aneb* signal is true, then we continue to s2, where the larger of the two numbers is selected by using the other computed flag from s1. This *agtb* flag will either send us into s3 or s4, subtracting the appropriate values from one another and re-storing the result into the appropriate *a* or *b* register, again with the help of one of the supplied merge elements.

As mentioned earlier, the designer of control-driven NCL circuits using Uncle has greater responsibility in that he or she must connect the control signals his- or herself, as well as have a complete understanding of how the acknowledge network will connect those signals.

The acknowledge network generation step of Uncle differs slightly when using the control-driven design style as opposed to using the data-driven design style. When generating acknowledge networks for data-driven systems, the Uncle tool needs to generate a two-way communication to and from source and destination registers and ports. When generating the acknowledge networks for control-driven systems, a three-way communication protocol is realized. First, the control path of the circuit asserts a read signal on some set of elements in the design, then these elements (and only these elements) provide data to their destinations, and finally these destinations provide acknowledge signals back to the control path of the system. This realizes the three-way communication cycle that continues (and replaces the traditional two-way DATA/NULL waves).

Table 14.1 Design characteristics of NCL circuits designed using Uncle

Design	**Data-driven**			**Control-driven**		
	Transistors	**Cycle time (ns)**	**Energy (pJ)**	**Transistors**	**Cycle time (ns)**	**Energy (pJ)**
16-bit GCD (this example)	16,192	105.7	32.4	8,658	75.7	10.2
Viterbi decoder PMU	20,184	13.4	5.1	18,838	13.3	4.6

14.4 Conclusion

Uncle is a hardware mapping software suite that is used for creating NCL systems. The beauty of Uncle is that it allows both data-driven and control-driven NCL design styles to be used at the discretion of the designer. By fully understanding the benefits and drawbacks of both design styles, the designer can make better choices of which methodology to use when designing NCL systems. As discussed in this chapter, it is typically more beneficial to use the data-driven NCL design style unless the system can benefit from conditionally read and written registers. Complex controlled systems, like microprocessors, typically fall into this category, but the distinction should be made by the designer on a case-by-case basis.

Table 14.1 shows the final design characteristics of two circuits designed using both data-driven and control-driven NCL design styles. The first is the design of the 16-bit GCD described in this chapter. The second is the path metric unit submodule of a Viterbi decoder, which has registers that are read and written every cycle. As can be seen, significant area and energy saving are realized in the GCD implementation by using the control-driven NCL design style. However, for the Path Metric Unit (PMU), a design seemingly best suited for a data-driven NCL implementation, the savings are minimal.

In conclusion, Uncle can be used to assist asynchronous digital designers in converting traditionally synchronous designs to data-driven and control-driven NCL design styles. Both design styles offer their own benefits and drawbacks, leaving the responsibility for optimization of this selection best left to the designer. Built-in area and speed optimizations, along with a simulator, make Uncle a full end-to-end toolset when combined with the power of commercial synthesis tools, such as Cadence or Synopsys. Uncle can be downloaded for free,[†] and also includes a complete user manual and sample Verilog code for the example circuits presented in this chapter. There remain to be many areas where Uncle can be expanded upon, such as how to quantitatively determine whether a data-driven or control-driven design style is best suited for any given circuit, and automatic control path generation. Ongoing research continues to develop new features and further optimizations for this toolset.

[†]Download at https://sites.google.com/site/asynctools/

References

[1] J. Sparsø, J. Staunstrup, and M. Dantzer-Sorensen, "Design of Delay Insensitive Circuits Using Multi-Ring Structures," in *European Design Automation Conference, 1992. EURO-VHDL '92, EURO-DAC '92.* 1992, pp. 15–20.

[2] S.C. Smith and J. Di, "Designing Asynchronous Circuits using NULL Convention Logic (NCL)," *Synthesis Lectures on Digital Circuits and Systems*, Morgan & Claypool Publishers, San Rafael, CA, Vol. 4/1, July 2009.

[3] J. Cheoljoo and S.M. Nowick, "Optimization of Robust Asynchronous Circuits by Local Input Completeness Relaxation," in *Asia and South Pacific Design Automation Conference, 2007. ASP-DAC '07.* 2007, pp. 622–627.

[4] A. Kondratyev, L. Neukom, O. Roig, A. Taubin, and K. Fant, "Checking Delay-Insensitivity: 10^4 Gates and Beyond," *IEEE International Symposium on Asynchronous Circuits and Systems*, pp. 149–157, April 2012.

Chapter 15

Formal verification of NCL circuits

Ashiq Sakib[1], Son Le[2], Scott C. Smith[3] and Sudarshan Srinivasan[2]

Validation is a critical component of any commercial design cycle. Testing-based approaches have been predominantly used to ensure design correctness. Formal verification is an alternate approach to design validation, where correctness is established using mathematical proofs. Since a proof can correspond to a very large number of test cases, formal verification has been found to be extremely useful in establishing design correctness and finding corner-case errors that often escape traditional testing. Since the now infamous floating-point division (FDIV) bug (i.e., bug found in the floating-point unit of the Intel Pentium processor in 1994 after shipping, which cost Intel $500 million to correct), the semiconductor industry has aggressively incorporated formal verification into its design cycle for validation.

One of the more popular formal verification approaches that have been found to be extremely scalable and useful in semiconductor design is equivalence checking. Typically, a lot of time, money, and effort are invested into ensuring the correctness of a design. However, the design itself is never static, as it is continuously tinkered with and optimized. Equivalence checking technology can, with a high degree of automation and efficiency, check that the golden model (i.e., the design that has been extensively validated) and its derivate are functionally equivalent. Scalability is harnessed by exploiting the structural similarity of the golden model and its derivate. Examples of commercial equivalence checkers include IBM Sixth Sense, Jasper Gold Sequential Equivalence Checker, Calypto SLEC, Mishchenko EBCCS13, and Cadence Encounter Conformal Equivalence Checker.

In this chapter, we describe an equivalence checking methodology for NULL Convention Logic (NCL) circuits. Note that currently, there are no commercial equivalence checkers for quasi-delay insensitive (QDI) circuits. For commercial applications, NCL circuits, and QDI circuits in general, are often synthesized from synchronous intellectual property designs. The resulting NCL design may then be further optimized and tinkered with. Therefore, we have designed an equivalence checker that can be used in two ways: (1) to verify the functional equivalence of

[1]Department of Electrical and Computer Engineering, Florida Polytechnic University, Lakeland, FL, USA
[2]Department of Electrical and Computer Engineering, North Dakota State University, Fargo, ND, USA
[3]Department of Electrical Engineering and Computer Science, Texas A&M University Kingsville, Kingsville, TX, USA

two NCL designs and (2) to verify the equivalence between an NCL design and a synchronous design.

15.1 Overview of approach

Vidura *et al.* [1] have previously developed an approach for verifying the equivalence of an NCL circuit against a synchronous circuit. They use the theory of Well-Founded Equivalence Bisimulation (WEB) refinement [2] as the notion of equivalence. In WEB refinement, both the circuit to be verified (here the NCL circuit) and the specification circuit (here the synchronous circuit) are modeled as transition systems (TSs), which capture the behavior of the circuit as a set of states and transitions between the states. WEB refinement essentially defines what it means for two TSs to be functionally equivalent. Their approach performs symbolic simulation on both the NCL circuit and the synchronous circuit to generate the TSs corresponding to both circuits. A decision procedure is then used to verify that the two TSs satisfy the WEB refinement property.

In working with the above approach, we found that because NCL circuits exhibit highly nondeterministic behaviors, the corresponding TSs are very complex, even for relatively simple circuits. This complexity leads to two issues. First is state space explosion. Second, it becomes very difficult to compute the reachable states of the resulting TS. Computing reachable states is important because unreachable states often flag numerous spurious counterexamples, which makes verification intractable.

We have therefore developed an alternate approach to circumvent having to deal with the NCL TS. The high-level idea is to perform structural transformation on the NCL circuit netlist to convert the NCL circuit into an equivalent synchronous circuit. The converted synchronous circuit is then compared against the specification synchronous circuit, using WEB refinement as the notion of correctness. The converted synchronous circuit, specification synchronous circuit, and the WEB refinement property are then automatically encoded in the Satisfiability Modulo Theory Library (SMT-LIB) language [3]. The resulting equivalence property is then checked using an SMT solver. Additional checks need to be performed to ensure that the NCL circuit is live (i.e., deadlock free). Thus, the overall verification has three high-level steps: (1) conversion from NCL to synchronous; (2) verification of converted synchronous against specification synchronous; and (3) additional checks on original NCL circuit to ensure liveness. The methodology can also be used to check the equivalence of two NCL circuits by applying the conversion technique to both NCL circuits to obtain two corresponding synchronous circuits, verifying these two synchronous circuits against each other, and performing the additional liveness checks on both NCL circuits.

15.2 Related verification works for asynchronous paradigms

Several formal verification techniques have been implemented to verify the two major asynchronous design paradigms: bounded-delay and QDI. The bounded-delay model is based on the assumption that the delay in all circuit components and wires is bounded—i.e., worse case delay can be calculated. Because of these timing constraints, most of the verification schemes for timed asynchronous models involve trace theory, Signal Transition Graph [4], and timed Petri nets. Reference [5] illustrates a gate-level verification method based on trace theory where the circuit, as well as the correctness properties, is modeled as Petri nets. An approach based on time-driven unfolding of Petri nets is used to verify freedom from hazards in asynchronous circuits consisting of logic gates and micropipelines [6]. However, timed-model-based verification methods are not applicable to QDI circuits, which are based on exactly the opposite assumption that circuit delays are unbounded and therefore indeterminate.

There exist several verification schemes specific to QDI circuits as well. Verbeek and Schmaltz [7] illustrate a deadlock-verification scheme for QDI circuits based on the Click Library [8]. Circuits based on this primitive library are structurally different from other QDI paradigms, such as NCL. Moreover, this method does not verify the functional correctness (safety) of the circuit. Refinement-based formal methods have been successful in verifying both bounded-delay and QDI asynchronous models. Desynchronized circuits, which are based on a bounded-delay structure, can be verified by a refinement-based approach, as discussed in [9]. As mentioned in the previous section, reference [1] presents a method to check the functional equivalence of NCL circuits against their synchronous counterparts using WEB refinement; and a model-checking-based method that checks for safety and liveness of pre-charge half buffer (PCHB) circuits is presented in [10]. However, both of these techniques suffer from state space explosion, since they model the QDI circuits as TSs, which become very complex for large circuits due to the nondeterministic behavior of QDI paradigms. Using a conversion technique along with WEB refinement, similar to that presented herein but applied to PCHB circuits, we were able to verify equivalence of combinational PCHB circuits with their Boolean specification, which proves to be highly scalable and much faster than previous techniques [11]. That method is currently being extended to sequential PCHB circuits.

Along with safety and liveness, input-completeness and observability are two critical properties of NCL circuits, which must be verified in order to ensure delay insensitivity, since a circuit may function correctly under normal operating conditions while not being input-complete or observable, but may then malfunction under extreme timing scenarios, such as those caused by process, voltage, or temperature variations. A manual approach to checking input-completeness is outlined in [12], which requires an analysis of each output term. For example, in order for output Z to be input-complete with respect to input A, every product term in all rails of Z (in sum of products (SOP) format) must contain any rail of A. This ensures that Z cannot be DATA until A is DATA; and if Z is constructed solely out of NCL gates with hysteresis, the gate hysteresis ensures that Z cannot transition from DATA to NULL until A transitions from DATA to NULL. Hence, Z is input-complete with respect to A. However, this method cannot ensure input-completeness of relaxed NCL circuits [13], where not all gates contain hysteresis. Also, scalability is a problem with this approach, as the number of product terms that need to be verified grows exponentially as the number of inputs increase. Kondratyev *et al.* [14] provide a formal verification approach for observability verification, which entails determining all input combinations that assert $gate_i$, then forcing $gate_i$ to remain de-asserted while checking that none of those input combinations result in all circuit outputs becoming DATA. This check is performed for all gates to ensure circuit observability; and if also applied to each circuit input (i.e., replace $gate_i$ with $input_i$ in the observability check explanation), it will guarantee input-completeness. Our approach for observability checking, detailed in Section 15.3.5, is very similar to [14], while our approach checks input-completeness for all inputs simultaneously, as detailed in Section 15.3.4.

15.3 Equivalence verification for combinational NCL circuits

In industry, asynchronous NCL circuits are typically synthesized from their synchronous counterparts. Throughout the synthesis and optimization process, the synchronous specification undergoes several transformations, resulting in major structural differences between the implemented NCL circuit and its synchronous specification. For this kind of scenario, equivalence checking is a widely used formal verification method that checks for logical and functional equivalence between two different circuits.

NCL verification based on equivalence checking has proved to be a unified, fast, and scalable approach that eliminates most of the limiting factors of previous verification works in the field. The NCL equivalence verification method requires five steps, as described below and detailed in the following subsections:

Step 1: The netlist of an NCL circuit to be verified is converted into a corresponding Boolean/synchronous netlist, which is modeled in the SMT-LIB language using an automated script that we developed. The converted netlist is then checked against its corresponding Boolean/synchronous specification using an SMT solver to test for functional equivalence, as detailed in Section 15.3.1.

Step 2: Step 1 only checks the converted circuit's signals corresponding to the original NCL circuit's $rail^1$ signals, with their equivalent Boolean/synchronous specification external outputs or register outputs; hence, the original NCL circuit's $rail^0$ signals must also be ensured to be inverses of their respective $rail^1$ signals, through the invariant check detailed in Section 15.3.2, in order to guarantee safety after passing Step 1.

Step 3: The NCL netlist is then automatically converted into a graph structure, and information related to the handshaking control is gathered by traversing the graph. This information is utilized to analyze the handshaking correctness of the circuit in order to check for deadlock, as detailed in Section 15.3.3.

Steps 4 and 5: Once the NCL circuit passes Step 2, each combinational logic (C/L) block must be verified to be both input-complete (Step 4) and observable (Step 5) in order to guarantee liveness of the circuit under all timing scenarios, as detailed in Sections 15.3.4 and 15.3.5, respectively.

15.3.1 Functional equivalence check

A 3 × 3 NCL multiplier, shown in Figure 15.1(a), is used as an example to illustrate the equivalence verification procedure for combinational NCL circuits. NCL multipliers use input-incomplete NCL AND functions (denoted with an I inside the AND symbol), input-complete NCL AND functions (denoted with a C inside the AND symbol), NCL Half-Adders (HA), and NCL Full-Adders (FA), which all consist of a combination of NCL threshold gates, as shown in Figure 15.1(b), (c), (d), and (e), respectively. All signals in Figure 15.1(a) are dual-rail; and all registers are reset-to-NULL, denoted as REG_NULL. In addition to the I/O registers, the multiplier in Figure 15.1(a) includes one intermediate register stage to increase throughput.

The netlist of the NCL 3 × 3 multiplier is shown in Figure 15.2(a). The first two lines indicate all primary inputs and primary outputs, respectively. Lines 3–44 correspond to the NCL C/L threshold gates, where the first column is the type of gate, the second column lists the gate's inputs, in comma separated format starting with input A, and the last column is the gate's output. Lines 45–64 correspond to 1-bit NCL registers, where the first column is the reset type of the register (i.e., _NULL, _DATA0, or _DATA1, for reset to NULL, DATA0, or DATA1, respectively), the second column denotes the register's level (i.e., the depth of the path through registers without considering the C/L in-between. For the 3 × 3 multiplier example, there are three stages of registers, with levels 1, 2, and 3, starting from the input registers), the third and fourth columns are the register's $rail^0$ and $rail^1$ data inputs, respectively, the fifth and sixth columns are the register's *Ki* input and *Ko* output, respectively, and the seventh and eighth columns are the register's $rail^0$ and $rail^1$ data outputs, respectively. Lines 65–72 correspond to the C-elements (i.e., THnn gates) used in the handshaking control circuitry, where the first column is C*n*, with *n* indicating the number of inputs to the C-element, the second column lists the inputs in comma separated format, and the last column is the C-element's output. For example, C4 on line 65 is a four-input C-element.

The NCL netlist is input to a conversion algorithm that converts it into an equivalent Boolean netlist, as shown in Figure 15.2(b) for the Figure 15.2(a) example. Each NCL C/L gate is replaced with its corresponding Boolean gate that has the same set function, but no hysteresis; each internal dual-rail signal is already represented as two Boolean signals, the first for $rail^1$ and the second for $rail^0$, so no changes are needed for these; and each primary dual-rail input is replaced with that signal's $rail^1$, as this corresponds to the equivalent Boolean signal. The $rail^1$ primary inputs are then inverted to produce internal signals corresponding to what

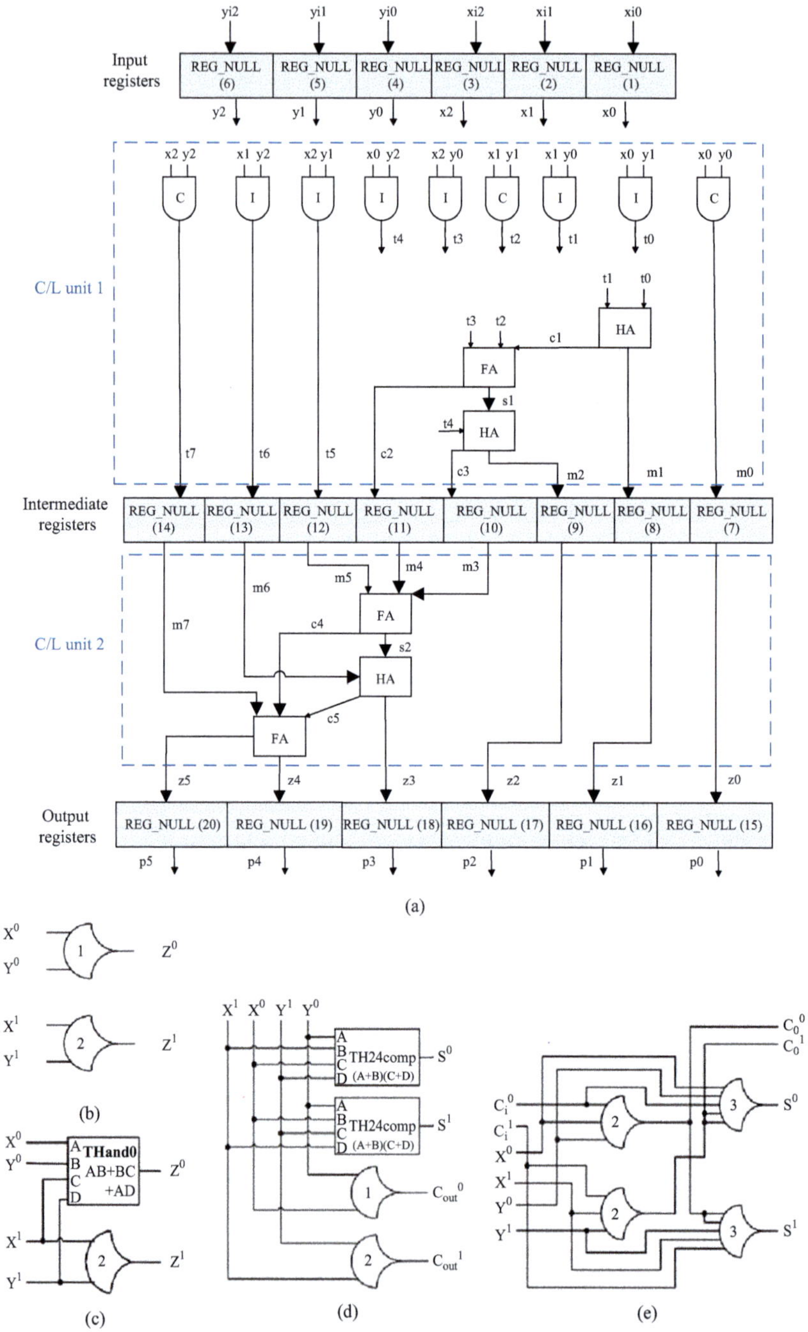

Figure 15.1 *(a) 3 × 3 NCL multiplier circuit. (b) Input-incomplete NCL AND. (c) Input-complete NCL AND. (d) NCL HA. (e) NCL FA*

```
1.  xi0_0,xi0_1,xi1_0,xi1_1,...,yi1_0,yi1_1,yi2_0,yi2_1
2.  p0_0,p0_1,p1_0,p1_1,...,p5_0,p5_1
3.  th22 x0_1,y0_1 m0_1
4.  thand0 y0_0,x0_0,y0_1,x0_1 m0_0
5.  th22 x0_1,y1_1 t0_1
6.  th12 x0_0,y1_0 t0_0
7.  th22 x0_1,y2_1 t4_1
8.  th12 x0_0,y2_0 t4_0
9.  th22 x1_1,y0_1 t1_1
10. th12 x1_0,y0_0 t1_0
11. th22 x1_1,y1_1 t2_1
12. thand0 y1_0,x1_0,y1_1,x1_1 t2_0
13. th22 x1_1,y2_1 t6_1
14. th12 x1_0,y2_0 t6_0
15. th22 x2_1,y0_1 t3_1
16. th12 x2_0,y0_0 t3_0
17. th22 x2_1,y1_1 t5_1
18. th12 x2_0,y1_0 t5_0
19. th22 x2_1,y2_1 t7_1
20. thand0 y2_0,x2_0,y2_1,x2_1 t7_0
21. th24comp t0_0,t1_0,t0_1,t1_1 m1_1
22. th24comp t0_0,t1_1,t1_0,t0_1 m1_0
23. th22 t0_1,t1_1 c1_1
24. th12 t0_0,t1_0 c1_0
25. th23 t3_0,t2_0,c1_0 c2_0
26. th23 t3_1,t2_1,c1_1 c2_1
27. th34w2 c2_0,t3_1,t2_1,c1_1 s1_1
28. th34w2 c2_1,t3_0,t2_0,c1_0 s1_0
29. th24comp s1_0,t4_0,s1_1,t4_1 m2_1
30. th24comp s1_0,t4_1,t4_0,s1_1 m2_0
31. th22 s1_1,t4_1 c3_1
32. th12 s1_0,t4_0 c3_0
33. th23 m5_0,m4_0,m3_0 c4_0
34. th23 m5_1,m4_1,m3_1 c4_1
35. th34w2 c4_0,m5_1,m4_1,m3_1 s2_1
36. th34w2 c4_1,m5_0,m4_0,m3_0 s2_0
37. th24comp s2_0,m6_0,s2_1,m6_1 z3_1
38. th24comp s2_0,m6_1,m6_0,s2_1 z3_0
39. th22 s2_1,m6_1 c5_1
40. th12 s2_0,m6_0 c5_0
41. th23 m7_0,c4_0,c5_0 z5_0
42. th23 m7_1,c4_1,c5_1 z5_1
43. th34w2 z5_0,m7_1,c4_1,c5_1 z4_1
44. th34w2 z5_1,m7_0,c4_0,c5_0 z4_0
45. Reg_NULL 1 xi0_0 xi0_1 KO3 ko1 x0_0 x0_1
46. Reg_NULL 1 xi1_0 xi1_1 KO3 ko2 x1_0 x1_1
47. Reg_NULL 1 xi2_0 xi2_1 KO3 ko3 x2_0 x2_1
48. Reg_NULL 1 yi0_0 yi0_1 KO3 ko4 y0_0 y0_1
49. Reg_NULL 1 yi1_0 yi1_1 KO3 ko5 y1_0 y1_1
50. Reg_NULL 1 yi2_0 yi2_1 KO3 ko6 y2_0 y2_1
51. Reg_NULL 2 m0_0 m0_1 ko15 ko7 z0_0 z0_1
52. Reg_NULL 2 m1_0 m1_1 ko16 ko8 z1_0 z1_1
53. Reg_NULL 2 m2_0 m2_1 ko17 ko9 z2_0 z2_1
54. Reg_NULL 2 c3_0 c3_1 KO4 ko10 m3_0 m3_1
55. Reg_NULL 2 c2_0 c2_1 KO4 ko11 m4_0 m4_1
56. Reg_NULL 2 t5_0 t5_1 KO4 ko12 m5_0 m5_1
57. Reg_NULL 2 t6_0 t6_1 KO4 ko13 m6_0 m6_1
58. Reg_NULL 2 t7_0 t7_1 KO5 ko14 m7_0 m7_1
59. Reg_NULL 3 z0_0 z0_1 Ki ko15 p0_0 p0_1
60. Reg_NULL 3 z1_0 z1_1 Ki ko16 p1_0 p1_1
61. Reg_NULL 3 z2_0 z2_1 Ki ko17 p2_0 p2_1
62. Reg_NULL 3 z3_0 z3_1 Ki ko18 p3_0 p3_1
63. Reg_NULL 3 z4_0 z4_1 Ki ko19 p4_0 p4_1
64. Reg_NULL 3 z5_0 z5_1 Ki ko20 p5_0 p5_1
65. C4 ko7,ko8,ko9,ko10 KO1
66. C4 ko11,ko12,ko13,ko14 KO2
67. C2 KO1,KO2 KO3
68. C3 ko18,ko19,ko20 KO4
69. C2 ko19,ko20 KO5
70. C3 ko4,ko5,ko6 KO6
71. C3 ko1,ko2,ko3 KO7
72. C2 KO7,KO6 KO
```

(a)

```
1.  xi0_1,xi1_1,xi2_1,yi0_1,yi1_1,yi2_1
2.  p0_0,p0_1,p1_0,p1_1,...,p5_0,p5_1
3.  not xi0_1 xi0_0
4.  not xi1_1 xi1_0
5.  not xi2_1 xi2_0
6.  not yi0_1 yi0_0
7.  not yi1_1 yi1_0
8.  not yi2_1 yi2_0
9.  th22 xi0_1,yi0_1 p0_1
10. thand0 yi0_0,xi0_0,yi0_1,xi0_1 p0_0
11. th22 xi0_1,yi1_1 t0_1
12. th12 xi0_0,yi1_0 t0_0
13. th22 xi0_1,yi2_1 t4_1
14. th12 xi0_0,yi2_0 t4_0
15. th22 xi1_1,yi0_1 t1_1
16. th12 xi1_0,yi0_0 t1_0
17. th22 xi1_1,yi1_1 t2_1
18. thand0 yi1_0,xi1_0,yi1_1,xi1_1 t2_0
19. th22 xi1_1,yi2_1 t6_1
20. th12 xi1_0,yi2_0 t6_0
21. th22 xi2_1,yi0_1 t3_1
22. th12 xi2_0,yi0_0 t3_0
23. th22 xi2_1,yi1_1 t5_1
24. th12 xi2_0,yi1_0 t5_0
25. th22 xi2_1,yi2_1 t7_1
26. thand0 yi2_0,xi2_0,yi2_1,xi2_1 t7_0
27. th24comp t0_0,t1_0,t0_1,t1_1 p1_1
28. th24comp t0_0,t1_1,t1_0,t0_1 p1_0
29. th22 t0_1,t1_1 c1_1
30. th12 t0_0,t1_0 c1_0
31. th23 t3_0,t2_0,c1_0 c2_0
32. th23 t3_1,t2_1,c1_1 c2_1
33. th34w2 c2_0,t3_1,t2_1,c1_1 s1_1
34. th34w2 c2_1,t3_0,t2_0,c1_0 s1_0
35. th24comp s1_0,t4_0,s1_1,t4_1 p2_1
36. th24comp s1_0,t4_1,t4_0,s1_1 p2_0
37. th22 s1_1,t4_1 c3_1
38. th12 s1_0,t4_0 c3_0
39. th23 t5_0,c2_0,c3_0 c4_0
40. th23 t5_1,c2_1,c3_1 c4_1
41. th34w2 c4_0,t5_1,c2_1,c3_1 s2_1
42. th34w2 c4_1,t5_0,c2_0,c3_0 s2_0
43. th24comp s2_0,t6_0,s2_1,t6_1 p3_1
44. th24comp s2_0,t6_1,t6_0,s2_1 p3_0
45. th22 s2_1,t6_1 c5_1
46. th12 s2_0,t6_0 c5_0
47. th23 t7_0,c4_0,c5_0 p5_0
48. th23 t7_1,c4_1,c5_1 p5_1
49. th34w2 p5_0,t7_1,c4_1,c5_1 p4_1
50. th34w2 p5_1,t7_0,c4_0,c5_0 p4_0
```

(b)

Figure 15.2 (a) 3 × 3 NCL multiplier netlist. (b) Converted Boolean netlist

used to be the rail^0 primary inputs, as these are utilized in the internal logic. The first two lines in the converted netlist are the list of primary inputs and outputs, respectively, where the inputs correspond to the original NCL netlist's rail^1 inputs, and the outputs include both rail^0 and rail^1 outputs. Lines 3–8 in the converted netlist are the added inverters used to produce the equivalent signals to the original rail^0 inputs, as these were removed in the conversion. The format of each gate is the same as explained above for the NCL netlist. All *Reg_NULL* components are removed during conversion by setting their data outputs equal to their data inputs, since these have no corresponding functionality in the equivalent Boolean circuit. Purely C/L circuits will not include *Reg_DATA* components, as these correspond to synchronous registers; these will be discussed in Section 15.4.

The converted Boolean netlist is automatically encoded in the SMT-LIB language [3], using a conversion tool we developed, which is then input to an SMT solver to check for functional equivalence with the corresponding specification. For the 3 $\times$ 3 multiplier example, the SMT solver checks for the following safety property: $F_{\text{NCL\_Bool\_Equiv.}}$ $(x2\_1, x1\_1, x0\_1, y2\_1, y1\_1, y0\_1) = \text{MUL}\ (x, y)$, where $(x2\_1, x1\_1, x0\_1)$ and $(y2\_1, y1\_1, y0\_1)$ are the x and y rail^1 inputs, respectively, starting with the most significant bit (MSB). We use the Z3 SMT solver [15] to check for equivalence, but any combinational equivalence checker could be used. Note that only the rail^1 outputs need to be checked here, as these correspond to the Boolean specification circuit outputs. The rail^0 outputs will be utilized for the invariant check, described next.

15.3.2 Invariant check

Since only the rail^1 outputs are utilized for the functional equivalence check, the rail^0 outputs must also be checked to ensure safety. To address correctness of the rail^0 outputs, an additional SMT invariant proof obligation is required for the original NCL circuit, which states that in any reachable NCL circuit state where the outputs are all DATA, every rail^0 output must be the inverse of its corresponding rail^1 output.

One way to achieve this is to initialize all registers to NULL, all C/L gate outputs to 0, and all register *Ki* inputs to *rfd* (i.e., logic 1), then make all the primary inputs DATA (i.e., represented in SMT as all combinations of valid DATA) and step the circuit. This will allow the input DATA to flow through all stages of the circuit, generating all possible combinations of valid DATA at the primary outputs. For each primary dual-rail output, the invariant is then checked to ensure that the rail^0 output is the inverse of its corresponding rail^1 output. For a C/L circuit with j registers $r^1, \ldots, r^j$, k C/L threshold gates $g^1, \ldots, g^k$, q dual-rail inputs $i^1, \ldots, i^q$, and l dual-rail outputs $o^1\langle R^0, R^1\rangle, \ldots, o^l\langle R^0, R^1\rangle$, where R^0 and R^1 are the output's rail^0 and rail^1, respectively, the proof obligation for this invariant check is shown below as Proof Obligation 1. Predicate *P1* indicates that all registers in are reset-to-NULL. *P2* and *P3* state that all threshold gates and *Ki* register inputs are initialized to logic 0 and 1, respectively. *P4* indicates that all inputs are DATA. *P5* represents the symbolic step of the circuit with all threshold gates set to 0 and all inputs set to DATA, with the new values of the threshold gates stored in $(g_B^{1}, \ldots, g_B^{k})$. *P6* states that the rails of each dual-rail output are complements of each other. The proof obligation, *PO1*, indicates that if DATA is allowed to flow from the primary inputs to the primary outputs, then for all possible valid DATA inputs, each output's rail^0, R^0, is always the inverse of its respective rail^1 output, R^1.

Proof Obligation 1:
P1: $\wedge_{n=1}^{j}(r_A{}^n = 0b00)$
P2: $\wedge_{n=1}^{k}(g_A{}^n = 0)$
P3: $\wedge_{n=1}^{j}(Ki_A{}^n = 1)$
P4: $\wedge_{n=1}^{q}(i_A{}^n = 0b01) \vee (i_A{}^n = 0b10)$
P5: $(g_B{}^1, \ldots, g_B{}^k) = NCLStep(i_A{}^1, \ldots, i_A{}^q)$
P6: $\wedge_{n=1}^{n} o_B^n < R^0 > = \neg o_B{}^n < R^1 >$
PO1: $\{P1 \wedge P2 \wedge P3 \wedge P4 \wedge P5 \Rightarrow P6\}$

An alternative, faster method to check invariants is to check each NCL circuit stage independently. To do this, we developed an algorithm that reads the original NCL circuit netlist and separately extracts each circuit stage. Then, for each extracted stage, we set all gate outputs to 0, all stage inputs to DATA, and step the circuit, such that the stage's outputs become all possible combinations of valid DATA. Finally, the invariant is checked for each of the stage's dual-rail outputs to ensure that its rail^0 is the inverse of its corresponding rail^1. The proof obligation for this second invariant check method is shown below as Proof Obligation 2, where the extracted stage has j dual-rail inputs $i^1, \ldots, i^j$, m threshold gates $g^1, \ldots, g^m$, and k dual-rail outputs $o^1<R^0, R^1>, \ldots, o^k<R^0, R^1>$, where R^0 and R^1 are the output's rail^0 and rail^1, respectively. Predicate *P1* indicates that all stage inputs are valid DATA; *P2* indicates that all NCL threshold gates in the stage are initialized to 0; *P3* corresponds to a NULL to DATA transition of the stage; and *P4* states that the rails of each dual-rail output are complements of each other. The Proof Obligation, *PO2*, states that after a NULL to DATA transition of the stage with all possible valid DATA inputs, that each output's rail^0, R^0, is always the inverse of its respective rail^1 output, R^1.

Proof Obligation 2:
P1: $\wedge_{n=1}^{j}(i_A{}^n = 0b01) \vee (i_A{}^n = 0b10)$
P2: $\wedge_{n=1}^{m}(g_A{}^n = 0)$
P3: $(g_B{}^1, \ldots, g_B{}^m) = NCLStep(i_A{}^1, \ldots, i_A{}^j)$
P4: $\wedge_{n=1}^{k} o_B{}^n < R^0 > = \neg o_B{}^n < R^1 >$
PO2: $\{P1 \wedge P2 \wedge P3 \Rightarrow P4\}$

This second invariant check method is much faster than the first, since it breaks the problem into a set of smaller invariant checks (i.e., one per stage), whereas the first method checks the invariant for the entire circuit all at once. For example,

Method 2 is 38% faster for a two-stage 10 × 10 multiplier, and becomes even faster when the circuit includes additional stages. Note that for both invariant check methods, the NCL gates are modeled in SMT as Boolean functions (i.e., no hysteresis), since invariant checking only requires the NULL to DATA transition, which only utilizes each gate's set function, that is, the same for the Boolean and NCL state-holding gate implementations. This optimization reduces the invariant check time by approximately half (e.g., 377 vs. 192 s for a non-pipelined 10-bit × 10-bit unsigned multiplier).

15.3.3 Handshaking check

Liveness means absence of deadlock in a circuit. For combinational NCL circuits, proper connections between handshaking signals, along with observable and input-complete C/L, ensures liveness. The same NCL netlist shown in Figure 15.2(a), used as input for the functional equivalence and invariant checks, is also utilized as input for the liveness checks. For the handshaking check, the NCL netlist is automatically converted into a graph structure, and the handshaking paths and C-element connections are traced back to verify proper handshaking, ensuring that every register acknowledges all preceding stage register outputs that took part in calculating its input. For each NCL register, *i*, its dual-rail input is traced back through its preceding C/L to identify every NCL register's dual-rail output that took part in its calculation, generating a fan-in list, *reg_fanin(i)*. For example, referring to Figure 15.1(a), *reg_fanin(8)* would be *1, 2, 4, 5*, since *x0*, *x1*, *y0*, and *y1* are all used to generate *m1*. Also, for each NCL register, *i*, its *Ko* output is traced through the C-element handshaking circuitry to identify every NCL register's *Ki* input that register$_i$'s *Ko* output took part in calculating, generating a *Ko* fanout list, *ko_fanout(i)*. For example, referring to Figure 15.3, which shows the handshaking circuitry for the 3 × 3 multiplier example, *ko_fanout(8)* would be *1, 2, 3, 4, 5, 6*, since *ko8* takes part in the generation of the *Ki* input for all of the preceding stage's registers (i.e., 1–6).

After *reg_fanin* and *ko_fanout* for each NCL register is calculated, as shown in Figure 15.4 for the 3 × 3 multiplier example, *reg_fanin(k)* is checked to ensure that it is a subset of *ko_fanout(k)*, for all NCL registers. Note that 0 in *reg_fanin* denotes a primary data input; and 0 in *ko_fanout* denotes the external *Ko* output. Bit-wise completion results in *reg_fanin(k)* being equal to *ko_fanout(k)*, while full-word completion results in *reg_fanin(k)* being a proper subset of *ko_fanout(k)*, with the restriction that each register that is in *ko_fanout(k)* and not in *reg_fanin(k)* must be from the immediate preceding register stage of register *k*. *reg_fanin(k)* not being a subset of *ko_fanout(k)* could result in deadlock, while *reg_fanin(k)* being a proper subset of *ko_fanout(k)* but violating the stage restriction described above, could either result in deadlock or may just decrease circuit performance. Hence, if *reg_fanin(k)* is a proper subset of *ko_fanout(k)*, then each register that is in *ko_fanout(k)* and not in *reg_fanin(k)* is automatically inspected to ensure that it meets this stage restriction. If not, a warning message is generated denoting the

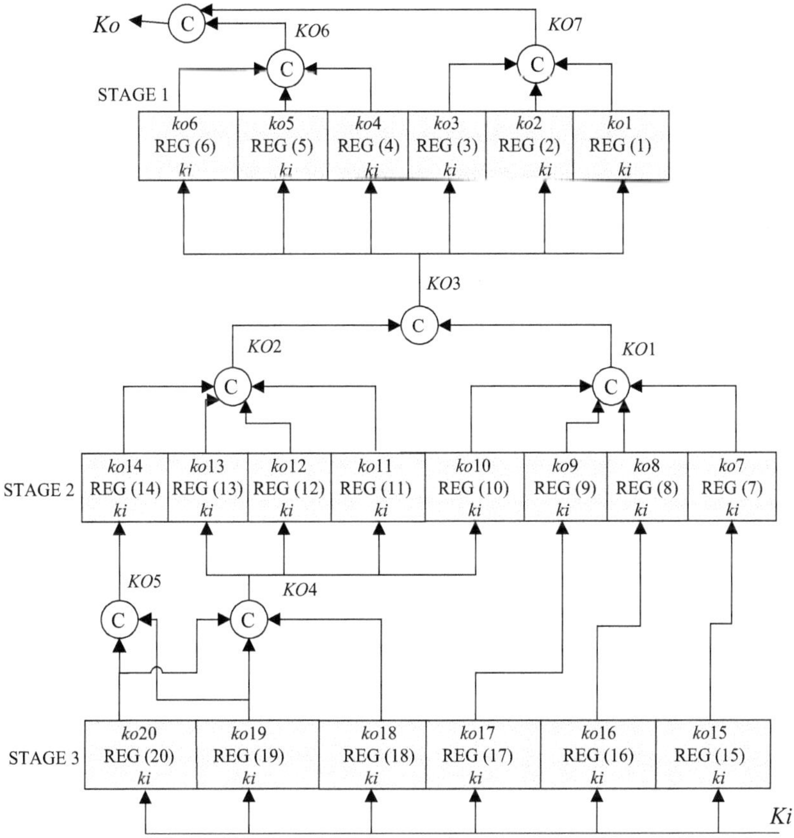

Figure 15.3 Handshaking connections for the 3 × 3 NCL multiplier

extra register in that particular register's *ko_fanout* list, to allow for easier manual inspection. For the Figure 15.3 example, the first stage utilizes full-word completion, while the second stage uses bit-wise completion.

An additional check is needed to ensure correct connection of the external *Ki* input, namely that the external *Ki* should be the *Ki* input to every register that produces a primary data output. The developed algorithm generates an appropriate descriptive error message in case the NCL circuit fails to satisfy any of these handshaking checks. Furthermore, it checks to ensure that no data signal is part of the handshaking circuitry, and that no handshaking signal is part of a data signal.

The methodology has been demonstrated on several multipliers and ISCAS-85 [16] combinational circuit benchmarks, as shown in Table 15.1. *umultN* represents a non-pipelined *N*-bit × *N*-bit unsigned multiplier. The NCL-to-Boolean netlist conversion time was negligible compared to the safety and invariant checks

1: reg_fanin: 0	ko_fanout: 0
2: reg_fanin: 0	ko_fanout: 0
3: reg_fanin: 0	ko_fanout: 0
4: reg_fanin: 0	ko_fanout: 0
5: reg_fanin: 0	ko_fanout: 0
6: reg_fanin: 0	ko_fanout: 0
7: reg_fanin: [1, 4]	ko_fanout: [1, 2, 3, 4, 5, 6]
8: reg_fanin: [1, 2, 4, 5]	ko_fanout: [1, 2, 3, 4, 5, 6]
9: reg_fanin: [1, 2, 3, 4, 5, 6]	ko_fanout: [1, 2, 3, 4, 5, 6]
10: reg_fanin: [1, 2, 3, 4, 5, 6]	ko_fanout: [1, 2, 3, 4, 5, 6]
11: reg_fanin: [1, 2, 3, 4, 5]	ko_fanout: [1, 2, 3, 4, 5, 6]
12: reg_fanin: [3, 5]	ko_fanout: [1, 2, 3, 4, 5, 6]
13: reg_fanin: [2, 6]	ko_fanout: [1, 2, 3, 4, 5, 6]
14: reg_fanin: [3, 6]	ko_fanout: [1, 2, 3, 4, 5, 6]
15: reg_fanin: [7]	ko_fanout: [7]
16: reg_fanin: [8]	ko_fanout: [8]
17: reg_fanin: [9]	ko_fanout: [9]
18: reg_fanin: [10, 11, 12, 13]	ko_fanout: [10, 11, 12, 13]
19: reg_fanin: [10, 11, 12, 13, 14]	ko_fanout: [10, 11, 12, 13, 14]
20: reg_fanin: [10, 11, 12, 13, 14]	ko_fanout: [10, 11, 12, 13, 14]

Figure 15.4 reg_fanin and ko_fanout lists for the 3 × 3 NCL multiplier

Table 15.1 Verification results of various C/L NCL circuits

Circuits	Functional check (s)	Invariant check (s)	Handshaking check (s)	Total time (s)
ISCAS c17	0.01	0.01	0.0020	0.0220
***umult*2**	0.02	0.01	0.0997	0.1297
***umult*3**	0.04	0.02	0.1087	0.1687
***umult*6**	0.32	0.33	0.8238	1.4738
***umult*8**	10.62	6.79	9.3090	26.719
***umult*10**	683.49	192.39	70.370	946.25
ISCAS c432	1.03	1.06	3.0111	5.1011
umult*10-*B1	0.08 (B)	0.10 (B)	70.370	70.550
umult*10-*B2	0.06 (B)	192.39	70.370	262.82
umult*10-*B3	683.49	192.39	69.1538 (B)	945.034
umult*10-*B4	683.49	0.08 (B)	72.0235 (B)	755.5935
umult*10-*B5	0.1 (B)	0.09 (B)	70.37	71.37

performed by the Z3 SMT solver [15] on an Intel® Core™ i7-4790 CPU with 32GB of RAM, running at 3.60 GHz. To test the methodology, we injected several bugs. The *umult*10-Bn multipliers are circuits with n different kinds of bugs, and the (B) in either the Functional Check, Invariant check, or Handshaking Check column denotes which check detected the bug. The –B1 bug incorrectly swaps rails of a dual-rail signal. –B2 represents a faulty data connection. For example, the F output of NCL $gate_i$ should be connected to the X input of NCL $gate_j$; however, X is instead connected to the output of NCL $gate_k$, which results in a logical error. –B3 corresponds to an incorrect handshaking connection; and external *Ki* and *Ko* bugs are represented by –B4. –B5 denotes a rail-duplication error, where rail^0 and rail^1 of a particular signal are the same wire. Z3 reported all functional and invariant bugs along with a counter example; and our handshaking check tool identified and reported the location of all inserted completion logic bugs.

15.3.4 Input-completeness check

Input-completeness requires that all outputs of a combinational circuit may not transition from NULL to DATA until all inputs have transitioned from NULL to DATA, and that all outputs of a combinational circuit may not transition from DATA to NULL until all inputs have transitioned from DATA to NULL [12]. In circuits with multiple outputs, it is acceptable according to Seitz's "weak conditions" of delay-insensitive signaling, for some of the outputs to transition without having a complete input set present, as long as all outputs cannot transition before all inputs arrive [17]. Input-completeness of every C/L stage is required for NCL circuits to be QDI; an input-incomplete stage may cause the circuit to deadlock under some timing scenarios.

There are two proof obligations required for verification of input-completeness. These two proof obligations have been developed to accommodate two scenarios, the first for when the circuit transitions from NULL to DATA, and the second for the transition from DATA to NULL. Both proof obligations have been generalized so that they apply to all NCL combinational circuits. The proof obligations have been encoded in a decidable fragment of first-order logic, and are automatically checked using an SMT solver.

15.3.4.1 Input-completeness proof obligation: NULL to DATA

Assume an NCL circuit has m threshold gates, p dual-rail-inputs, and q dual-rail outputs. Let $g_A{}^1, \ldots, g_A{}^m$ represent Boolean variables that correspond to the current state of the NCL threshold gates before *step A*, and $g_B{}^1, \ldots, g_B{}^m$ represent the same threshold gates' state after *step A*. Let $i_A{}^1, \ldots, i_A{}^p$ represent the circuit inputs for *step A*, and $i_B{}^1, \ldots, i_B{}^p$ for *step B*. Let $o_A{}^1, \ldots, o_A{}^q$ be the circuit output values after symbolically stepping the circuit using inputs $i_A{}^1, \ldots, i_A{}^p$ and threshold gate states $g_A{}^1, \ldots, g_A{}^m$. Let $o_B{}^1, \ldots, o_B{}^q$ be the circuit output values after symbolically stepping the circuit using inputs $i_B{}^1, \ldots, i_B{}^p$ and threshold gate states $g_B{}^1, \ldots, g_B{}^m$. The predicates used in the proof obligations for input-completeness are given in Table 15.2.

p_0 indicates that no dual-rail inputs are in an illegal state. p_1 states that all the threshold gate's current output values are 0, which indicates that the circuit is in the NULL state before a DATA transition. p_2 indicates that at least one of the dual rail inputs is NULL, and p_3 indicates that at least one of the dual-rail outputs is NULL. Proof Obligation *PO3*, below, is used to check input-completeness of the NULL to DATA transition of the circuit. *PO3* essentially states that if none of the inputs are ILLEGAL, all current threshold gate outputs are 0, and at least one of the dual-rail

Table 15.2 Proof obligation predicates for input-completeness

p_n	Predicate
p_0	$\bigwedge_{n=1}^{n=p} \sim ({i_A}^n = 0b11)$
p_1	$\bigwedge_{n=1}^{n=m} ({g_A}^n = 0)$
p_2	$\bigvee_{n=1}^{n=p} ({i_A}^n = 0b00)$
p_3	$\bigvee_{n=1}^{n=q} ({o_A}^n = 0b00)$
p_4	$\bigwedge_{n=1}^{n=p} (({i_A}^n = 0b01) \vee ({i_A}^n = 0b10))$
p_5	$({g_B}^1, \ldots, {g_B}^m) = NCLStep({i_A}^1, \ldots, {i_A}^p)$
p_6	$\bigwedge_{n=1}^{n=p} (({i_B}^n = {i_A}^n) \vee ({i_B}^n = 0b00))$
p_7	$\bigvee_{n=1}^{n=p} ({i_B}^n = {i_A}^n)$
p_8	$\bigvee_{n=1}^{n=q} (({o_B}^n = 0b01) \vee ({o_B}^n = 0b10))$

inputs is NULL, then at least one of the dual-rail outputs must be NULL. Since the dual-rail inputs in the proof obligation are symbolic, the SMT solver checks this property for all possible input combinations.

$PO3$: $\{p_0 \wedge p_1 \wedge p_2\} \rightarrow p_3$

15.3.4.2 Input-completeness proof obligation: DATA to NULL

When NCL circuits are constructed using only threshold gates with hysteresis, ensuring input-completeness of the NULL to DATA transition guarantees input-completeness of the DATA to NULL transition, since gate hysteresis ensures that a gate output cannot transition to 0 until all its inputs transition to 0. However, this is not the case for relaxed NCL circuits [13], which are comprised of both threshold gates with hysteresis and Boolean gates. Hence, for relaxed NCL circuits, input-completeness of the DATA to NULL transition must also be checked.

To formulate the DATA to NULL proof obligation, the circuit must first be symbolically initialized with all possible threshold gate outputs after a transition from NULL to DATA. This is done by first initializing the circuit to the NULL state (i.e., all threshold gates are set to 0) and then stepping the circuit with valid symbolic DATA (i.e., not NULL and not illegal) inputs, identified as *step A*.

The symbolic values of the threshold gates from *step A* are retained, and the circuit is symbolically stepped again with new inputs, identified as *step B*, which represents the DATA to NULL transition.

p_1 initializes all threshold gate outputs to 0 before *step A*. p_4 indicates that all *step A* inputs are DATA. p_5 represents the symbolic step of the circuit with all threshold gates set to 0 and all inputs set to DATA, with the new values of the threshold gates stored in $(g_B{}^1, \ldots, g_B{}^m)$. p_6 indicates that each input for *step B* is either the same DATA value it was for *step A*, or has transitioned to NULL. p_7 indicates that at least one of the inputs for *step B* is still DATA; and p_8 indicates that at least one of the outputs of *step B* remains DATA. The final proof obligation for input-completeness of the DATA to NULL transition is given below as *PO4*. It states that after initializing the circuit to the NULL state and symbolically stepping the circuit with all possible DATA inputs to generate all possible DATA states, that if at least one dual-rail input remains DATA while other inputs may transition to NULL, then at least one of the outputs must remain DATA, meaning that the circuit has not fully transitioned to the NULL state, because all inputs have not yet transitioned to NULL. Like the NULL to DATA proof obligation, all inputs are symbolic, so the SMT solver checks all combinations.

$$PO4: \{p_1 \wedge p_4 \wedge p_5 \wedge p_6 \wedge p_7\} \rightarrow p_8$$

15.3.4.3 Input-completeness results

Verification of the proof obligations for input-completeness can be performed using any SMT solver. To perform input-completeness verification, we developed a tool to automatically generate the circuit model and proof obligation specifications, encoded in SMT-LIB format, from the original circuit netlist, such as the one shown in Figure 15.2(a) for the 3×3 multiplier. For the verification results presented here, N-bit $\times$ N-bit unsigned dual-rail NCL multipliers were used as benchmarks, where $3 \leq N \leq 15$. The ISCAS-85 C432 27-channel interrupt controller circuit was also used as a benchmark [18]. The verification proof obligations were checked using the Z3 SMT solver on an Intel® Core™ i7-4790 CPU with 32GB of RAM, running at 3.60 GHz.

The verification results are listed in Table 15.3, where the first column is the Circuit Name, the second column is the verification time for the NULL to DATA proof obligation of a correct input-complete implementation, the third column is the verification time for the NULL to DATA proof obligation of an incorrect input-incomplete implementation, and columns four and five report the verification times for the DATA to NULL proof obligations for input-complete and input-incomplete implementations, respectively. *umultN* represents an N-bit $\times$ N-bit unsigned multiplier constructed using only NCL gates with hysteresis, while $r - umultN$ represents a relaxed version of the N-bit $\times$ N-bit multiplier, where NCL gates are replaced with Boolean gates when hysteresis is not required for input-completeness. Timeout (TO) is listed in the verification results when the verification time exceeded 1 day.

Table 15.3 Input-completeness verification times (s)

Circuit	N to D	Buggy N to D	D to N	Buggy D to N
*umult*3	0.02	0.01	0.03	0.04
*umult*4	0.02	0.05	0.06	0.06
*umult*5	0.09	0.05	0.12	0.11
*umult*6	0.11	0.15	0.38	0.24
*umult*7	0.38	0.27	1.49	1.23
*umult*8	1.44	0.49	5.47	3.60
*umult*9	5.30	2.37	22.38	1.28
*umult*10	20.22	8.92	102.42	18.45
*umult*11	54.09	2.99	430.29	22.81
*umult*12	236.00	8.21	1,909.44	23.17
*umult*13	885.30	3.85	7,401.11	15.11
*umult*14	3,424.89	114.41	34,961.6	8.26
*umult*15	9,663.01	19.41	***TO***	112.55
r − *umult*3	0.02	0.02	0.04	0.07
r − *umult*4	0.02	0.02	0.06	0.07
r − *umult*5	0.05	0.04	0.10	0.08
r − *umult*6	0.15	0.12	0.42	0.07
r − *umult*7	0.39	0.12	1.48	0.11
r − *umult*8	1.38	1.43	6.38	0.17
r − *umult*9	4.74	5.17	28.03	0.20
r − *umult*10	16.26	19.02	146.95	0.20
r − *umult*11	58.04	46.53	642.80	0.31
r − *umult*12	215.75	228.47	3,635.01	0.35
r − *umult*13	729.11	34.97	15,663.24	0.40
r − *umult*14	3,045.99	4,104.45	80,213.90	0.68
r − *umult*15	10,561.11	9,974.39	***TO***	0.308
*C*432	0.062	0.068	0.074	0.94

The benchmark multipliers were designed as shown for the 3 × 3 version in Figure 15.1, with input-complete AND functions to generate the X_iY_i partial products and input-incomplete AND functions for the X_iY_j partial products, where $i \neq j$, but without the intermediate NCL register (i.e., a single stage with only input and output registers [19]). To create the buggy non-relaxed versions, $1 \leq k \leq N$ was chosen at random and the input-complete AND function used to generate the X_kY_k partial product was replaced with an input-incomplete version. NCL HAs and FAs are inherently input-complete and therefore cannot be made input-incomplete when constructed only using NCL gates with hysteresis. The relaxed version of each multiplier was constructed by taking the non-relaxed version and replacing the TH22 gate within the input-incomplete AND functions and HAs with a Boolean AND gate. Buggy relaxed circuits were constructed by relaxing one of the following: either the TH22 or THand0 gate in a X_iY_i partial product AND function, a TH24comp gate in a HA, or either a TH34w2 or TH23 gate in a FA. The ISCAS-85 C432 circuit was designed using input-incomplete functions when possible while maintaining input-completeness. The buggy version was obtained by replacing one of the input-complete 3-input NAND functions that calculate RC, in Module M3 [20], with an input-incomplete version. Z3 reported all bugs along with a counter example.

15.3.5 Observability check

Observability requires every gate transition to be observable at the output, which means that every gate that transitions is necessary to transition at least one output. Observability of every gate in every C/L stage is required for NCL circuits to be QDI; an unobservable gate in any stage may cause the circuit to deadlock under some timing scenarios. Observability can be proven in a similar fashion to input-completeness. Two proof obligations are needed for each C/L gate, one for the NULL to DATA transition, and the other for the DATA to NULL transition. The proof obligations, like those for input-completeness, have been encoded in a decidable fragment of first order logic and are automatically checked using an SMT solver.

15.3.5.1 Observability proof obligation: NULL to DATA

To verify observability, a check must be performed on each C/L gate. For each gate $g^1, \ldots, g^m$, assertion of that gate is first computed, denoted as $f_1, \ldots, f_m = 1$, respectively. During the NULL to DATA observability verification of g^n, where $1 \leq n \leq m$, the output of g^n is forced to 0. Simulation of a circuit with g^n forced to 0 is called a Gn0 simulation, and the resulting function is $nclcktGn0(i^1, \ldots, i^p)$. To formulate the DATA to NULL observability proof obligation, the circuit must first be symbolically initialized with all possible threshold gate outputs that assert g^n after a transition from NULL to DATA. This is done by first initializing the circuit to the NULL state (i.e., all threshold gates are set to 0) and then stepping the circuit with valid symbolic DATA (i.e., not NULL and not illegal) inputs, identified as *step A*. The symbolic values of the threshold gates from *step A* are retained as $g_B{}^1, \ldots, g_B{}^m$, and the circuit is symbolically stepped again with new inputs, identified as *step B*, which represents the DATA to NULL transition. During the verification of g^n, where $1 \leq n \leq m$, the output of g^n is forced to 1. Simulation of a circuit with g^n forced to 1 is called a Gn1 simulation, and the resulting function is $nclcktGn1(i^1, \ldots, i^p)$. Additional predicates used in the proof obligations for observability are given in Table 15.4.

p_1 states that all the threshold gates' current output value is 0, which indicates that the circuit is in the NULL state before a DATA transition. p_4 indicates that every circuit input is valid DATA. p_9 assigns the outputs of the NCL circuit for a Gn0 simulation, where the output of g^n, the gate under test, is forced to 0. p_{10} enables only valid input combinations that would assert g^n to be used to step the circuit in p_9. Finally, p_{11} ensures that at least one of the outputs is NULL. The proof obligation to test observability of the NULL to DATA transition is given below as *PO5*, which tests observability of all gates, $g^1, \ldots, g^m$. If true for g^n, this ensures that there is at least one output that will not be asserted if g^n is not asserted, for all

Table 15.4 Additional proof obligation predicates for observability

p_n	Predicate
p_9	$(o_A{}^1, \ldots, o_A{}^q) = nclcktGn0(i_A{}^1, \ldots, i_A{}^p)$
p_{10}	$f_n = 1$
p_{11}	$\bigvee_{n=1}^{n=q} (o_B{}^n = 0b00)$
p_{12}	$\bigwedge_{n=1}^{n=p} (i_B{}^n = 0b00)$
p_{13}	$(o_B{}^1, \ldots, o_B{}^q) = nclcktGn1(i_B{}^1, \ldots, i_B{}^p)$
p_{14}	$\sim \bigwedge_{n=1}^{n=q} (o_B{}^n = 0b00)$

sets of inputs in which g^n should be asserted, therefore proving that g^n is observable for the NULL to DATA transition.

$$PO5: \bigwedge_{n=1}^{n=m} (\{p_1 \wedge p_4 \wedge p_9 \wedge p_{10}\} \rightarrow p_{11})$$

15.3.5.2 Observability proof obligation: DATA to NULL

Like input-completeness, NCL circuits consisting only of NCL gates with hysteresis are inherently observable for the DATA to NULL transition if observable for the NULL to DATA transition, since gate hysteresis ensures that a gate output cannot transition to 0 until all its preceding gates' outputs transition to 0. However, this is not the case for relaxed NCL circuits, which are comprised of both threshold gates with hysteresis and Boolean gates. Hence, for relaxed NCL circuits, observability of the DATA to NULL transition must also be checked.

p_1 initializes all threshold gate outputs to 0 before *step A*. p_4 indicates that all *step A* inputs are DATA. p_5 represents the symbolic step of the circuit with all threshold gates set to 0 and all inputs set to DATA, with the new values of the threshold gates stored in $(g_B{}^1, \ldots, g_B{}^m)$. p_{10} enables only valid input combinations that would assert g^n to be used to step the circuit in p_5. p_{12} indicates that all inputs for *step B* have transitioned to NULL. p_{13} assigns the outputs of the NCL circuit for a Gn1 simulation, where the output of g^n, the gate under test, is forced to 1. Finally, p_{14} ensures that all outputs are not NULL. The proof obligation to test observability of the DATA to NULL transition is given below as *PO6*, which tests observability of all gates, $g^1, \ldots, g^m$. If true for g^n, this ensures that following a NULL to DATA transition that asserts g^n, there is at least one output that will not become NULL during the subsequent DATA to NULL transition while g^n remains asserted, therefore proving that g^n is observable for the DATA to NULL transition.

$$PO6: \bigwedge_{n=1}^{n=m} (\{p_1 \wedge p_4 \wedge p_5 \wedge p_{10} \wedge p_{12} \wedge p_{13}\} \rightarrow p_{14})$$

Table 15.5 Observability verification times (s)

Circuit	N to D	D to N
*umult*4	0.001	0.001
*umult*5	8.203	8.944
*umult*6	13.7599	16.1921
*umult*7	27.8229	36.528
*umult*8	54.062	105.4979
*umult*9	138.3139	412.605
*umult*10	363.7079	1,968.434
*umult*11	902.046	9,657.475
*umult*12	2,384.504	52,093.64
*umult*13	5,797.037	***TO***
*C*432*M*1	1.53	3.882

15.3.5.3 Observability results

Verification of the proof obligations for observability can be performed using any SMT solver. To perform observability verification, we developed a tool to automatically generate the circuit model and proof obligation specifications, encoded in SMT-LIB format, from the original circuit netlist, such as the one shown in Figure 15.2(a) for the 3 $\times$ 3 multiplier. For the verification results presented here, N-bit $\times$ N-bit unsigned dual-rail NCL multipliers were used as benchmarks, where $3 \leq N \leq 13$. The ISCAS-85 C432 27-channel interrupt controller circuit was also used as a benchmark [18]. The verification proof obligations were checked using the Z3 SMT solver on an Intel® Core™ i7-4790 CPU with 32GB of RAM, running at 3.60 GHz.

The verification results are listed in Table 15.5, where the first column is the Circuit name, the second column is the verification time for the NULL to DATA proof obligation, and the third column is the verification time for the DATA to NULL proof obligation. *umultN* represents an N-bit $\times$ N-bit unsigned multiplier constructed using only NCL gates with hysteresis. TO is listed in the verification results when the verification time exceeded 1 day.

The test multipliers were designed exactly the same as the ones used for testing input-completeness (i.e., input-complete AND functions generate the X_iY_i partial products, and input-incomplete AND functions generate the X_iY_j partial products, where $i \neq j$). To create buggy multipliers that were input-complete but not observable, an HA was chosen at random and the XOR function to generate its sum (i.e., the two TH24comp gates in Figure 15.1(d)) was replaced with the unobservable XOR function, shown in Figure 15.5. To check observability of relaxed circuits, the M1 module of the ISCAS-85 C432 benchmark [21] was used, where the nine-input NAND function that generates PA was composed of two relaxed input-incomplete four-input AND functions, followed by an input-complete two-input AND function, and then an input-complete two-input NAND function, as shown in Figure 15.6. To create a buggy version that was

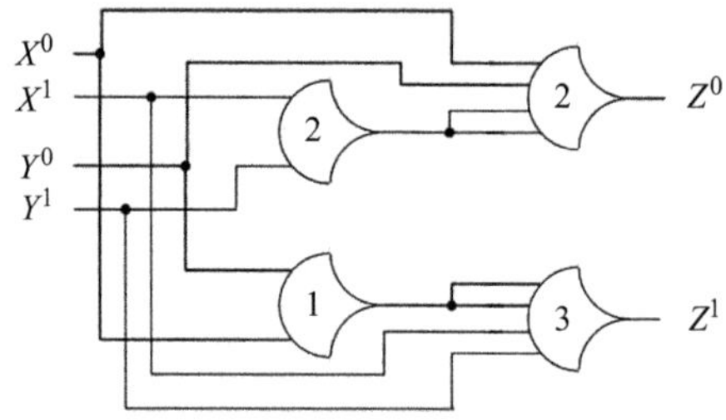

Figure 15.5 Unobservable NCL XOR

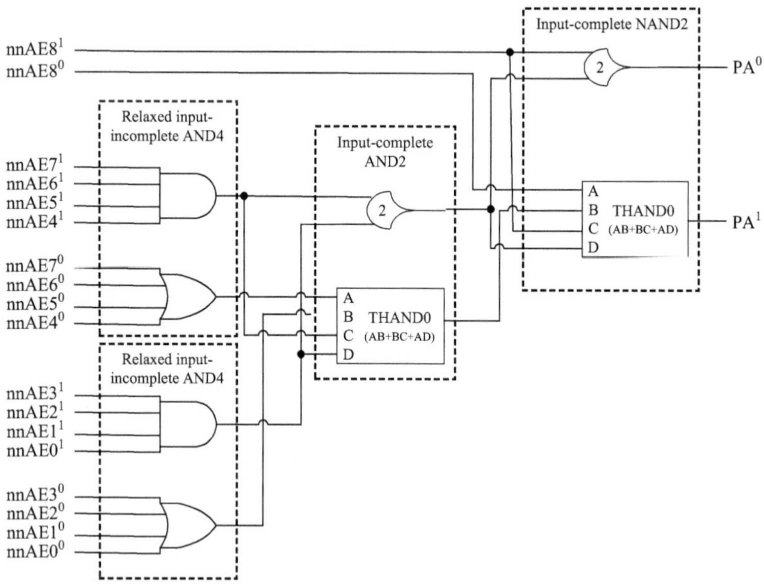

Figure 15.6 ISCAS-85 C432 M1 module nine-input NCL NAND that generates PA

input-complete but not observable, any of the four gates comprising the two-input AND function or the two-input NAND function in Figure 15.6 could be relaxed. The test times reported for the circuits are for testing every single gate for observability, even if a previous gate was found to be unobservable. Therefore, the time to detect a buggy circuit will be less than or equal to the reported times, since the rest of the gates would no longer need to be tested once an unobservable gate was identified. Z3 reported all bugs along with a counter example.

15.4 Equivalence verification for sequential NCL circuits

As described in Section 15.3.1, our equivalence verification methodology proved to be a fast and scalable approach for C/L NCL circuits. Hence, in this section we extend that approach to verify both safety and liveness of sequential NCL circuits, which is more complex due to datapath feedback.

To describe our methodology, we'll use an unsigned Multiply and Accumulate (MAC) unit as an example circuit. Figure 15.7(a) shows a synchronous MAC, where $A' = A + X \times Y$; and Figure 15.7(b) shows the equivalent NCL version. The four-phase QDI handshaking protocol utilized for NCL circuits requires at least $2N + 1$ NCL registers in a feedback loop that contains N DATA tokens, in order to avoid deadlock [12].

Hence, at least three NCL registers are needed in the MAC feedback loop to avoid deadlock, as shown in Figure 15.7(b). Although the synchronous and NCL MACs seem similar, they are structurally very different. Synchronous registers are clocked, whereas alternating DATA/NULL transitions in NCL are maintained via C-elements and a well-defined handshaking scheme. *Ki* and *Ko* are the external *request* input and *acknowledge* output, respectively.

Figure 15.8 shows the datapath diagram for a $4 + 2 \times 2$ NCL MAC with two C/L stages and four registers in the feedback loop (note that including a 4th register

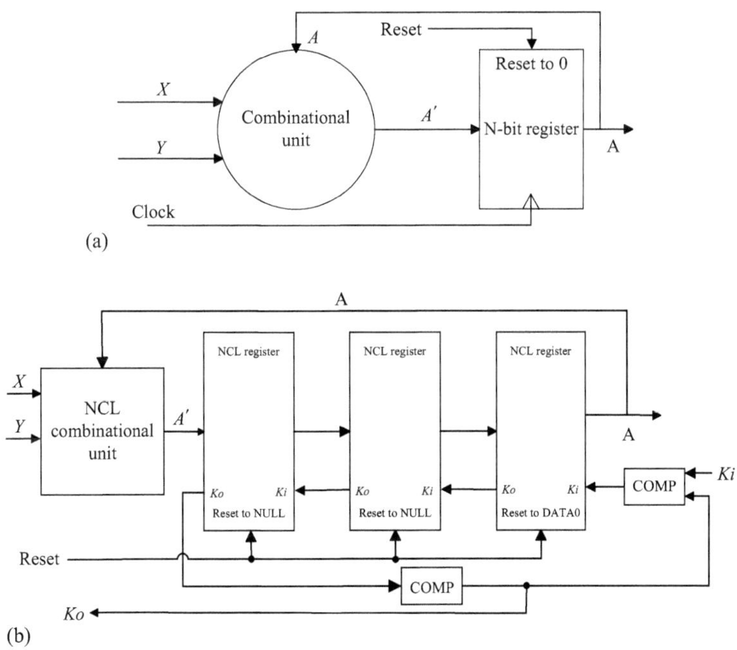

Figure 15.7 MAC circuit: (a) synchronous; (b) NCL

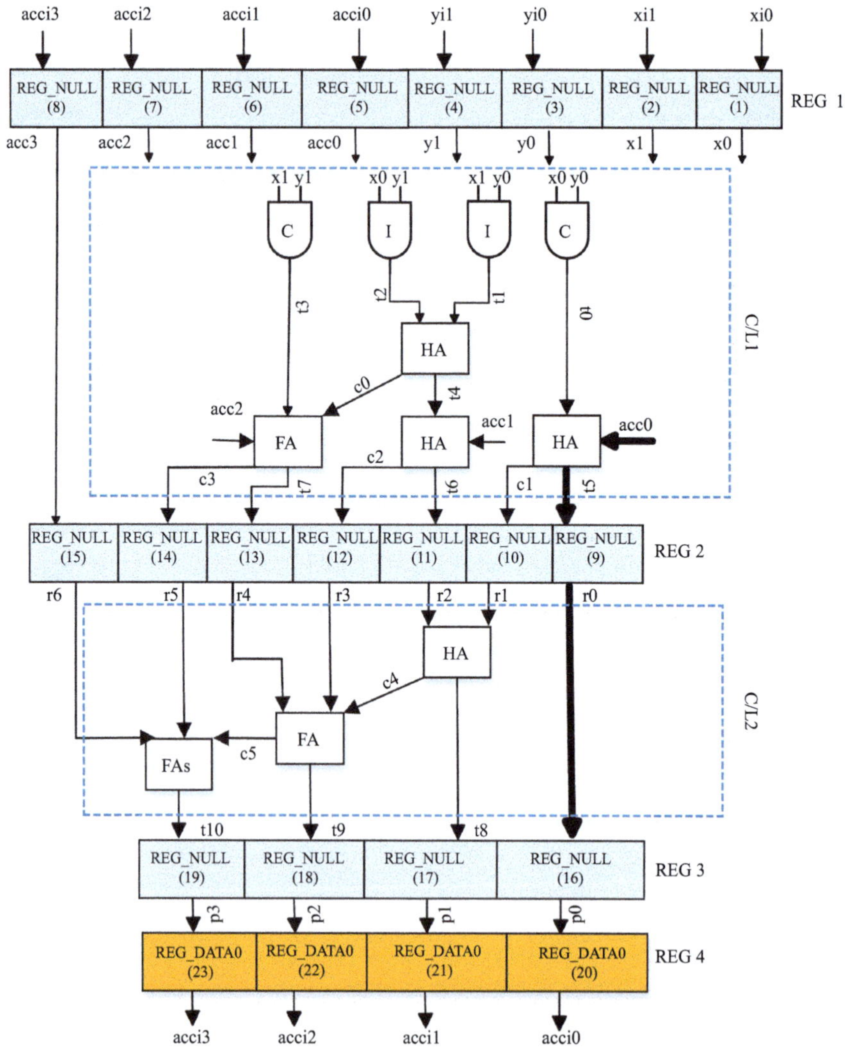

Figure 15.8 4 + 2 × 2 NCL MAC datapath

in the feedback loop increases throughput compared to using the minimum required three registers, since this allows the DATA and NULL wavefronts to flow more independently [12]). (Xi_1, Xi_0) and (Yi_1, Yi_0) are the two bits of inputs *Xi* and *Yi*, respectively. The product of *Xi* and *Yi* is added with the 4-bit accumulator output, *Acci*, where $Acci_3$ and $Acci_0$ are the MSB and least significant bit (LSB), respectively. All signals shown in Figure 15.8 are dual-rail signals. *HA* and *FA* are the NCL half-adder and full-adder components, shown in Figure 15.1(d) and (e), respectively; and *FA* is a full-adder component without a *carry* output; hence, it utilizes two two-input XOR functions, each comprised of two TH24comp gates (same as the HA *sum* output shown in Figure 15.1(d)), to compute its *sum* output. The highlighted components in Figure 15.8 are the NCL registers.

Figure 15.9(a) shows the netlist of the NCL 4 + 2 × 2 MAC, following the same structure as described in Section 15.3.1. The first two lines are the circuit inputs and outputs, respectively; lines 3–38 are the NCL threshold gates; lines 39–61 are the NCL registers; and lines 62–69 are C-elements used in the handshaking network.

15.4.1 Safety

Safety verification requires two steps. In the first step, we take a sequential NCL circuit and convert it to an equivalent synchronous circuit. We utilize the theory of WEB refinement [2] to compare the synchronous netlist generated from the NCL circuit with the original synchronous specification, as the notion of correctness. The major advantage of applying WEB refinement to the generated equivalent synchronous circuit instead of the actual NCL circuit is that a synchronous circuit is much more deterministic compared to its NCL equivalent, which makes the verification time much faster. The generated synchronous circuit, specification synchronous circuit, and the WEB-refinement property are automatically encoded in the SMT-LIB language. The resulting equivalence property is then checked using an SMT solver. In the second step, we check the invariant for each C/L stage, the same as previously discussed in Section 15.3.2.

The converted netlist (NCL-SYNC) is depicted in Figure 15.9(b). The conversion algorithm for sequential NCL circuits is slightly different than for C/L NCL circuits, described in Section 15.3.1, since the sequential NCL circuit contains reset-to-DATA registers, which are replaced with a 2-bit resettable synchronous register, 1 bit for each rail of the corresponding NCL dual-rail register. Like for C/L NCL circuits, all reset-to-NULL registers, handshaking signals, and C-elements are eliminated; and all C/L NCL gates are replaced with their corresponding relaxed (i.e., Boolean) gate.

The NCL-SYNC netlist must next be checked against the synchronous specification (SPEC-SYNC) netlist for equivalence. When verifying C/L NCL circuits, the circuit functionality could be specified as a Boolean function. However, since sequential circuits involve states and transitions, we use TSs as the formal model to capture the behaviors of both the NCL-SYNC netlist as well as the SPEC-SYNC netlist. The theory of WEB refinement [2] defines what it means for an implementation TS to be functionally equivalent to a specification TS. Therefore, we use the theory of WEB refinement for checking equivalence for sequential circuits.

The theory of WEB refinement allows for stutter between the implementation TS and the specification TS. What this means is that multiple but finite transitions of the implementation can match to a single specification transition. Rank functions (functions that map circuit states to natural numbers) are used to distinguish finite stutter from deadlock (infinite stutter). Another characteristic of WEB refinement is the use of refinement maps, which are functions that map implementation states to

(a)

```
1. xi0_0,xi0_1,xi1_0,xi1_1,yi0_0,yi0_1,yi1_0,yi1_1
2. acci0_0,acci0_1,acci1_0,acci1_1,…,acci3_0,acci3_1
3. th22 x0_1,y0_1 t0_1
4. thand0 y0_0,x0_0,y0_1,x0_1 t0_0
5. th12 x1_0,y0_0 t1_0
6. th22 x1_1,y0_1 t1_1
7. th12 x0_0,y1_0 t2_0
8. th22 x0_1,y1_1 t2_1
9. th12 x1_0,y1_0 t3_0
10. th22 x1_1,y1_1 t3_1
11. th24comp t2_0,t1_1,t1_0,t2_1 t4_0
12. th24comp t2_0,t1_0,t2_1,t1_1 t4_1
13. th12 t2_0,t1_0 c0_0
14. th22 t1_1,t2_1 c0_1
15. th24comp acc0_0,t0_1,t0_0,acc0_1 t5_0
16. th24comp acc0_0,t0_0,acc0_1,t0_1 t5_1
17. th12 acc0_0,t0_0 c1_0
18. th22 t0_1,acc0_1 c1_1
19. th24comp acc1_0,t4_1,t4_0,acc1_1 t6_0
20. th24comp acc1_0,t4_0,acc1_1,t4_1 t6_1
21. th12 acc1_0,t4_0 c2_0
22. th22 t4_1,acc1_1 c2_1
23. th23 t3_0,acc2_0,c0_0 c3_0
24. th23 t3_1,acc2_1,c0_1 c3_1
25. th34w2 c3_1,t3_0,acc2_0,c0_0 t7_0
26. th34w2 c3_0,t3_1,acc2_1,c0_1 t7_1
27. th24comp r1_0,r2_1,r2_0,r1_1 t8_0
28. th24comp r1_0,r2_0,r1_1,r2_1 t8_1
29. th12 r1_0,r2_0 c4_0
30. th22 r2_1,r1_1 c4_1
31. th23 r4_0,r3_0,c4_0 c5_0
32. th23 r4_1,r3_1,c4_1 c5_1
33. th34w2 c5_1,r4_0,r3_0,c4_0 t9_0
34. th34w2 c5_0,r4_1,r3_1,c4_1 t9_1
35. th24comp r5_0,r6_1,r6_0,r5_1 c6_0
36. th24comp r5_0,r6_0,r5_1,r6_1 c6_1
37. th24comp c5_0,c6_1,c6_0,c5_1 t10_0
38. th24comp c5_0,c6_0,c5_1,c6_1 t10_1
39. Reg_NULL 1 xi0_0,xi0_1 KO2 ko1 x0_0,x0_1
40. Reg_NULL 1 xi1_0,xi1_1 KO2 ko2 x1_0,x1_1
41. Reg_NULL 1 yi0_0 yi0_1 KO2 ko3 y0_0 y0_1
42. Reg_NULL 1 yi1_0 yi1_1 KO2 ko4 y1_0 y1_1
43. Reg_NULL 1 acci0_0 acci0_1 KO2 ko5 acc0_0 acc0_1
44. Reg_NULL 1 acci1_0 acci1_1 KO2 ko6 acc1_0 acc1_1
45. Reg_NULL 1 acci2_0 acci2_1 KO2 ko7 acc2_0 acc2_1
46. Reg_NULL 1 acci3_0 acci3_1 KO2 ko8 acc3_0 acc3_1
47. Reg_NULL 2 t5_0 t5_1 ko16 ko9 r0_0 r0_1
48. Reg_NULL 2 c1_0 c1_1 KO3 ko10 r1_0 r1_1
49. Reg_NULL 2 t6_0 t6_1 KO3 ko11 r2_0 r2_1
50. Reg_NULL 2 c2_0 c2_1 KO3 ko12 r3_0 r3_1
51. Reg_NULL 2 t7_0 t7_1 KO3 ko13 r4_0 r4_1
52. Reg_NULL 2 c3_0 c3_1 KO3 ko14 r5_0 r5_1
53. Reg_NULL 2 acc3_0 acc3_1 KO3 ko15 r6_0 r6_1
54. Reg_NULL 3 r0_0 r0_1 ko20 ko16 p0_0 p0_1
55. Reg_NULL 3 t8_0 t8_1 ko21 ko17 p1_0 p1_1
56. Reg_NULL 3 t9_0 t9_1 ko22 ko18 p2_0 p2_1
57. Reg_NULL 3 t10_0 t10_1 ko23 ko19 p3_0 p3_1
58. Reg_DATA0 4 p0_0 p0_1 KO4 ko20 acci0_0 acci0_1
59. Reg_DATA0 4 p1_0 p1_1 KO5 ko21 acci1_0 acci1_1
60. Reg_DATA0 4 p2_0 p2_1 KO6 ko22 acci2_0 acci2_1
61. Reg_DATA0 4 p3_0 p3_1 KO7 ko23 acci3_0 acci3_1
62. C4 ko9,ko10,ko11,ko12 KO1
63. C4 ko13,ko14,ko15,KO1 KO2
64. C3 ko17,ko18,ko19 KO3
65. C2 Ki,ko5 KO4
66. C2 Ki,ko6 KO5
67. C2 Ki,ko7 KO6
68. C2 Ki,ko8 KO7
69. C4 ko1,ko2,ko3,ko4 KO
```

(b)

```
1. xi0_1,xi1_1, yi0_1,yi1_1
2. acci0_0,acci0_1,acci1_0,acci1_1,…,acci3_0,acci3_1
3. not xi0_1 xi0_0
4. not yi0_1 yi0_0
5. not xi1_1 xi1_0
6. not yi1_1 yi1_0
7. th12 xi0_0,yi0_0 t0_0
8. th22 xi0_1,yi0_1 t0_1
9. th12 xi1_0,yi0_0 t1_0
10. th22 xi1_1,yi0_1 t1_1
11. th12 xi0_0,yi1_0 t2_0
12. th22 xi0_1,yi1_1 t2_1
13. th12 x1_0,y1_0 t3_0
14. th22 x1_1,y1_1 t3_1
15. th24comp t2_0,t1_1,t1_0,t2_1 t4_0
16. th24comp t2_0,t1_0,t2_1,t1_1 t4_1
17. th12 t2_0,t1_0 c0_0
18. th22 t1_1,t2_1 c0_1
19. th24comp acci0_0,t0_1,t0_0,acci0_1 t5_0
20. th24comp acci0_0,t0_0,acci0_1,t0_1 t5_1
21. th12 acci0_0,t0_0 c1_0
22. th22 t0_1,acci0_1 c1_1
23. th24comp acci1_0,t4_1,t4_0,acci1_1 t6_0
24. th24comp acci1_0,t4_0,acci1_1,t4_1 t6_1
25. th12 acci1_0,t4_0 c2_0
26. th22 t4_1,acci1_1 c2_1
27. th23 t3_0,acci2_0,c0_0 c3_0
28. th23 t3_1,acci2_1,c0_1 c3_1
29. th34w2 c3_1,t3_0,acci2_0,c0_0 t7_0
30. th34w2 c3_0,t3_1,acci2_1,c0_1 t7_1
31. th24comp c1_0,t6_1,t6_0,c1_1 t8_0
32. th24comp c1_0,t6_0,c1_1,t6_1 t8_1
33. th12 c1_0,t6_0 c4_0
34. th22 t6_1,c1_1 c4_1
35. th23 t7_0,c2_0,c4_0 c5_0
36. th23 t7_1,c2_1,c4_1 c5_1
37. th34w2 c5_1,t7_0,c2_0,c4_0 t9_0
38. th34w2 c5_0,t7_1,c2_1,c4_1 t9_1
39. th24comp c3_0,acci3_1,acci3_0,c3_1 c6_0
40. th24comp c3_0,acci3_0,c3_1,acci3_1 c6_1
41. th24comp c5_0,c6_1,c6_0,c5_1 t10_0
42. th24comp c5_0,c6_0,c5_1,c6_1 t10_1
43. Reg_0 t5_0 t5_1 acci0_0 acci0_1
44. Reg_0 t8_0 t8_1 acci1_0 acci1_1
45. Reg_0 t9_0 t9_1 acci2_0 acci2_1
46. Reg_0 t10_0 t10_1 acci3_0 acci3_1
```

Figure 15.9 (a) 4 + 2 × 2 NCL MAC netlist. (b) Converted synchronous equivalent netlist

specification states. Refinement maps allow for the implementation and specification to be specified at significantly different abstraction levels. However, since the rail^1 registers of NCL-SYNC and the registers of SPEC-SYNC have a one-to-one mapping, there is no stutter between these two TSs, and the refinement is simply a projection of the rail^1 registers of the implementation state to the registers of the specification state. Therefore, the correctness proof obligations required for verifying WEB refinement can be reduced to the proof obligation depicted in Figure 15.10, where s is a state of NCL-SYNC; u is a SPEC-SYNC state obtained by projecting the values of the rail^1 registers from state s; $Step_{SYNC\_NCL}$ and $Step_{SYNC\_SPEC}$ are the functions that correspond to a single step of the NCL-SYNC circuit and the SPEC-SYNC circuit, respectively; w is the state obtained by stepping NCL-SYNC from state s; and v is the state obtained by stepping SPEC-SYNC from state u. The proof obligation states that the two circuits are functionally equivalent if for every state s of NCL-SYNC, the corresponding projection of values from the rail^1 registers of the w state are equivalent to the values of the corresponding registers in the v state. This proof obligation can be encoded in the SMT-LIB language, as shown below in *PO7*, and checked using an SMT solver.

$$PO7: \{\forall s :: s \in S_{SYNC\_NCL} :: \\ [u = Reg\_Proj(s) \wedge w = Step_{SYNC\_NCL}(s) \wedge v = Step_{SYNC\_SPEC}(u)] \\ \Rightarrow Reg\_Proj(w) = v\}.$$

After verifying function equivalence, the rail^0 outputs of each C/L stage must also be checked to ensure safety, as detailed in Section 15.3.2. Note that for sequential circuits, which include datapath feedback, the first invariant check method that checks the entire circuit simultaneously won't work; hence, the second, much faster method that performs the invariant check independently for each stage is utilized.

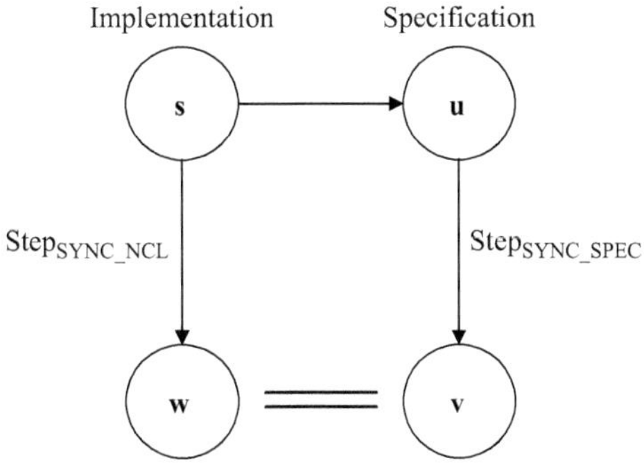

Figure 15.10 Depiction of proof obligation to check equivalence of NCL-SYNC and SPEC-SYNC netlists

15.4.2 Liveness

Figure 15.11 shows the handshaking connections for the 4 + 2 × 2 NCL MAC. Full-word completion is used by the input register, Reg 1, to generate a single *Ko*. Full-word completion is also utilized between Reg 1 and Reg 2, since bit-wise completion would have the same delay and require more area. Partial bit-wise

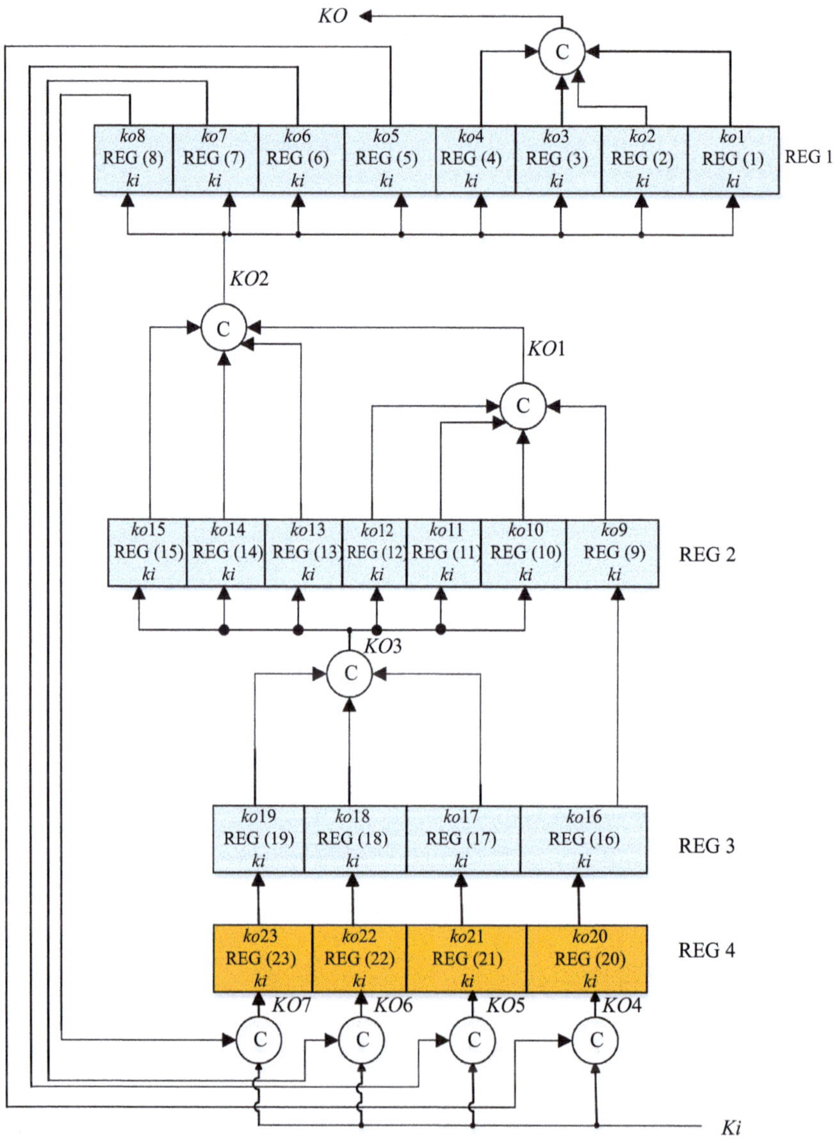

Figure 15.11 Handshaking connections for the 4 + 2 × 2 NCL MAC

completion is utilized between Reg 2 and Reg 3, since full bit-wise completion would have the same delay and require more area. Bit-wise completion is utilized between Reg 3 and Reg 4, and for the output register, Reg 4. The handshaking check for sequential NCL circuits is essentially the same as that for C/L NCL circuits, described in Section 15.3.3. The only addition is calculating a feedback register's level, which should be assigned the same level as other registers that share its *Ki* input signal, or one level more than its previous register, if its *Ki* input signal is not shared with another register already assigned a level. For the MAC example in Figure 15.11, feedback registers 5–8 would be assigned level 1, since they share their *Ki* input with the other level 1 registers, 1–4; and feedback register 15 would be assigned level 2, since it shares its *Ki* input with other level 2 registers, 9–14. Figure 15.12 shows the *reg_fanin* and *ko_fanout* lists for each register in the $4 + 2 \times 2$ NCL MAC example.

After verifying handshaking correctness, each stage's C/L must also be checked for input-completeness and observability, utilizing the methods detailed in Sections 15.3.4 and 15.3.5, respectively, to guarantee liveness.

1: *reg_fanin*: 0	*ko_fanout*: 0
2: *reg_fanin*: 0	*ko_fanout*: 0
3: *reg_fanin*: 0	*ko_fanout*: 0
4: *reg_fanin*: 0	*ko_fanout*: 0
5: *reg_fanin*: [20]	*ko_fanout*: [20]
6: *reg_fanin*: [21]	*ko_fanout*: [21]
7: *reg_fanin*: [22]	*ko_fanout*: [22]
8: *reg_fanin*: [23]	*ko_fanout*: [23]
9: *reg_fanin*: [1, 3, 5]	*ko_fanout*: [1, 2, 3, 4, 5, 6, 7, 8]
10: *reg_fanin*: [1, 3, 5]	*ko_fanout*: [1, 2, 3, 4, 5, 6, 7, 8]
11: *reg_fanin*: [1, 2, 3, 4, 6]	*ko_fanout*: [1, 2, 3, 4, 5, 6, 7, 8]
12: *reg_fanin*: [1, 2, 3, 4, 6]	*ko_fanout*: [1, 2, 3, 4, 5, 6, 7, 8]
13: *reg_fanin*: [1, 2, 3, 4, 7]	*ko_fanout*: [1, 2, 3, 4, 5, 6, 7, 8]
14: *reg_fanin*: [1, 2, 3, 4, 7]	*ko_fanout*: [1, 2, 3, 4, 5, 6, 7, 8]
15: *reg_fanin*: [8]	*ko_fanout*: [1, 2, 3, 4, 5, 6, 7, 8]
16: *reg_fanin*: [9]	*ko_fanout*: [9]
17: *reg_fanin*: [10, 11]	*ko_fanout*: [10, 11, 12, 13, 14, 15]
18: *reg_fanin*: [10, 11, 12, 13]	*ko_fanout*: [10, 11, 12, 13, 14, 15]
19: *reg_fanin*: [10, 11, 12, 13, 14, 15]	*ko_fanout*: [10, 11, 12, 13, 14, 15]
20: *reg_fanin*: [16]	*ko_fanout*: [16]
21: *reg_fanin*: [17]	*ko_fanout*: [17]
22: *reg_fanin*: [18]	*ko_fanout*: [18]
23: *reg_fanin*: [19]	*ko_fanout*: [19]

Figure 15.12 reg_fanin and ko_fanout lists for the $4 + 2 \times 2$ NCL MAC

Table 15.6 Verification results for sequential NCL circuits

Circuits	**Functional check (s)**	**Handshaking check (s)**	**Total time (s)**
ISCAS s27	0.01	0.0019	0.0119
$4 + 2 \times 2$ *MAC*	0.01	0.0045	0.0145
$8 + 4 \times 4$ *MAC*	0.05	0.7852	0.8352
$12 + 6 \times 6$ *MAC*	0.77	2.331	3.101
$16 + 8 \times 8$ *MAC*	47.55	21.7411	69.2911
$20 + 10 \times 10$ *MAC*	2,643.99	163.6463	2,807.6363
$20 + 10 \times 10$ *MAC-B1*	0.11 (B)	163.6463	163.7563
$20 + 10 \times 10$ *MAC-B2*	0.13 (B)	163.6463	163.7763
$20 + 10 \times 10$ *MAC-B3*	2,643.99	169.8422 (B)	2,813.8322
$20 + 10 \times 10$ *MAC-B4*	2,643.99	159.3253 (B)	2,803.3153
$20 + 10 \times 10$ *MAC-B5*	0.20 (B)	163.6463	163.8463

15.4.3 Sequential NCL circuit results

The verification results for sequential NCL circuits, including functional equivalence and handshaking checks, are shown in Table 15.6. Since the invariant, input-completeness, and observability checks are exactly the same for combinational and sequential NCL circuits, these results are not included in Table 15.6. Test circuits include multiple MAC units and one ISCAS-89 benchmark, s27 [22]. The MAC units are represented as $A + M \times N$, where A, M, and N represent the length of the accumulator, multiplicand, and multiplier, respectively. The same types of bugs were tested for the MACs as tested for the multipliers, and the same machine was used to perform the sequential circuit verification, both as described at the end of Section 15.3.3. Z3 reported all functional bugs along with a counter example, and our handshaking check tool identified and reported the location of all inserted completion logic bugs.

15.5 Conclusions and future work

This chapter presents a novel methodology for formally verifying the correctness (both safety and liveness) of combinational and sequential NCL circuits. The approach includes methods for ensuring handshaking correctness, and functional correctness of both $rail^1$ and $rail^0$ outputs, and methods to ensure that NCL C/L circuits, or pipeline stages, are both input-complete and observable, which is required for correct operation under all timing scenarios. The presented methodology is applicable to both NCL circuits designed using only NCL gates with hysteresis and relaxed NCL circuits, where NCL gates with hysteresis are replaced with their Boolean equivalent gate when hysteresis is not required for input-completeness and/or observability.

The framework of this verification methodology can also be applied to other QDI paradigms, such as MTNCL and PCHB. For Multi-Threshold NULL Convention Logic (MTNCL), the functional checking and invariant checking methods are essentially the same as for NCL, but the handshaking check is slightly different [23]. Additionally, MTNCL circuits do not require input-completeness or observability, so these checks are not needed. For PCHB, the handshaking check is essentially the same as for NCL, but the functional checking method is a bit different [11]. Since PCHB gates consist of dual-rail input(s) and output(s), invariant, input-completeness, and observability checking are not required, as these are ensured within the primitive PCHB gates themselves.

References

[1] V. Wijayasekara, S.K. Srinivasan, and S.C. Smith, "Equivalence verification for NULL Convention Logic (NCL) circuits," *32nd IEEE International Conference on Computer Design* (ICCD), pp. 195–201, October 2014.

[2] P. Manolios, "Correctness of pipelined machines," in *FMCAD 2000*, ser. LNCS, W.A. Hunt, Jr. and S.D. Johnson, Eds., vol. 1954, Springer, New York, 2000, pp. 161–178.

[3] D. Monniaux, "A survey of Satisfiability Modulo Theory" [online]. Available: https://hal.archives-ouvertes.fr/hal-01332051/document [Accessed: May 5, 2019].

[4] C.W. Moon, P.R. Stephan, and R.K. Brayton, "Specification, synthesis, and verification of hazard-free asynchronous circuits," *Journal of VLSI Signal Processing* (1994) 7(1): 85–100.

[5] C.J. Myers, *Asynchronous Circuit Design*. New York: Wiley, 2001.

[6] A. Semenov and A. Yakovlev, "Verification of asynchronous circuits using time Petri net unfolding," *33rd Design Automation Conference*, Las Vegas, NV, USA, 1996, pp. 59–62.

[7] F. Verbeek and J. Schmaltz, "Verification of building blocks for asynchronous circuits," in *ACL2*, ser. EPTCS, R. Gamboa and J. Davis, Eds., vol. 114, 2013, pp. 70–84.

[8] A. Peeters, F. te Beest, M. de Wit, and W. Mallon, "Click elements: An implementation style for data-driven compilation," *IEEE Symposium on Asynchronous Circuits and Systems*, Grenoble, France, 2010, pp. 3–14.

[9] S.K. Srinivasan and R.S. Katti, "Desynchronization: design for verification," *FMCAD*, Austin, TX, 2011, pp. 215–222.

[10] A.A. Sakib, S.C. Smith, and S.K. Srinivasan, "Formal modeling and verification for pre-charge half buffer gates and circuits," *60th IEEE International Midwest Symposium on Circuits and Systems*, Boston, MA, 2017, pp. 519–522.

[11] A.A. Sakib, S.C. Smith, and S.K. Srinivasan, "An equivalence verification methodology for combinational asynchronous PCHB circuits," *61st IEEE International Midwest Symposium on Circuits and Systems*, Windsor, ON, Canada, 2018, pp. 767–770.

[12] S.C. Smith and J. Di, "Designing asynchronous circuits using NULL Convention Logic (NCL)," in *Synthesis Lectures on Digital Circuits and*

Systems, Morgan & Claypool Publishers, San Rafael, CA, vol. 4/1, July 2009.

[13] C. Jeong and S.M. Nowick, "Optimization of robust asynchronous circuits by local input completeness relaxation," *Asia and South Pacific Design Automation Conference*, Jan. 2007, pp. 622–627.

[14] A. Kondratyev, L. Neukom, O. Roig, A. Taubin, and K. Fant, "Checking delay-insensitivity: 10^4 gates and beyond," *8th International Symposium on Asynchronous Circuits and Systems*, April 2002, pp. 149–157.

[15] L.M. de Moura and N. Bjørner, "Z3: An efficient SMT solver," *in TACAS*, ser. Lecture Notes in Computer Science, C.R. Ramakrishnan and J. Rehof, Eds., vol. 4963, Springer, New York, 2008, pp. 337–340.

[16] D. Bryan, "The ISCAS '85 benchmark circuits and netlist format" [online]. Available: https://ddd.fit.cvut.cz/prj/Benchmarks/iscas85.pdf [Accessed: May 5, 2019].

[17] C.L. Seitz, "System timing," in *Introduction to VLSI Systems*, Boston, MA, USA: Addison Wesley Longman Publishing Co., Inc., 1979, pp. 218–262.

[18] "ISCAS-85 c432 27-channel interrupt controller" [online]. Available: http://web.eecs.umich.edu/~jhayes/iscas.restore/c432.html [Accessed: May 5, 2019].

[19] S.C. Smith, R.F. DeMara, J.S. Yuan, D. Ferguson, and D. Lamb, "Optimization of null convention self-timed circuits," *Integr. VLSI J.* (2004) 37(3): 135–165.

[20] "ISCAS-85 c432 27-channel interrupt controller Module M3" [online]. Available: http://web.eecs.umich.edu/~jhayes/iscas.restore/c432m3.html [Accessed: May 5, 2019].

[21] "ISCAS-85 c432 27-channel interrupt controller Module M1" [online]. Available: http://web.eecs.umich.edu/~jhayes/iscas.restore/c432m1.html [Accessed: May 5, 2019].

[22] [Online] http://www.pld.ttu.ee/~maksim/benchmarks/iscas89/bench/, [Accessed: May 5, 2019].

[23] M. Hossain, A.A. Sakib, S.K. Srinivasan, and S.C. Smith, "An equivalence verification methodology for asynchronous sleep convention logic circuits," *IEEE International Symposium on Circuits and Systems*, May 2019.

Chapter 16

Conclusion

Jia Di[1] *and Scott C. Smith*[2]

Asynchronous logic has been around for the past 50+ years; but until recently, synchronous circuits have been good enough to meet industry needs, so asynchronous circuits were primarily utilized for niche markets and in the research domain. Recently, however, the continued advances in semiconductor technology and newly emerging commercial demands have induced resurgence in investments in asynchronous circuits. This book highlights a set of applications where asynchronous circuits outperform their synchronous counterparts due to one or more of their advantages, such as no clock tree, flexible timing requirement, robust operation, improved performance, high energy efficiency, high modularity and scalability, and low noise and emission. Note that this book is by no means a complete list of suitable applications for asynchronous circuits; in addition to the topics covered herein, new applications are being investigated by academia and industry, including robust digital circuits designed using emerging devices.

In order to promote the development and widespread industry adoption of asynchronous circuits, a series of challenges will need to be addressed. A commercial grade automated design flow, including logic synthesis, formal verification, and testing, is amongst the most urgent needs. With such a design flow in place, asynchronous ICs could be easily generated from RTL and other high-level descriptions, the same as synchronous circuits currently are, thereby releasing the burden of asynchronous design details and optimization from the designers, reducing design effort, shortening time-to-market, and enhancing design quality/reliability. A related challenge is asynchronous design workforce development. Increasing industry usage demands more IC design engineers who understand asynchronous logic, including its advantages and disadvantages, and when and where it could potentially be utilized to benefit their products.

The asynchronous logic research community has been actively seeking to address these challenges. Many efforts in developing automated asynchronous circuit design tools have been carried out, some of which are introduced in this book; and

[1]Computer Science and Computer Engineering Department, University of Arkansas, Fayetteville, AR, USA
[2]Department of Electrical Engineering and Computer Science, Texas A&M University-Kingsville, Kingsville, TX, USA

asynchronous IC design has been incorporated into undergraduate and graduate courses taught at a variety of universities. One example is the EENG426/CPSC459/ENAS876 Silicon Compilation course at Yale University, which is "an upper-level course on compiling computations into digital circuits using asynchronous design techniques. Emphasis is placed on the synthesis of circuits that are robust to uncertainties in gate and wire delays by the process of program transformations. Topics include circuits as concurrent programs, delay-insensitive design techniques, synthesis of circuits from programs, timing analysis and performance optimization, pipelining, and case studies of complex asynchronous designs." Utilizing the course's asynchronous IC design flow, students are able to go from RTL code all the way to asynchronous chip tape-out in a combination of QDI and BD logic. Other examples include EE 552 Asynchronous VLSI Design at University of Southern California, 02204 Design of Asynchronous Circuits at Denmark Technical University, ECE/CS 5960/6960 Relative Timed Asynchronous Design at University of Utah, ECE 475/773 Advanced Digital Design at North Dakota State University, and CSCE4233 Low Power Digital Systems at University of Arkansas.

As observed from the semiconductor industry trend over the past 15 years, and as predicted by the International Technology Roadmap for Semiconductors, asynchronous circuit usage will continue to increase in the multibillion dollar semiconductor industry. While the academic research community continues to develop new asynchronous design techniques and a workforce familiar with both synchronous and asynchronous design, increased investment from industry, especially EDA companies, will further speed up the rate at which asynchronous logic is being incorporated into commercial ICs, which will benefit the variety of applications that are more suited to an asynchronous logic implementation.

Index